CRYSTAL CLEAR

CRYSTAL CLEAR

The Autobiographies of Sir Lawrence and Lady Bragg

Edited by
A. M. Glazer and
Patience Thomson

UNIVERSITY PRESS

Great Clarendon Street, Oxford, OX2 6DP,
United Kingdom

Oxford University Press is a department of the University of Oxford.
It furthers the University's objective of excellence in research, scholarship,
and education by publishing worldwide. Oxford is a registered trade mark of
Oxford University Press in the UK and in certain other countries

First Edition published in 2015

Impression: 1

Published in the United States of America by Oxford University Press
198 Madison Avenue, New York, NY 10016, United States of America

British Library Cataloguing in Publication Data
Data available

Library of Congress Control Number: 2015934148

ISBN 978–0–19–874430–6

Printed in Great Britain by
Clays Ltd, St Ives plc

Dedicated to the memories of
Stephen Lawrence Bragg (1923–2014) and
David William Bragg (1926–2005)

Foreword

On 14 November 1915 in the midst of the carnage of the Great War, a young officer at the front in northern France received a telegram informing him that he had been awarded that year's Nobel Prize in Physics, to be shared with his father. The officer was the twenty-five-year old William Lawrence Bragg (WLB), and his father was William Henry Bragg (WHB). This made WLB the youngest Nobel laureate ever and, apart from the Peace Prize awarded in 2014, that remains true. Moreover, WHB and WLB are to date the only father and son team to be jointly awarded the Nobel Prize. Their discoveries made between 1912 and the outbreak of the war in 1914 can truly be said to have transformed all our lives, for they enabled us to understand and study for the first time the atomic structures of crystalline solids.

This has led to many of the most important scientific achievements of the last hundred years, continuing to the present day and no doubt beyond. WLB and WHB, in the late summer of 1912, were the first to show how to interpret the patterns of spots seen on a film or detector when X-rays are incident on a crystal. The spots arose from the scattering of the X-rays by the atoms in the crystal, a process known to physicists as diffraction. From the positions and intensities of these spots, WLB and WHB showed how to derive the arrangements of atoms in crystals. They thus jointly ushered in an entirely new scientific discipline, known as X-ray crystallography. It is a subject that has enabled scientists to determine the structures of thousands of crystals, starting from the very simple to the most complex materials. In molecular biology, the structures of proteins, viruses and, famously, DNA have been solved. One of the greatest triumphs of this kind was exemplified by the Nobel Award in 2009 to Venki Ramakrishnan, Thomas Steitz and Ada Yonath for the determination of the structure and function of the ribosome, the molecular machine in the cell that is responsible for the synthesis of proteins. This work over many years resulted in the location of some hundreds of thousands of atoms, a feat that the two Braggs could never have envisaged in their time. In addition, our knowledge of the atomic structures and properties of metals, electronic materials, pharmaceuticals and inorganic and organic materials, including polymers, comes directly from the two Braggs' seminal research. More than twenty-six Nobel Prizes have been awarded since for research that has built upon the work of the two Braggs. Today crystallographic techniques are used throughout the world to study solid materials in universities, industries and research institutes worldwide. The World Directory of Crystallographers published by the International Union of Crystallography(IUCr) lists today 12,922 persons in 114 countries (2,119 in the USA, 865 in India, 824 in Germany, 730 in the UK, 718 in the Russian Federation, 574 in France, 573 in Japan, 493 in Italy, 369 in China. There are many more who use crystallographic techniques but whom are not listed, with a total estimated to be in excess of 30,000.

It all began with the discovery in Würzburg, Germany, of X-rays by Wilhelm Conrad Roentgen (1845–1923) in the late afternoon of 8 November 1895, for which he received the first Nobel Prize in Physics in 1901. Like most great scientific discoveries, this was made by accident, when Roentgen was experimenting with an electrical discharge tube covered with cardboard in a darkened room and happened to notice a shimmering glow from a fluorescent screen made from barium platinocyanide on a distant bench.[1] The rays were mysterious at the time, and hence the name given to them, at least outside Germany, was X-rays, signifying unknown rays. In the following years up to 1912, one of the burning questions in physics was whether these mysterious rays should be described in terms of waves or in terms of particles of some kind (the corpuscular theory). The world of physics was divided on this, with, it is true to say, the majority believing that X-rays were wave-like, as suggested by some experiments. However, WHB was one of those who argued in favour of the corpuscular theory, suggesting the X-rays could be described by the flow of neutral (uncharged) particles. WHB had begun his study of X-rays while in Adelaide, Australia, before coming to England to take up a professorial post in Leeds and could point to experiments that strongly suggested a corpuscular model (for instance, the observation of particle tracks seen in the cloud chamber invented by C. T. R. Wilson in Cambridge).

[1] There is a famous and pertinent quotation from the great French scientist, Louis Pasteur (1822–1895) who had observed many years earlier, '*Dans les champs de l'observation le hasard ne favorise que les esprits préparés*,' that is, 'In the fields of observation chance favours only the prepared mind!'

In the spring of 1912, according to the most widely accepted story, a *Privatdozent* in Arnold Sommerfeld's Institute of Theoretical Physics in Munich, Max Laue, was in conversation with a student of Sommerfeld, Paul Ewald, discussing the nature of crystals. The idea that crystals consisted of repeating groups of atoms, rather like the pattern in a wallpaper, had been a matter of conjecture for the previous two to three centuries. Laue asked Ewald that, if this was so, what would be the typical distances between the repeating units? When Ewald said that this would be very small, of the order of angstroms (1 Å = 10^{-8} cm), Laue was said to have had a flash of inspiration. He thought that if X-rays were waves, they would have wavelengths of that order, and so in a stroke of genius he thought that it should be possible to use crystals to diffract X-rays, rather like the way in which visible light can be diffracted by a series of fine slits. He initially approached Sommerfeld for permission to use the institute's personnel to help to try out the experiment, but this was refused in the belief that they had more important things to do. However, not to be put off, Laue approached Paul Knipping (a student of Roentgen, who by this time had moved to Munich) and Walter Friedrich (an assistant of Sommerfeld) to try the experiment out. After several abortive attempts, one night in April 1912 (reputed to be 21 April) they obtained on a photographic film a pattern of spots, thus showing that the X-rays had been scattered by the crystal, consistent with diffraction. (Ewald once told me that in fact Friedrich and Knipping had secretly stolen Roentgen's apparatus in order to carry out the experiments!). The pattern of spots then strongly suggested that X-rays had wave-like properties. As Max von Laue (he acquired the 'von' in 1913 when his father was ennobled), he was awarded the 1914 Nobel Prize in Physics for this important discovery. However, in spite of producing the correct equations to explain the diffraction process, he made several incorrect assumptions and failed to understand fully the origin of the patterns of spots.

In the summer of 1912, WHB received a letter from a colleague telling him about Laue's work, but he was still convinced that X-rays were particle-like in nature. So, with his son WLB, he set about showing how Laue's patterns could be explained by particles travelling along 'avenues' within the crystals. Their experiments, however, were far from conclusive. Shortly after, WLB became convinced that X-rays in fact did consist of waves, and that Laue's experiment was indeed evidence of diffraction. It was on 11 November

1912 that WLB had his famous paper read to the Cambridge Philosophical Society in which he showed how to interpret Laue's diffraction patterns in terms of the positions of atoms in a crystal. He also gave a remarkably simple, yet brilliant, explanation by thinking of crystal diffraction as a kind of reflection of X-rays from parallel planes of atoms in the crystal. This was like the reflection of light from mirrors, except that the reflected waves, coming from different depths of the crystal, combined together, sometimes in phase and sometimes out of phase, thus creating the observed spots. To this day, these spots are called *reflections* by crystallographers. WLB had acquired a deep knowledge of optical physics while a student at Cambridge, and he applied this to the case of X-ray diffraction from a crystal. This led him to the very simple formula $\lambda = 2d \sin \theta$, which today is known to all scientists as Bragg's Law. The variable λ is the wavelength of the X-rays, d is the distance between successive planes of atoms and θ is the angle between the beam and the planes. This is one of the most important equations in all of science. With this, WLB had solved a problem that others had failed to do. Not bad for a mere twenty-two-year-old!

WHB modified his view of X-rays in the belief that they could be treated as *both* particles and waves, depending on the experiment employed and here he was well ahead of the then current thinking, as it was in the 1920s that wave–particle duality was indeed shown to be true.[2] WHB immediately realised the importance of WLB's insight, and WLB recognised the value of WHB's new X-ray spectrometer. This was a most important development, as this spectrometer can be regarded as the ancestor of the modern 'diffractometer' used by crystallographers throughout the world today.[3] Father and son then collaborated to study the structures of several crystals. In 1913 WLB solved the structure of common salt (sodium chloride) and together with his father that of a crystal of diamond. The sodium chloride

[2] WHB said in his Robert Boyle lecture in Oxford in 1921: 'On Mondays, Wednesdays, and Fridays we use the wave theory; on Tuesdays, Thursdays, and Saturdays we think in terms of flying energy quanta or corpuscles.'

[3] Incidentally, WLB had originally in 1912 written his law as $\lambda = 2d \cos \theta$, where the angle θ is the angle between the X-ray beam and the <u>normal</u> to the crystal planes being considered. I suggest that it was recast into its more familiar sine form in 1913 by changing the angle θ by 90°, as a result of using the spectrometer, where it is natural to measure angles from the direction of the incident X-ray beam.

structure was shown to consist of a chessboard-like pattern of alternating, equally spaced sodium and chlorine atoms. This structure was initially not accepted by many chemists who argued that the sodium and chlorine atoms should form molecules rather than equally spaced atoms. For instance, it is said that Professor Arthur Smithells begged WLB to 'find that one sodium atom was just a tiny bit nearer to one chlorine than it was to the others'![4] Of course, WLB's salt structure was finally accepted. It even appeared on a British postage stamp in 1977 (although I note that the artist actually made a tiny error in the drawing!).

At the time, WHB was a well-established scientist, while the young WLB was completely unknown. This meant that it was WHB who tended to receive the attention of the scientific community and that of the public. It was WHB who was invited to attend international meetings, such as the famous Solvay meeting in 1913. For a long time, WLB felt in his father's shadow and this upset him. Despite this, WHB, a genial, kind, humble and shy man, always went out of his way to credit his son with the initial discovery wherever possible, even in his own publications (WHB must have explained to those attending the Solvay meeting that it was his son who had made the initial breakthrough, as all the attendees signed a postcard to send back to WLB—see Figure 17). In truth, both of them were very close and in constant contact, and they collaborated on many important contributions to the subject. One way in which the two Braggs later decided to alleviate the problem of overlapping research was for WHB to concentrate on organic crystals while WLB would work on metals and inorganic crystals.

Throughout 1913 and into 1914, father and son 'plundered' the field of crystallography, determining the structures of many crystals with no competition. WLB once said, 'It was like discovering an alluvial gold field with nuggets lying all round waiting to be picked up'. Unfortunately, the two Braggs' research was interrupted by the Great War of 1914–1918, during which time WLB served as an officer in a sound ranging unit in northern France, while WHB conducted studies of hydrophones to detect enemy submarines. WLB then went to Manchester and set up an important crystallography research group.

[4] Arthur Smithells (1873–1960) was Professor of Chemistry in the University of Leeds and author of scientific papers on flame and spectrum analysis.

Many famous and important discoveries continued to be made. After working in Manchester, followed by a brief spell at the National Physical Laboratory (NPL), WLB was appointed Head of the world-famous Cavendish Laboratory in Cambridge, where he worked from 1938 until 1953. In this position, he stimulated the research of many great scientists, including the Nobel Prize winners Max Perutz, John Kendrew, Francis Crick and James Watson. In 1954 he was appointed to the Royal Institution of Great Britain, where he remained until his retirement in 1966. WLB died in 1971.

Another important aspect of the two Braggs is that they both, unusually for the period, strongly encouraged female students into crystallography. For instance, of eighteen of WHB's students, eleven were women! It is sometimes claimed that the large numbers of female scientists working in crystallography today owes its origin to the two Braggs.

If I may inject a personal note, I first encountered WLB while a teenager, when I had the good luck to attend his 'Schools Lectures' at the Royal Institution. I well remember his wonderful demonstrations: sometimes these were accompanied by loud whizzes and bangs followed by his boyish grin! He particularly liked teaching (or rather, in his own words, 'showing') science to children. It is believed that his Schools Lectures were attended by up to 20,000 school students per year over a ten-year period! Later, when I carried out my Ph.D. research in crystallography, I had the great pleasure to meet him in person, when he used to visit my supervisor Professor Dame Kathleen Lonsdale (who incidentally had been a student of WHB, thus making him my scientific grandfather!).

In later years, WLB came to be regarded as one of the great old men of science, much loved and respected, especially within the crystallographic community. I recall that many years ago our crystallography meetings in the UK used to take place at the Institute of Electrical Engineers in Savoy Place, London. I think by this time WLB had retired; but suddenly, during one such conference, attended by perhaps 300 to 400 people, he appeared at the back door of the auditorium. The whole audience rose en masse to its feet as he walked in.

Another example of the respect in which WLB was held is illustrated by the following note that I received in 2013 while I was planning, at the University of Warwick, an exhibition devoted to the work

Two screenshots taken from a film made of a Schools Lecture by WLB, probably in the late 1950s at the Royal Institution, in which he illustrates the principle of the Paget speech synthesiser—this uses compressed air, a tube and pieces of modelling clay to create human voices. He was in his element when demonstrating science to children. The top picture shows some of the audience of school children, and I am the teenager at the top left!

of the two Braggs; the note was from Michael Fuller, MBE, who had been a laboratory assistant at the Cavendish Laboratory under WLB:

> One day, after school in Cambridge, I picked up a book in the library about the Cavendish Laboratory. An article about high voltage physics really excited me, so when I was due to leave school, a very naive 15 year old, without any qualifications, wrote direct to Sir Lawrence asking if I could come and work at the Cavendish Laboratory. Two days later, I received a very nice hand-written letter from him asking if I would like to come for an interview with a Dr Max Perutz, who was looking for an apprentice technician to work with Tony Broad on the development of a Rotating Anode X-ray set being set up in the basement of the Austin Wing.[5] After a short interview, I was offered a post and started work on Jan 2nd 1952 in the MRC unit, learning all the techniques from glass-blowing to electronics and vacuum engineering.
>
> As Sir Lawrence always left his bicycle in the basement corridor of the Austin Wing, where our x-ray generator set was being developed, he would often come and ask how I was getting on, and was the X-ray set working? Sometimes, after leaving his bicycle, he would call into the electrical maintenance workshop and have a chat with Bert [the electrician], and over a cup of tea catch up with all the gossip about what was happening in the lab.
>
> Sir Lawrence was a classic English gentleman, who insisted on everybody being polite and well-dressed with lab coats buttoned up at all times, and tidy clean corridors etc. The only time I ever saw him cross and lose his temper was when a Dr Brandenberger, a nuclear physicist, came to work, one hot summer day in his 'Lederhosen' German outfit, walking along the corridor with his brown lab coat, his bare legs showing. Sir Lawrence, standing at the end of the ground floor corridor went very red faced and shouted to Dr Brandenberger to go home at once and come back to the lab properly dressed, stating that this is the Cavendish Laboratory and that certain standards of dress are expected.
>
> He was a very talented painter, and presented two pen and ink sketches of Max Perutz and John Kendrew to the MRC unit. These hung in the corridor for many years, until a visit from an exhibition curator spotted them and pointed out that they were very valuable and should be kept under lock and key.
>
> I have fond memories of Sir Lawrence, who helped me on my career, and would always find time to have a chat and provide advice.

[5] The so-called Broad rotating-anode set was designed by the laboratory technician Tony Broad, a technician in the Cavendish Laboratory.

So how did this book come about? Several years ago I was invited to a conference in Madrid to talk about the two Braggs. I made contact with Lady Heath (Margaret), elder daughter of WLB. Amazingly, and rashly as I thought at the time, she let me take away the family photograph album (this is now in the Bragg archives at the Royal Institution). I also obtained from her a copy of WLB's autobiography, which had never been published.

In August 2013, to celebrate the centenary of the Braggs' seminal work, I mounted an exhibition entitled 'The Two Braggs' and held at the University of Warwick (see <http://www.amg122.com/twobraggs>). In order to borrow the exhibits, much of them hitherto unseen in public, I contacted other members of the Bragg family. These included Stephen Bragg (WLB's son), Patience Thomson (younger daughter of WLB) and the Lady Adrian (Lucy, niece of WLB, and a member of the family who was particularly close to her grandfather WHB). I learnt that WHB, his wife Gwendoline, WLB and his sister Gwendy were all competent amateur artists, and so we had, in addition to the historic items of equipment and letters, a splendid display of their paintings and sketches. I think that it was this artistic tendency that enabled WLB, at such a young age, to see the solution to Laue's crystal diffraction problem, where the more formal approach of the German scientists struggled. Crystallography is by its nature both a highly mathematical and a visual subject!

I also was given a copy of a manuscript by Lady Bragg (née Hopkinson), WLB's wife, telling her side of the story. On reading through the manuscripts, I realised that these accounts may be of interest not just to crystallographers like me but perhaps to a more general audience. Publication of these accounts is especially apt, as 2015 is one hundred years since the two Braggs' Nobel Prize. WLB writes about his experience from his early childhood in Australia through to his work in England. His account of his work during the First World War is most interesting. It is not appreciated, I think, that he and his team of sound rangers actually were credited for significantly shortening the war. With current interest in this war, it seems to me that WLB's account is timely. Alice's account presents her life with WLB as seen from her point of view. Therefore, some anecdotes are repeated in both accounts, but there are also many other observations made about the many well-known people whom they both encountered. Alice admitted that she had no head for science, but she made her mark through public service. For example, she was the Mayor of Cambridge just after the Second World War and later served on government commissions and as a magistrate. In spite of having a completely different personality and background from her husband, she was a great support to WLB throughout their lives together.

The Bragg family originated from the area around Wigton, in what was then known as Cumberland. They were mainly middle class, involved in areas such as shipping and farming. Alice's family, the Hopkinsons and Cunliffe-Owens, on the other hand, was very different, as it was well connected to the establishment, and even to royalty. It was a large family with many talented and distinguished members. Alice was a very outgoing person with strong opinions. On the other hand WLB was by nature, according to Alice in the 1965 BBC programme '50 Years a Winner', which celebrated the fiftieth anniversary of the Braggs' Nobel Prize, 'a shy, private, family man, whose mind was partly child-like in its ability to see behind the complex and to relate so well to children'. He never felt that he belonged to the establishment. He was an emotional man, but he kept his feelings under control. In fact, he was so private that he addressed almost everyone he talked to only by their surname, reserving first-name terms only for his closest friends and family. He also hated controversy, preferring to avoid it wherever possible. I can record here a story that was related to me many years ago when I worked at the Cavendish Laboratory. My chief technician in the Crystallography Laboratory, R. A. (Sam) Cole, had been a young technician there at the time that WLB was Cavendish Professor. The brilliant metallurgist–crystallographer A. J. Bradley was there too, having come from Manchester to Cambridge with WLB, but had developed a serious mental condition. Sam Cole was shocked when Bradley suddenly started to shout uncontrollably at WLB, with the result that WLB fled immediately from the scene!

I have attempted to edit with the lightest of touches, letting the Braggs speak for themselves in the language and style of their times. Apart from reorganising a few passages that had inadvertently been repeated, my main contribution to this work has been to supply brief explanatory footnotes. I assume that WLB had intended at some time to go back and flesh out and revise his manuscript but unfortunately he died in 1971 before doing it. That he never finished his account may be because he was so busy with organising the Royal Institution. He may also have become discouraged: a clue may be found in a letter to him in 1968 from C. P. Snow (in the Archive at the Royal Institution London), whom WLB had consulted about the manuscript. In it Snow says he is not egocentric enough and that 'nice men find it hard to write an honest autobiography. . . The interest of Jim Watson's book [referring to the famous *The Double Helix*] lies very largely in the fact that he is not at all a nice man'! The autobiography finishes before his move to the Royal Institution, although Lady Bragg's account does cover that period. I have removed some references to a very small number of people who are not relevant to the central story and whose present families might be disappointed to see what was said in private many years ago.

I have included some of WLB's fine sketches too and many other illustrations, some of which have never been seen in public before. It is fascinating to see how many famous people, including many Nobel Prize winners, were known personally to the Braggs, their names sprinkled casually throughout the manuscripts. As WLB's autobiography was an unfinished piece of work, it contains many names of people without explanation as to who they were. Some were obvious to me, but many others not. Thank goodness for the internet and Wikipedia for helping me to identify most of them!

In planning this book, I have worked closely with Patience Thomson, who has added her own account of her parents and their relationship to each other. This is told through a number of delightful anecdotes and vignettes which I think will give an additional, personal, insight into Sir Lawrence and Lady Bragg.

I am grateful to Professor Frank James and Charlotte New at the Royal Institution for permission and help to use the 'Braggiana' from the archives, and especially to members of the Bragg family, including Stephen,[6] Patience, Margaret and Lucy. I am especially indebted also to John Jenkin, biographer of the two Braggs, for his comments and help with the manuscript, and Corinna Dahnke of the Clarendon Laboratory for initial typing. I thank also Ian Butson for help with identifying members of the Hopkinson and Cunliffe-Owen families. The following figures are shown by courtesy of the Royal Institution London: Figures 4, 5, 7–9, 13, 14, 16, 17, 20, 21, 25, 28, 29, 34–40, 41, 42, 46, 47, 58 and 69.

All royalties for this book will be donated to support the Royal Institution of Great Britain.

Mike Glazer
Emeritus Professor of Physics Oxford University
Visiting Professor at the University of Warwick

[6] Sadly, Stephen Bragg passed away in November 2014.

Contents

1

Meet my Mother and Father (by Patience Thomson)

1.1 Introduction

My contribution to this book is, I feel, to bring alive my parents as real people and vibrant personalities. Because I am recording the memories of what I observed and what I was told as I grew up, my account is naturally weighted towards the years when I was living at home and seeing my parents on a daily basis. My late brothers Stephen and David and my sister Margaret would surely have had their own stories and their own perspectives on the Bragg family life, our upbringing and experiences. It is not for me to tell their story, but Margaret has been very helpful in reminding me of some of the classic Bragg anecdotes.

What I have written about my parents is intentionally not a chronical of their lives. Others have written biographies. My offering is a series of sketches illustrating different facets of their characters, their lifestyle, their attitudes and their feelings.

1.2 Personalities

A Double Act

My mother, my father and I are sitting in a café having tea and cake. My mother bends over to whisper something in my father's ear, which he then starts to rub ruefully.

'You're blowing cake crumbs into my ear,' my father complains, 'and I can't hear you.'

'Do you see that lady in the corner with the extraordinary purple hat?' asks my mother, a little louder.

My father rears back in his seat, looks all around and bellows, 'Lady in an extraordinary purple hat? Where?' and I nearly die of embarrassment.

Variants on this theme were not unusual. My parents played a double act, egging each other on and laughing together. They were a totally committed team, yet they had very little in common. 'Life is flavourless without you,' wrote my father to my mother in a letter from Canada in the war. Their marriage was a lifelong romance. They adored each other.

1.3 Meet my Mother

My mother was beautiful—not just pretty, but beautiful. Her only disadvantage was her prominent front teeth. She had thick dark hair, a lovely skin, a slim figure, boundless vivacity and a flirtatious nature. She was a people-person, happiest with a crowd around her. On her deathbed she unexpectedly made a statement that explained a lot to me. 'I have acted my way through life,' she said, 'and I have played many parts.' One might add that she had played them all with great confidence and skill.

Mother got away with being mildly eccentric and even outrageous. I remember her greeting my father on the doorstep. 'You have a letter from the War Office marked Strictly Private and Confidential, but it's very dull.' For a dare, she once ate an entire meal in a restaurant backwards. She started with coffee and ended with tomato soup.

Only my mother could write a poem to the porters at King's College, Cambridge, asking permission to wheel my pram through their grounds, as this would save her a long journey round to reach the centre of town. Alas, her poem does not survive, but the reply does.

Dear Madam,

It has lately come to our knowledge,

That you were not allowed to come through the College.

We deeply regret we caused you pain,

And hasten the veto to explain.

We should never have thought your way home to dam,

Had you not been accompanied by a pram.

Our pathways were not made to cater

For the wheels of a perambulator.

If one came through the rest would follow,

Marmot and Tansad, Dunkley and Swallow.

And after the prams the bikes would race
Raleigh and Humber and Hercules.
You understand we are not to blame!
If only you could establish a claim . . .
You are not one of our corporate life
In as much as you are not a Fellow's wife.
Our ruling cannot be any other
Because you are not a Fellow's mother.
We cannot presume the rule to alter
Since you say you are not a Fellow's daughter.
We could settle at once this unfortunate bother,
If you could claim a Fellow as brother.
You could sail triumphantly through the College
Wheeling your pram over sacred grass
Through Jumbo's arch, to the best of my knowledge
And no one could say 'You may not pass'.
But as you affirm you are none of these things,
. . . I remain, yours sincerely
(Not) the Porter of King's

My Mother's Roots

My mother's father, Dr Albert Hopkinson, was a saintly man; unfortunately, saintliness is a quality which children do not always appreciate. My cousin, Margaret Barton, wrote of Albert's parents in her book *John and Alice Hopkinson* that they 'could have come out of no other environment than the protestant north in the great days of individualism'.

Albert was the fifth son and tenth child in a family of thirteen. There was no money left in the kitty for him to train as a consultant, so he became a general practitioner in Manchester and practised there for many years while my mother was growing up. He was deeply committed to helping the poor and often did not send in a bill. My children and grandchildren have played endlessly with a huge box of wooden bricks which had been presented by a carpenter whose large family had been successfully attended by my grandfather during a scarlet fever epidemic.

His work took its toll and he moved to Cambridge, where he became a demonstrator in anatomy. I used to look on with fascination as he carved the Sunday joint with impeccable precision.

However, Albert Hopkinson was no pushover. Once when travelling with his family, he was informed by a station master that the children's perambulator must go in the luggage van. He picked up the offending article and whirled it round his head. 'Hand luggage,' he said triumphantly.

Incidentally, Albert's father, John Hopkinson, was illegitimate. In spite of this disadvantage in Victorian Britain, he made good and became Lord Mayor of Manchester. All letters and references concerning his illegitimacy were destroyed by his daughter, Mary.

Albert Hopkinson, my mother's father, married Olga Cunliffe-Owen. Her family were spirited, talented and eccentric. Sir Philip Cunliffe-Owen spent some years in the navy but left on account of ill health. There followed several years of idleness, travelling abroad with his father. I find it extraordinary that Sir Philip Cunliffe-Owen ended up as the second-ever Director of the South Kensington Museum, later the Victoria and Albert, from 1874 to 1893. One cannot but question his credentials for the job. He was a navy man and a traveller. However, he was a close friend of the Royal Family and this might be part of the explanation. He lived in the residence within the grounds of the museum and brought up his nine children there. They played in the galleries on Sunday, when the museum was closed. Olga's account of her girlhood is beautifully written and gives fascinating insight into upper-class life in the nineteenth century. My grandmother was often at court and had an idyllic existence with continental travel, pleasure and parties. I still have and wear her mother's deep blue velvet ball dress, made in Paris.

In 1853, while staying in Vevey, Philip met Baroness Jenny von Reitzenstein, his future wife. According to family tradition, he courted her barefoot in the snow, singing under her window. Jenny brought a strong German influence into the family. She was certainly an aristocrat. Her father, Baron von Reitzenstein, commanded the Prussian Garde du Corps and she was brought up at court with the future Kaiser. She wrote two diaries covering the period from 1849 to 1853. There is no mention in them of the political upheavals of the time. It is Jenny's romantic story. She fell in love with a French officer of no means. They were discovered and separated and Jenny was married off to Sir Philip.

In the diaries I found a cartoon, drawn in pencil but signed in ink. The signature was that of Crown Prince Friedrich Wilhelm. Years later, on visits to London, he would send his bandsmen to play outside the residence in the museum.

When I was sixteen, my mother took me to stay with a certain Princess Wittenstein in Westphalia. We were introduced all round in the drawing room and my mother was immediately forthcoming on her German relatives, all 'von' or 'zu' something. Someone whipped the *Almanach de Gotha* out of his pocket to test her authenticity. She passed with flying colours.

Mother was much influenced by her German heritage. On Christmas Eve, celebrated in Germanic style, she always sang 'Stille Nacht, heilige Nacht'. She had a sentimental and emotional streak and a passionate and intensely loyal love of family. 'Patience is perfect,' she wrote on my holiday report for school. 'And even if she isn't, I wouldn't tell you.'

Delving a little further back in the Cunliffe-Owen history I discovered a strain of derring-do that was obviously a dominant trait in the family. A slim green volume, with no print on the cover or spine, fell out of a suitcase. On opening it, I discovered the title *The Descendants of the Elder Branch of the Cunliffes of Wycoller*. The Cunliffes, it transpired, from whom my grandmother, Olga, descended, had been lords of the manor in the small village of Wycoller, in Lancashire, for many hundreds of years. The manor, now a ruin, was finally lost to them through the extravagance of the last family owner. Incidentally, it was his son, Charles, who added 'Owen' to his surname.

I knew from my mother that Wycoller Hall was frequently visited by the Brontës, who lived nearby, and that, damp and dismal, it was reckoned by some to be the model for Ferndean Manor in *Jane Eyre*. What I did not know, until I visited Wikipedia, was that the Cunliffe forebears were a racy lot.

One of the Cunliffe squires was hunting during the reign of Charles II. The fox that he was pursuing ran into the manor, up the stairs and into the women's chamber, hotly followed by the hounds and my Cunliffe ancestor, who did not bother to dismount from his horse. Arriving with raised hunting whip, he startled his wife to such an extent that she died from fright.

Once a year he returns, in the hours of darkness, to haunt the manor, dressed in a costume of the Stuart era. His visit is proclaimed by the clattering of his horse's hooves. His wife's ghost put a curse on the Cunliffes, predicting their downfall and the ruin of the manor.

Another Cunliffe is said to have married a West Indian woman. Realising on his way back to England that he had made a huge mistake, he threw her overboard. She drowned. Her ghost also haunts the manor.

Obviously these two stories must be taken with a pinch of salt but do reflect the Cunliffes' reputation, which was not the image of the usual English country gentleman. There was definitely dramatic blood in their veins, a trait which I think my mother inherited.

Mother's Upbringing and Education

My mother was an accomplished writer; here is a typical example taken from an undated draft for *The Manchester Guardian* or *The Times*. It was entitled 'Memories of Mancunians'.

> My grandparents lived at Grove house, Rusholme, where the Whitworth Gallery now stands. There they had their thirteen children and an Alderney cow in the paddock towards their support. Those who survived childhood spent most of their lives in Manchester so that it was natural that we knew some of the city's 'characters'.
>
> One of my earliest memories is that of Dendy (Miss Mary Dendy),[1] a prim figure in a jet-trimmed bonnet, and cloak, pioneer in mental deficiency work. She was indeed runner up for the place of first woman Commissioner in Lunacy when Dame Ellen Pinsent was appointed.[2] Dendy founded the Sandlebridge Homes for the feeble-minded in Cheshire, and we had a curious rake-off from this venture. She bought the children large spotted handkerchiefs wholesale, and included our needs in this purchase. Except for parties the five of us always had 'feeble-minded' hankies as we called them. My sister and I had tea with her occasionally, (gentlemen's relish and meringues) and were allowed to use her typewriter for our stories, as long as they had moral endings. Had we but known it, I am sure now that she was studying us as a control group for her mental health activities.

[1] Mary Dendy (1859–1933) was the secretary (and later president) of the Mary Dendy Society, which was involved in agitation for the reform of provision for the 'mentally subnormal'.

[2] Ellen Frances Pinsent, née Parker (1866–1949), was a British mental health worker.

Then there were the 'Manchester Guardian families', the Scotts and the Montagues. Mr C. P. Scott was a legend to me, but later, as a bride, my husband and I had dinner with him in Fallowfield, and had to return his hospitality.[3] The prospect of this alarmed me and I rang up his doctor for advice. 'Get on to Lloyd George with the soup,' he said. I did, and we hardly left the subject till the famous editor left for his office at 9.30 in a hansom cab. His grandchildren were all at school with us, and the Montagues, though young devils, were a fascinating crowd. They were fond of pouring small bags of sand from behind their wall on people they did not like; and selling their grandfather's soft fruit in grubby handfuls at his gate. C. E. Montague, their father, was on the literary side of The Guardian.[4] With his rugged red face and bristling white hair, he was athletic, and always won the fathers' race at our school, though as he was the official starter, he had the advantage of gym shoes. In the first war, he served in the Sportsman's Battalion, which in fact was raised by my Aunt Emma.[5] Emma Cunliffe-Owen had decided that in the hunting, shooting and fishing, but over-age male, there was an untapped source of recruitment for the army. She put this personally to Lord Kitchener at the war office.[6] Everything would be done by Aunt Emma if she could have a suite at the Hotel Cecil. Kitchener finally said that she could have what she wanted as long as he need never see her again. C. E Montague reported that my Aunt weighed and measured her recruits herself, and equipped them economically and thoroughly. He said the toothbrushes especially were the best in the army. When the battalion was finally disbanded, my aunt took the salute from a bath-chair, in Hyde Park, the drum on her lap, her decoration round her neck, wearing ospreys and an ermine tippet. Had Mr Montague not told us this, even knowing my Aunt Emma, I should hardly have believed it.

The University personalities were of course generally only Manchester's by adoption, but many stayed for years. Any south-going tram would have a selection. Mrs Perkins, for instance, wife of the Professor of

[3] Charles Prestwich Scott (1846–1932) was a British journalist, publisher and politician. He was editor of *The Manchester Guardian* (now *The Guardian*) from 1872 until 1929 and its owner from 1907 onwards.

[4] Charles Edward Montague (1867–1928) was an English journalist, known also as a writer of novels and essays.

[5] Emma Paulina Cunliffe-Owen (b. 1863).

[6] Horatio Herbert Kitchener (1850–1916), the first Earl Kitchener, was a senior British Army officer and colonial administrator who won fame for his imperial campaigns and later played a central role in the early part of the First World War.

Chemistry, would sit, supporting on her tray-like bosom a spray of carnations weirdly coloured by her husband's latest discovery in aniline dye.[7] Professor Herford might be there, having corrected exam papers placed on top of the pillar box at his stop.[8] Professor Alexander always bicycled.[9] A curious figure, a cross between Jehovah and Father Christmas, with his great beard, and twinkling eyes, he generally dismounted and walked with us to school, he was very fond of children. Rutherford was a bit frightening, he had such a loud, gruff voice, and was so large.[10] I remember going to a fancy dress party at his house in Withington. We were playing hide and seek, but told that there was one room into which we must not go. I suppose I forgot which it was, as I crept into an empty room and was just going behind a curtain, when Rutherford loomed up from his desk, and shouted 'Who are you?' 'A violet,' I said, and added quickly, 'A modest violet.' 'Ho! Ho! Ho!' he roared, 'you can hide here with me.' Alas! None of course dared look for me there, and I nearly missed tea.

One of the most colourful of all was Miss Anna Phillips of Prestwick Park, a loud-spoken cheerful woman in elastic-sided boots and tweeds. Soon after I was married, I had to second a vote of thanks in the Town Hall, and simple though this should have been, I was very bad. I had never seen Miss Phillips before, but at the end she bustled up and shouted, 'Take the advice of an old woman, my dear, and never speak in public. It's clearly painful to you and much more painful for everyone else.' I went home in tears, and my father-in-law urged me to speak again next day, but there was no opportunity. Miss Phillips was really the soul of kindness. She had lunch parties for new arrivals to Manchester, and especially young clergy. On one such occasion, I remember her nephew G. M. Trevelyan was staying with her just after he had been awarded the Order of Merit.[11] 'George,' said his

[7] William Henry Perkin, Jr (1860–1929) was an English organic chemist who was primarily known for his groundbreaking research work on the degradation of naturally occurring organic compounds. He was the son of Sir William Henry Perkin (1838–1907), who founded the aniline dye industry.

[8] Charles Harold Herford (1853–1931) was an English literary scholar and critic.

[9] Samuel Alexander (1859–1938) was an Australian-born British philosopher. He was the first Jewish Fellow of an Oxbridge college.

[10] Ernest Rutherford (1871–1937), born in New Zealand, has been called the father of nuclear physics and was awarded the Nobel Prize in Chemistry in 1908, primarily for his research into radioactivity. He also proved that atoms had most of their mass concentrated within a very small nucleus and that alpha particles were helium atoms.

[11] George Macaulay Trevelyan (1876–1962) was a British historian. He was Master of Trinity College, Cambridge, from 1940 to 1951.

> aunt, after lunch, 'run up and fetch your nice medal. We should all like to see it.' The modest Trevelyan demurred, but he was no match for his aunt. He returned and gave it to her. She clapped her hands, 'No, no, George, put it on and walk round slowly.' I shall never forget that Gandhi-like figure, in acute embarrassment, walking round in a circle.
>
> These are pictures of just a few of the great figures. My cousin, Katherine Chorley, wrote an admirable book, 'Manchester made them'. If I ever wrote down my memories, I would call the book 'They made Manchester'.

The decision was taken to send my mother to boarding school. St Leonard's School in St Andrews was a highly traditional establishment. Mother was there during the First World War, which was a tragic time. Girls lost fathers and brothers. Mother was a rebel. When she was a fag, which entailed running errands and doing useful small jobs for a senior girl, she was so bad at it that she was made to continue for a second year. She played atrocious cricket—an obligatory exercise—and actually managed to bowl backwards over her shoulder.

Tragedy erupted into her life when her beloved brother was killed near Ypres in June 1915. His letters to her from the front survive. In these letters to his sister, he tries to reassure her and writes about life on the front as if it were some kind of adventure. He stresses anything positive he can invent to allay anxieties.

> 19 April 1915
> I have just finished my second four days in the trenches and am still safe and sound. I move off in about an hour with dugouts. You must understand that all our movements here are nocturnal. It is never safe to go about behind the line in daylight as snipers are always on the lookout!
>
> 1st June 1915
> Meanwhile we are very comfortable. At one end of the trench is a farm not inhabited now by anything but rats. It is surrounded by a moat of black inky water which, sad to say, is full of weeds. We found an old sort of make-shift punt on it and have made various perilous journeys. I have also shot at the rats with my revolver but with very little success.
>
> I must close now with much love from Eric

He went missing a couple of days later. His body was never found. Eric's great friends were Bob Bragg and Cecil Hopkinson.[12, 13] When Cecil died of his wounds, back in England, the last of the three to go, my Hopkinson grandfather said sadly, 'Tonight Cecil will be with Bob and Eric in Paradise.'

Mother studied history at Newnham College in Cambridge, where I followed her many years later. Cambridge after the First World War was full of gaiety, parties, tea dances, trips to point-to-points and, of course, balls. There were so few female undergraduates, with only Newnham and Girton to supply them, that the girls were much in demand. The men were, of course, older and more mature, having returned from war, and such qualities added to their attraction. It is surprising that my mother managed to do any work at all; but my grandfather arranged coaching for her and she came out with a respectable second-class degree. Incidentally, she and her great friend Cecily Carter scandalised their fellow undergraduates by wearing low-cut black satin nightdresses.

Mum had always told us that she had had thirteen proposals of marriage before she took her degree. My father always countered that these had constituted quantity, not quality. One proposal from a Wills cousin took place in a summerhouse. My aunt Enid sneaked up and lay on the roof taking notes, which she entitled 'Through the eyes of a neutral'.

Mother was having too good a time to tie herself down. Even my father was rejected first time round and told to go on his glorious way alone. It was two years later that they finally got engaged.

Mother and our Education

There was never a teaching element in my mother's involvement in our education. She never helped with history projects or with our homework in other subjects. What she did give us was a command of

[12] Robert Charles Bragg, born in 1892, sadly was killed in 1915 at Gallipoli during the Great War.

[13] Rudolf Cecil Hopkinson (1891–1917) became a close friend of WLB. He died from wounds during the First World War. A full account of his life can be downloaded from the 1918 book *Rudolph Cecil Hopkinson, Memoirs and Letters* (available at <https://archive.org/details/memoirletters00hopkiala>), in which, on pages 21–24, WLB has written about him.

language, written and spoken, and a love of words. I remember her reading me *The Mill on the Floss* in the garden one hot summer. At one level, it was totally unsuitable for a seven-year-old child but I have remembered the content to this day. She brought things alive.

I always felt it was weird that my mother made all the decisions on our education in a rather haphazard way, when it was my dad who was the born teacher. My mother believed in education for girls but marriage was the ultimate aim for us. She called the shots on how and where we were educated. I never quite forgave her for sending me protesting to boarding school when she wanted to go to the USA for a semester with my father.

In our gap year before university, Margaret and I were both, in our turn, presented to the Queen and went to some of the parties of the debutante season. My father was more interested in the architecture of Buckingham Palace, and the question of whether the pillars were made of fake or genuine marble.

When my mother visited me at Newnham, she never asked about my work but quizzed me on my social life. My supervisor had been a contemporary of my mother and invited us both to tea. 'You are working Patience too hard,' said Mum. 'She needs to get out and enjoy herself more.'

'But you want her to do better than you did,' was the immediate sharp response.

Incidentally, in contrast to Mum, my father had read most of the French works on my list but dismissed Racine and Corneille as 'tripe', and much of the work of modern French writers as 'sordid'. Dad told me that, sometime after his marriage, he had been extremely upset and had fallen out badly with his parents on the issue of whether his sister, Gwendy, should go to Cambridge. He was violently in favour—'Look at Alice,' he said. His parents looked at my mother, who had quite a reputation, and shook their heads.

My Mother in Public

Mum was an accomplished amateur actress on stage from the days when she played the lead role in *Pinafore* at St Leonard's School. She acted out her many roles in real life too. She spoke well in public, with a delightfully dry sense of humour.

In the First World War my mother and her younger sister Enid entertained the wounded soldiers in the local hospital in Manchester.

Both had good voices and my mother was an accomplished pianist. She used to play songs like 'There's a long, long road a-winding' and 'Keep the home fires burning'. Her favourite was the music hall hit 'Which switch is the switch miss for Ipswich'. She still sang this for her grandchildren at top speed and word perfect in her eighties. Alas, Dad was tone deaf and appreciated none of this. He only recognised the national anthem when everyone stood up.

Once Mum came to my school, the Perse Girls in Cambridge, and gave out the prizes. My father fell into one of his rare rages when I pointed out after we returned that her petticoat had been showing one inch below her skirt.

As Mayor of Cambridge, she reviewed the Cambridge Regiment on their return after the war. In her heavy robes, standing on the balcony of the town hall above the teaming town square, she whispered urgently to her army minder, 'I'm going to faint.' Without turning his head he whispered back, 'Keep wiggling your toes.' Incidentally, my father did not wish to play the part of Lady Mayoress, so my sister and I shared it between us when appropriate. I was twelve, and the gold chain banged against my knees as I opened the Midsummer Fair.

When I went to observe her as Chairman of the Magistrates Court in Cambridge, I saw the clerk unobtrusively creep up to the table and tactfully reverse the vital map which explained the cause of the accident. Mum could not read maps. She could and did frequently get lost in her car during the war, when all the signposts had been removed.

Mother's Hobbies and Interests

Her great strength was her social confidence and personality. She was the life and soul of parties, people fell for her and she made deep and enduring friendships.

She did not share in my father's hobbies except for walking, travelling and perhaps tennis. She found birdwatching chilly and boring and she could not distinguish between a whimbrel and a curlew. Ostensibly, she loved wild flowers but could not remember their names. This applied to the butterflies and shells which my father collected and preserved. WLB loved to paint on holiday, while Mum lay on the beach and read *Vogue*. When uncommitted elsewhere, I was taken along to keep her company. I would much rather have painted with

Dad. But when my father died, a curious thing happened. My mother took herself off to painting classes. Instead of the soft and gentle watercolours my father loved, she produced portraits, sometimes copied, in the style of Matisse, a painter my father particularly disliked.

WLB's love of sailing was not shared by my mother. She used to strap a hot-water bottle to her stomach if it was the least bit chilly. Oddly enough, she never worked a rope or took the tiller, even though her childhood holidays, often spent in the Lake District, had been full of adventurous climbing and sailing opportunities.

Both parents loved the garden, but mother only tended a narrow strip outside the drawing-room window, where she grew *Iris stylosa* and other favourites. She did not plant, weed, dig or sow but she appreciated the results of my father's labour and loved visiting other people's gardens.

Dad was very keen on photography. However, mother never took photos or carried a camera. She herself was very photogenic and loved to pose.

Figure 1 Painting by Lady Bragg after the death of WLB. (Courtesy: Patience Thomson)

So what were my mother's hobbies? She read a lot and wrote a lot: informal letters, and articles for magazines and newspapers. She produced long letters, she collected friends and she knitted. While fire-watching in the war, she hand-stitched me exquisite doll's clothes from scraps of silk. She kept diaries. She had a flair for decorating and designing the interior of her various homes, with the help of her great friend Joan Worthington,[14] and she loved clothes. However, because she did so much voluntary work and entertained for my father, she had little spare time.

Opinions and Values

Alice Bragg's broadcast for 8 February 1961:

> What I feel strongly
>
> What a chance you are giving me, a chance to let a stream of bees out of my bonnet, ride hobby horses out of my stable, yes, and probably let cats out of my personal bag. Strong feelings can be roused against things and for things—I think I will let a few of the anti-bees out first, and in any old order—I worry that we can't prevent so many women doing too much and getting overtired, especially young women with small children, and we let them get lonely and house-bound, too.
>
> I am against so many people taking so many pills for anything and everything which might in time change us out of all recognition. I'm roused by people who drink and then drive vehicles, by over dramatized society divorces and the sex life of film stars recorded in newspapers. I get fierce about things that won't work properly in the home unless you remember to put your foot against something and one finger on another thing and untie a bit of string first. For me, at least, they must be fool-proof. I dislike intensely the expressions 'couldn't care less', 'can't be bothered', 'and 'it'll have to do', any kind of slap-dashery in fact. I get hot about people who insist on forcing food down small children against their will. The average child is much too sensible to starve.
>
> Now you'll be getting a rough idea of the inside of my bonnet with these few examples. But one must have strong feelings the other way, of pleasure and satisfaction. Well here come some of mine—of course I have them about art, travel, music and books and all that, but my odd

[14] Probably Sophie Joan Worthington, née Banham (b. 1905); English draughtswoman and architect.

ones are more amusing for you. I've got very warm feelings for taxi-drivers. When I push my grandchildren across Piccadilly (and by the way those curbs are far too high for prams) it's they who always stop for me, under any conditions, and bless the ones who shouted 'Stick to it, Grandma.' I'll throw in lorry drivers who unfailingly give the right signals when I'm driving. I am thrilled when I've made something. Oh! The sight of a perfect cake when I open the oven door (that happens infrequently alas!). I love very old possessions, for instance, my pre-war bicycle. When it was stolen, borrowed I mean, for a day, on its return I literally took it to my bedroom with me. I get a tremendous kick, if, and when, my married children say, as to a trusty dog, 'Good old Mother, what should we do without you'. I love hearing the younger generation talking, especially if it's about ours. Altogether I've a strong feeling that there's a lot of fun about in life.

Now, please, look at my Queen Bee, and here I get on to my big hobby horse. That's family life in all its aspects, and how to preserve it.

My mother goes on to talk about the role of the Marriage Guidance Council (she was their president) and ends with the following conclusive paragraph:

And so to sum up—what I feel deeply is that we should all matter more to each other, find out how to love more, if you like. In my mind I see that young man in the dock, convicted of larceny, that lonely deserted wife, that rejected adolescent girl, that fumbly, neglected old man, and I want conditions changed so that we do not hear that sad cry from them—nobody really cares.

Where Christian belief was concerned, my mother was a devoted member of the Anglican High Church community and deeply religious. She took us to church every Sunday, where the incense made us feel sick. My father was a 'blue sky worshipper', which meant that he stayed at home and worked in the garden on Sundays. However, he sometimes went to the early morning communion with my mother. Interestingly, when I asked him about the afterlife, he told me in a very relaxed voice that the only certainty he had was that any arrangements made by God would prove to be satisfactory. He wrote the following to my mother when he was in Canada:

June 2nd 1941
I went to early morning church which I always feel is a way of communing with you because it brings back so many memories—going with you

> in Didsbury and on those lovely sunny mornings in Alderley Edge, and at Kingston and bicycling along to St Benets in Cambridge and countless holidays before breakfast walks to unknown little churches. It is something I always associate particularly with you.

1.4 Meet my Father

My Father's Roots

My father's forebears were very different from my mother's. Some were farmers and sailors in the north of England. Apart from his father, Sir William Bragg, the most famous was the entrepreneur and explorer, Sir Charles Todd, who set up the Overland Telegraph link between Adelaide and Darwin.

My father spent a lot of time with Grandfather Todd at the Adelaide Observatory. He was not only Postmaster General but Astronomer Royal for South Australia. For many years, we had a large black umbrella in our hall. When opened, it revealed all the stars in the night sky of the Southern Hemisphere.

I do not think it is fanciful to suggest that there was a family gene that led to thinking in a visuospatial manner, in three dimensions, in fact, and which encouraged lateral thinking.

My Mother's View

I have my mother's preparatory notes on how she wished to present WLB in her autobiography; the notes are scrappy but worth recording and a good starting point.

> Great shyness, did not know ordinary things.
>
> Hated Senate and Committees unless he was chairman.
>
> Disconcerted people by not listening. Faraway look came over him. Sometimes had been listening, something said to him triggered off train of thought—his own—sometimes not interested . . .
>
> Often missed things because he had not been attending.
>
> Used often to drop off at the Whitworth [PT: Art Gallery] on the way home.
>
> Always tore up first letter. [PT: This would have been an angry response to some 'trigger'. He would write a second more moderate one after.]
>
> Not a club man. Did not like Conferences, e.g. the Brit. Ass [PT: British Association]

Always had blind dog, Nuttall,[15] Scott Dickson,[16] James[17] [PT: someone to lead him and guide him]

Blessed secretaries Mair Jones, Brenda Smith

Always someone like Bell—disloyal, critical [PT: who was a target for his annoyance]

Our trips to the RI [PT: Royal Institution]

Mind on agenda. Things worked v. slowly—he 'had not put his case well.' Certain people very fond of him and helped him through. . .

'Mortified' a great word. [PT: He used it a lot about his own feelings]

Great friend GPT [PT: George Paget Thomson[18]]

Both overshadowed by famous fathers in same line.

Anxious at Trinity—so many College things he did not know—often queried their importance.

Alarmed by Blackett and Fowler,[19, 20] tho' both fond of him—always pleased to see Bernal.[21]

Passion for books—reread them, really knew them.

Happiness in old age. Loved parties.

A natural worrier (Brushed off on me).

My father changed character during his lifetime. His childhood and adolescence were not always easy. He faced the trauma of the First

[15] John Mitchell Nuttall (1890–1958); famous for the Geiger–Nuttall rule showing that short-lived isotopes emit more energetic alpha particles than long-lived ones.

[16] Ernest Scott Dickson was Senior Lecturer at the University of Manchester. He died in 1944.

[17] Reginald William James (1891–1964) was a crystallographer and became Professor of Physics at Cape Town University.

[18] George Paget Thomson (1892–1975), son of J. J. Thomson, shared the 1937 Nobel Prize in Physics with Clinton Davisson for the discovery of electron diffraction. He remained a close friend of WLB.

[19] Patrick Maynard Stuart Blackett (1897–1974) was a British physicist known for work on cloud chambers, cosmic rays and palaeomagnetism. He was awarded the 1948 Nobel Prize in Physics.

[20] Ralph Howard Fowler (1889–1944) was known for his work on statistical physics and for the study of aerodynamics of aircraft spins during the First World War.

[21] John Desmond Bernal (1901–1971), known as 'Sage', had been a student of WHB and built up a formidable team of crystallographers while at Cambridge in the 1930s. He was a controversial figure because of his political views and sympathy for the Soviet Union. One of his students, Dorothy Hodgkin, née Crowfoot (1910–1994), went on to win the Nobel Prize in Chemistry in 1964 for the study of the structures of penicillin and vitamin B_{12}.

World War. His great friend Cecil Hopkinson died of his wounds, and his brother Bob was killed at Gallipoli. Afterwards he had the problem of being such a young Professor at Manchester University. He had little experience of lecturing, and none of running a department. There was the awkward element of rivalry between father and son. At one stage he had a breakdown while in Manchester and had to take time off. He was often overstretched and unprepared. But all this happened before I was born. I appeared on the scene in 1935 and by the time I can remember him, much of this early angst and underconfidence had waned. He had a firm grip on his priorities in life.

The self-effacement that had characterised him as a young man vanished with time and was replaced by his pleasure at the recognition he was receiving. At Cambridge and the Royal Institution he had very specific jobs to do and was doing them well. After the war ended he had time to travel, to pursue his hobbies and to enjoy his family. Alice was almost always by his side. These were golden years and

Figure 2 WLB with Patience in 1936. (Courtesy: Patience Thomson)

even the life-threatening cancer, from which he only just recovered, did not dent his enthusiasm or zest for life. The father I knew was a man who had come to terms with the world around him, basically a happy man, fulfilled. He would read us Browning's poem 'How good is man's life, the mere living'.

I have given my mother priority in this introduction, because people know far less about her. She has had no biography, no lengthy obituaries and no lectures in her name. However there are a few points I would like to make about my father.

Physically a very fit man, Dad had excelled at school in the 100 yards in athletics and still played a threatening role at net in tennis. He loved skiing and walking. I used to ride on his shoulders on family picnic expeditions. He enjoyed physical activities and had a very healthy appetite.

Dad was not part of the Establishment. He had not been to public or grammar school and so had no English friends from school days and did not automatically fall into a category; he was a colonial from Australia. He had two invitations to be Master of a College, at Cambridge and at Oxford, respectively, and accepted neither invitation. He delegated administration whenever he could. He was not clubbable and did not particularly relish dinners at the High Table in Trinity. My mother would pointedly put out the dyspepsia tablets on a tray in the hall and go to bed.

Modest and self-deprecating, my father was quick to acknowledge mistakes and quick to apologise, but he could still get very angry, so much so that he would go red in the face and start to stammer. His family, and especially his father, seldom expressed their personal inner feelings, either to each other within the family, or to the outside world. My mother thawed WLB out emotionally.

My father adored children. He was very unusual for his time and generation in that he loved them even when their heads were still wobbly. He could and did change nappies.

A walk with him on summer Sundays often involved the identification of birds, butterflies and wild flowers. Where I could not follow was in his vast knowledge about insects, as most of them seemed to me either dull or threatening. He would point out a praying mantis in Corfu or explain the singular behaviour of the cuckoo bee.

Mutual Attraction

So what did my parents have in common? First, a sense of humour. Then, a sense of proportion in identifying what was important to them in life. Complete loyalty and fidelity. Then, there was their driving energy and enthusiasm. They grasped opportunities—they had a sense of commitment. With their mutual admiration, support and encouragement they increased each other's confidence. But they still gave each other space to go off and do their own thing. Finally they were deeply, passionately in love.

They needed each other. By the end of his life, my father relished being lionised at parties. Mother helped him to enjoy the social scene by letting him be a guest at his own parties and briefing him with lists of those attending and their interests. She attended his lectures, even though she could not understand the scientific principles involved, and encouraged him to enjoy public speaking and acclaim. In turn, he would say how happy he was to move in the kind of prestigious circles, both national and international, that my mother so loved.

He orchestrated a popular TV series where he demonstrated dramatic experiments. People accosted him in shops and in the street to tell him how much they were appreciating his programmes. By the sixties he had become a 'grand old man of science' and loved it. In return, he gave my mother a solid and serious grip on life and encouraged her to take on major public roles.

WLB backed my mother up wholeheartedly in all her enterprises. They were, in a positive sense, a mutual admiration society. She distracted my father and calmed him down when he was angry. Dad cherished my mother.

Family

They had four children: Stephen (b. 1923), David (b. 1926), Margaret (b. 1931) and me (b. 1935). We were widely spaced and so only shared a limited amount of time in our childhood. They shared the pleasure of having children around them, and both were excellent grandparents.

My brother Stephen became an engineer, chief scientist of Rolls Royce and later Vice-Chancellor of Brunel University.

I need to say more about Dave because his problems affected my parents so acutely. David was a sensitive and gentle soul. He was for

Figure 3 Patience in 1951, finishing a thriller. (Sketch by WLB; Courtesy: Patience Thomson)

many years a gardener for Caius College, Cambridge. He was musical and artistic, but alas also schizophrenic. He spent considerable periods in hospital. Before his death, he gave me, for safe keeping, the despairing notes he made about his shock treatment in the fifties, and his feelings of inferiority, frustration and uncertainty. His treatment was scandalous. He had unsuitable boyfriends. One took his savings, stole his clothes and embarked on a tour of the country borrowing money from our parents' friends. This caused my parents great distress. He was treated in the fifties at a time when consultants insisted that they must treat their patient in confidence and in isolation and that the family should not be informed or become involved in any way. It was a cruel concept which thoroughly upset my parents. My mother felt responsible for Dave's condition and used to agonise trying to work out what she had done wrong. My father tried to convince her that it was all in the chemistry of the brain. Mental health was ill understood. There would be angry scenes when my brother was late for lunch or left his new raincoat on the train. He was a constant worry to my parents, and all solutions seemed only to be temporary. Mother said that this misfortune was the worst experience of her life.

My sister Margaret went to Oxford, married into the Foreign Office and spent much of her time abroad. She was my teacher and mentor throughout my childhood, and I owe her a great debt of gratitude.

I myself went to Cambridge. Subsequently, I worked on the translation and publication of the Nazi party papers, which the allies had confiscated after the war. The Foreign Office had undertaken to edit and publish a selection. My father was very interested in this work and I would discuss it with him at dinner.

After the birth of my four children, I trained as a special-needs teacher, working in St Bartholomew's Hospital Psychology Department, in a young offenders' unit and in various schools. I was Principal of Fairley House School in London for eight years and subsequently helped to co-found a publishing house, Barrington Stoke, which produces books for individuals, juvenile and adult, with reading difficulties. Both my parents had instilled in me a love and understanding of language that stood me in good stead.

Learning from daughters—BY ALICE BRAGG

June 8th 1966

'Nothing comes up to mother's gingerbread'—that's what husbands used to say to young wives. Times change; now mothers say: 'No one makes pâté like my daughters.' The young marrieds seem to me a complete women's magazine of tips for middle-aged mums. That is if one can take it. Either one preserves the 'I have forgotten more than you are ever likely to know attitude,' or one learns from married daughters.

They don't worry, as we used to do, at any rate not about the same things. This is not a matter of temperament; it's the difference in generation. They are quite happy to show the works. When people drop in, if they are in the kitchen they stay there, if dressing they continue to dress. If they aren't dressed it doesn't matter. I'm picking this up fast, and have been able to receive the postman without my shoes and stockings, and shout over the banisters to the plumber, in my vest. They are not too upset if they break, lose or burn things. We were trained to keep possessions (I've had my work basket all my life), with the view to passing them on to the children. But most of the things the children aren't going to want. They have not the room and they have different taste. They use what they have thoroughly, and lend freely. I am a bit slow in saying, as they do, 'have my flat, take my car, borrow my outfit for the party,' without adding 'You will be careful won't you?'

My generation were accustomed to order and method, so that it was not so easy to deal with the unexpected. It threw us out. But the daughters and their friends like it that way. If they have something better to do on Monday they don't wash; anyway of course they have washing machines and spin-driers, but they will wash at night, or any old time. If friends stay on, they feed them, any number, and when there's no bread (metaphorically speaking) they make it or do without. I still have to have the right tools and equipment, but I remember when one daughter's stove was temperamental she wrapped a presentation salmon in her trousseau nightdress and cooked it thus.

Out-of-date conventions

That brings me to the whole question of food. First lesson, why have a joint and apple pie on Sunday? Why cold supper on Sunday night? These are conventions all geared to having help, and as they have no help they've ditched them, and I'm going along with them here. The husbands can cook of course, and mine is learning, though he is best and safest with a frying pan. I notice they don't bother much about puddings, except for parties, that's all right for me, it is an insurance against middle-aged spread. They fancy French, Hungarian, Chinese dishes. British food has somewhat low priority. Well, I can just manage to make pizza, piperade and the like, but really only to be 'with it.'

As for the children, it is really too late to learn here for one's own benefit. One can only take an academic interest, but since at any moment a granny may have to function it is as well to watch the form. Basically, I note that children have not to be forced to eat what they don't like, or indeed to eat at all. Incidentally they exhibit surprising tastes, wanting pickles, chutney, pepper, and strong cheese. The same latitude goes for bed-time. If they are not tired, they stay up, watch something on television, or do a little gentle milling round till ready for sleep. It is a shock to realise that while still quite young they can cook breakfast for themselves, answer the telephone, and fly all over the world, carefully labelled. Even when these children are really small the daughters take them anywhere and everywhere, with little bags crammed with plastic feeders and pants, and plastic sheets. I thank God that they do, for I shall never be good at coping with wet mattresses, however hard I tell myself it's only nature.

Thus, I try and learn not to worry about appearances, to do a whole lot of things at once, to seize opportunities that may not be offered again, and never (hardly ever) say no to any new experience. Of course I shall not catch up, and sometimes my daughters' efficiency depresses me. But then I play my trump card. I tell myself that it could be just possible that they have inherited some of their qualities from me.

Down the other end of West Road lived old Lady Thomson and her daughter, who were sort of extended family. She was the widow of J. J. Thomson and mother of Dad's greatest friend, Sir George Thomson.[22] My father kept an eye on her and did odd jobs for her (how did he find the time?). In particular, he mowed her lawn. There was no petrol for the motor mower, so it was pulled by a pony wearing leather booties, so as not to make holes in the lawn.

Lady Thomson's eleven-year-old grandson, David, came to live with her during the war after his mother had died. He used to come to tea with us. Fifteen years later, I married him. When I announced our engagement to my parents, my father picked up the telephone to ring his friend George Thomson who had already received the news.

'Herumph, Willie'

'Herumph, George'

'Are you going to the Royal Society on Wednesday?'

And so it was settled.

1.5 Family Homes

Before my memories start, my parents had lived in Didsbury, Manchester, in Alderley Edge, Cheshire, and at Bushy House, the NPL (National Physical Laboratory) in Teddington outside London. I was three when we moved to Cambridge, and 3 West Road is the home of my childhood. It is there that I shall start this part of my Bragg story.

Life at 3 West Road, Cambridge

It was a great place to bring up a family. It was the first home that I can remember. The previous owner had been shot dead by one of his students, and his children left us a scary treasure hunt in the extensive cellars to welcome us in.

[22] John Joseph Thomson (1856–1940), discoverer of the electron, was awarded the Nobel Prize in Physics in 1906 for his research on the conductivity of electricity in gases.

We moved there in 1938. I lived there for fifteen years. It was extraordinary that on the salary of a professor of physics, even though topped up by a legacy to my mother from a childless aunt, one could live in a house of this size and character and employ numerous staff.

We had a nanny, a cook, a parlour maid and a housekeeper, who had her own private room behind the green baize door. There was also a sad-looking lady, with chilblained red hands, who did the washing and scrubbed the floors. Mrs B was given the old, cold breakfast toast and wolfed it down. There was a series of somewhat eccentric gardeners, like Mr Fishpool, who had changed his name by deed poll. They hid the surplus vegetables in their bike baskets under a sack and sold them off on the side. We never told our father.

Mr Fishpool was a real character. He came from a family of seven boys and seven girls, and 'only one died of drink'. He answered to my father exclusively. When mother asked if he had any cauliflowers for our lunch, he simply replied, 'Wouldn't you like to know, Lady Bragg.' He kept a pig in the backyard of his council house.

There was a green baize door, which divided the family living quarters from the kitchen and the maids' territory. The nanny, who lived on the family side of the door, was highly intelligent and read *The Times*. She was not welcome downstairs and could not follow us when we went to the kitchen. There we read *The News of the World* and *Gone with the Wind*. We even played cards for money (pennies), in complete defiance of nursery rules.

My mother had scheduled meetings with the cook. My father had long happy sessions with the gardener and was very hands-on himself. In fact, in his rough old clothes and cloth cap, he was sometimes mistaken for the gardener.

He had a large upstairs study with meticulously labelled cabinets, some full of tape and glue and carpentry tools, and some with papers. He used to retire to the window when he needed a break. There he would shoot the starlings, which were pecking at the ripe fruit in the pear tree, with his lethal homemade catapult.

At 3 West Road there was a carriage house, stables, a hot greenhouse, a cold greenhouse and a huge brick-walled manure heap. One lunch time the gardener rushed into the dining room to report to my father that there was a strange butterfly feeding on some rotten pears. Against all the mealtime rules we jumped up and rushed off to check it out. 'By Jove, it's a Camberwell Beauty,' shouted my father in great delight. It was duly caught to join the glass-topped butterfly collection on the landing.

My father was much involved in the two-and-a-half acre garden, with its herbaceous borders, croquet lawn, apple orchard and shrubbery. Local boys who took the liberty of looting apples had their ears boxed by WLB personally. One memory from childhood is lying in bed at 7 a.m. in the morning listening to the 'swish, swish' of my father's scythe in the long grass. One of his favourite scenes in all literature was the one in *Anna Karenina* when Levin joins the peasants to scythe the fields.

We lived in the nursery but ate breakfast and lunch in the dining room at a separate table. At breakfast we made walls of the cereal packets so that no one could see what we were doing.

My parents gave grand dinner parties—even during the war—with polished silver and magnificent floral arrangements constructed by my mother from flowers from the garden—blowsy lilac, sweet-smelling phlox and lupins.

I can remember a party where everybody was set a mathematical quiz called 'Pigley's Farm'. My sister reminded me of one of the questions, 'What was the square root of Mrs Pigley's age when her third child was born?' One worthy academic was sitting on the stairs, and my mother was offering him a second helping. Rudely, he shouted at her, 'Get away, damn you, I need to concentrate.' The standard of intellectual conversation was daunting, and the guests often distinguished. There were many famous scientific names recorded in the visitors book, including Ernest Rutherford (pre-West Road), Kathleen Lonsdale,[23] Charles Darwin,[24] John Bernal and Linus Pauling,[25] among others.

[23] Kathleen Lonsdale, née Yardley (1903–1971); British crystallographer who was a student of WHB. She was known mainly for her work on the structure of hexamethyl benzene and on disorder in crystals. Kathleen was one of the two first female Fellows of the Royal Society (the other being Marjorie Stephenson, a biochemist). She was head of the Crystallography Laboratory at University College London.

[24] Charles Galton Darwin (1887–1962); grandson of Charles Darwin of *Origin of Species* fame. He made many contributions to the theory of X-ray diffraction by crystals.

[25] Linus Carl Pauling (1901–1994) was an American scientist, said to have been one of the greatest chemists of the twentieth century. He was awarded the Nobel Prize in Chemistry in 1954, and the Nobel Peace Prize in 1962. During the quest to solve the structure of DNA, he came very close to the correct solution but was beaten to the final result by Crick and Watson working in WLB's department in Cambridge.

During the Second World War, the large house mopped up a series of visitors. They were not all scientists. There were actresses performing at the Arts Theatre. There were evacuee children with whom we had reluctantly to share our toys. The cook grumbled, but my mother welcomed them all.

It is breakfast time in the Bragg family some time in 1946. Dad is invisible behind *The Times*. Two young students have been billeted upon us because of pressure on college rooms. My father lowers his newspaper. 'And who are you?' he says to the students.

'We're your lodgers,' says one shyly. 'We've been here a month.'

'You can't have been', says my father, 'or I would have noticed you.'

And he went back to his newspaper.

My father had a workshop in the old hay loft. When he came home in the evening, he would take me up there. He would make me little wooden speedboats for the bath, powered by twisted elastic bands. Doll's-house furniture was beautifully finished and painted with two coats of shiny paint. Woe betide me if I did not sandpaper the tiny dolls beds between coats.

When I married, he gave me a wooden box which he had made and decorated. It had my new initials carved on each of the professional-looking tools inside. 'Let your husband use your toothbrush if you must, but never let him touch your tools,' he counselled me. I still use them.

Dad believed in buying us the best of everything: walking boots, gum boots and particularly bicycles, which had to be cleaned and maintained and carefully put away out of the rain. He invented all sorts of devices. There was an ingenious long-handled apple picker that worked a treat. He tried to patent the wire supports he designed for the taller flowers in the borders. He saved up all sorts of scraps of materials—just old bits and pieces. Fir cones were transformed into wise owls for the Christmas tree. I still have his boxes of raw construction materials. In the garden, he built us a log cabin with a stable behind it for our hobby horses. These were made from stuffed socks (of appropriate colours) attached to bamboo poles. He helped us to organise hobby-horse gymkhanas.

In the evenings, when I was older, there would be poetry readings and family 'kitchen' bridge. I remember my absent-minded father losing the ace of trumps down his trouser turn-ups. He spent a great deal of time with us. To illustrate one historical point, he made us a crossbow and a longbow so that we could compare their relative performances (there was no 'health and safety' back then).

For my school biology prep, he constructed me a beautiful spider's web, complete with a hairy spider made with blackened pipe cleaners. He pinned it to the cork mat in the children's bathroom. My mother was not pleased when she discovered that it had been taken to school.

When I was eleven, my father thought I was lonely. Much the youngest of the family, I was left behind at 3 West Road when my siblings had all left the nest. When my mother went off for a fortnight's holiday with a friend, my father and I bought a puppy. This was in contravention of a strict dictate that Mum would not allow a dog in the house. This mongrel terrier became an integral part of the West Road family, but my mother only relented at teatime, when he was allowed a saucer of milky tea. We also had endless rabbits, which we refused to eat even when meat was rationed in the war; chickens; ducks; and two cats who produced a large number of kittens, which were drowned by our gardener Fishpool in the water butt. We also had a strong-minded pony. We concluded that he had worked in a circus, because he spent quite a lot of time on his hind legs.

Dad cycled to work. I remember once his handle-bars dropped off in King's Parade. Another time he chained his bike to the leg of an undergraduate who was stooping down to do up his shoelaces, mistaking him for a handy lamp post in Senate House Passage. He could be very absent-minded when he was thinking of other things.

My mother delegated all household duties. When I was twelve or so, I came across the astonishing sight of my mother cleaning her bathroom. When she had finished, I went to use the basin to wash my hands and invoked a tirade, 'You can't do that, Patience. I've just cleaned it.'

Cooking was a blind spot for my mother, though she could instruct our cook to produce what was wanted. On a professor's salary, you didn't get anyone very good, and our cook, a jockey's daughter, produced rock cakes that lived up to their name. When Mum was sixty, and we were living in London, and staff were scarce, she had to learn. So she went to Cordon Bleu and had private lessons and was taught to produce ten dishes to perfection. She cooked these in rotation, keeping careful records of what she served to whom. On the strength of this arrangement, she achieved an amazingly high reputation.

Mum was not a natural mother. When our beloved nanny went on holiday, my mother found a suitable substitute. We locked her in the bathroom and went our feral ways. The poor substitute nanny was too scared that she would lose her job to tell my mother what had happened. We released her before Mum came home.

It was during the years at West Road that my mother developed her role in society independently from my father's fame. She ran the WVS (Women's Voluntary Service) in Cambridge, looking very smart in her dark-green uniform. She was of that generation of women who were well educated but simply did not take paid jobs after marriage. She was good at delegating and organisation, and this was recognised. She was asked to stand for the Town Council and was elected Mayor. She was only the third woman to hold this office in Cambridge. My father was immensely proud and wrote the following letter to her:

> November 1945
>
> My darling Alice
>
> My heart nearly burst with pride this morning.
>
> I could sense an admiration and affection for you in the room which was really wonderful. And after your speech several people came up to me and said how proud I must be, as indeed I was, and how much they liked what you said.
>
> It is a terrific job to take on, darling, but I think it would be a life opportunity missed if you had not done it. I am quite sure it will be a culminating success for you, I shall sometimes steal into the gallery just to hear and see you. I liked particularly your saying I would support you and by jove I mean to do it.
>
> My talk is just over, was in a theatre with about 3,000 people. . .[26]
>
> Your admiring and devoted husband
>
> WLB
>
> I would have put Mayor on the envelope but I don't know yet the proper form of address.

Later she would hold other prestigious positions. She was a member of Lord Denning's Royal Commission on Marriage and Divorce, Chairman of The National Marriage Guidance Council and Chairman of the Cambridge Magistrates.

[26] He then goes on to other matters.

My parents moved from West Road into a house in Madingley Road, Cambridge, shortly before the whole question of my father's going to the Royal Institution was raised. It was inconvenient, with half the ground floor let out as a flat, and cramped all our styles. It left very little impression on family life; we hardly got to know it.

Quietways, Waldringfield Suffolk

In the early fifties, shortly after their move to the Royal Institution, my parents acquired a large cottage near the Deben estuary. Built in the thirties, it had mock Tudor beams, pargetting on the plaster, and a very pretty garden. It was a perfect holiday retreat for them. My mother could easily drive over to Cambridge for her magistrate duties and it was an easy train ride from London. My family had a mobile home in the wood there.

It was the perfect setting for Dad's hobbies of gardening, sailing, sketching and birdwatching. He and I were often left alone together there, when my mother was performing her various duties. We lived on sausages and chocolate semolina pudding and only did the washing up when my mum came home. The light sandy soil was ideal for Dad to work as he grew older and physically less tough. Birdwatching on the mud flats of the Deben was highly rewarding. He built his own hide on the dyke. The Suffolk light and its wide horizons inspired many sketches. He took us out sailing in the *Tortoise*, a small wooden boat which 'taught us' how to sail. Grandchildren often came to stay.

He constructed a practice high jump for eight-year-old Ben; many years later, that same grandson would perform internationally in the Scottish athletics team. Dad was full of enthusiasm and encouragement.

At the end of one television series which he presented, Dad was told that one section would have to be reshot and asked to come to the studio in the same suit as he had worn throughout. Alas, it had gone to the local Waldringfield jumble sale. I was landed with the uncongenial task of discovering who had bought it and getting it back. I toured the village and at last spied the jacket hanging on a hook on the back door of a farm cottage. I pounced on it. 'And the trousers?' I asked. 'I didn't buy them, my husband didn't have no need of them,' the somewhat unhelpful housewife responded. So my father had to complete his TV performance while creeping around behind a table that obscured his lower half.

Once a year my parents would invite the Royal Institution staff down for a visit in the summer. I remember how my father devised a complicated treasure hunt all around the village. My mother provided magnificent refreshments. This event was an extension of my father's belief in communal coffee and tea breaks, first at the Cavendish, and then at the Royal Institution, where staff could meet informally and discuss their research. He often took me with him to participate. Now, we would call it bonding.

At Quietways, mother rather enjoyed acting out the helpless female, although she was naturally tough and resourceful. She and Dad were often alone together there, so there was no audience to impress. I didn't count. She would cry for help—'Willie, Willie, there's a spider in the bath' (or a wasp in the kitchen, or a mouse in the larder). My father would come pounding to the rescue. 'Willie, Willie, the toaster isn't working' (or the iron, or the kettle). 'Let me plug it in for you,' my father would chuckle. It was a game that they never tired of playing.

On a much more serious note, it was at Waldringfield, as an old man shortly before his death, that WLB talked for the first time about the First World War. One sunny afternoon, sitting on the garden bench of his beloved Waldringfield cottage, he opened up.

Yes, there had been horror and sickening sights. Yes, his only brother had been killed on a beach in Gallipoli. But subtly, those First World War years had been a relief from the stresses of life in England. He had a definite and worthwhile challenge, that of perfecting and implementing sound ranging. He had a team of loyal colleagues whom he liked and respected. Amidst the turmoil at the front, many of the minor irritants and problems of life in England faded into insignificance. Dad was insulated from his family, so that the rivalry with his father and the communal mourning over the death of his brother was happening somewhere else, remote and unreal.

WLB had a huge sense of the threatening world around him. Death was ever present. His response was to lose personal feelings in an intellectual task, that of locating the enemy guns, so they could be destroyed. His job was engrossing, time consuming and congenial and he would have been 100% focussed on it.

For him, the wartime years were not a nightmare, and this worried him in retrospect. At the end of our conversation, as I rose to make a mug of tea, he admitted that he had enjoyed his years at the front. The death of Bob would really only hit him hard when he returned to

England and keenly felt the empty gap in the family, and the fact that the happy, eager light had gone from his mother's eyes.

My mother could be really difficult when put out or overtired. She would resort to speaking in a whisper as if she had no strength left. My father and I referred to it as 'little voicing', which at least pinpointed the nature of the problem. We wrapped her up in her aunt's red carriage rug and put her to sleep on the sofa, returning thankfully to our various ploys. It was a help that it was two to one.

My parents went to South Africa when I was in my teens. To Margaret's and my great annoyance, Mum published an article entitled 'Difficult Daughters' in the Cape Times. Tongue in cheek, I wrote my first ever article in repartee and, to my astonishment, and with the connivance of my father, it was published.

Mothers—their upkeep and care

Every time I buy a pair of nylons I am amazed at the long list of instructions that are sold with them. I read them through most carefully. I avoid rubbing against chair legs and snagging them with rings or fingernails. I rinse them through every night with lukewarm water and the best soap flakes. Yet in spite of all my precautions they always seem to go unreasonably soon. Now with mothers it is quite the reverse. You can neglect them for days at a time, you can tease and bother them, rub them up the wrong way, shock and upset them and they show astonishing resilience. Yet it is marvellous how they will respond to a little extra care and attention, and I find the time and energy well spent if the result is a bright and cheerful mother.

If a mother looks tired and jaded there is nothing like breakfast in bed to freshen and restore her. Even a whole day in bed may sometimes be found useful, but this is rather a drastic cure and should only be resorted to occasionally when the situation is becoming acute. Dangerous symptoms to watch out for are nagging your father, tears when you leave the oven on all night and refusal to lend you any of her things because of what you did to them last time. Some mothers undergo treatment quite gladly; others have to be forced into it. However if you want a mother to stay in bed you must provide her with breakfast coffee that is really hot, take the telephone receiver off and organize meals that you are certain you are capable of cooking.

There is nothing that brings mothers bounding out of bed sooner than complaints from father that his food has not been agreeing with him. By the way though, keeping your mother in bed is the best way to deal with her when she is run down. If you want to keep her in good general repair see that she puts her feet up and has her forty winks regularly after lunch. You will find this prevents her showing signs of strain at about six o'clock.

A certain measure of criticism is a daughter's privilege and duty. There is, however, a right time for everything. It is an act of kindness to tell your mother her slip is showing before you go out anywhere, but very unwise to do so just as you are leaving the party. It is certainly helpful to remember that appreciation is also a form of criticism and tell your mother she looks nice whenever you really think so.

Do not tell her she looks perfectly marvellous in that pink feather hat because you know she herself is doubtful about it and you think her morale needs boosting. Much better to admit it looks awful, get her to see the funny side of it and go out together and buy another at the first opportunity. To tell your mother that something is too young for her because you think it would do nicely for you is definitely not playing fair.

A mother is there to help you to grow up and you are there to see she does not grow old. The warning sign is refusal to accept any innovations, having her new clothes modelled on the old ones and using the same old well-worn recipes that father liked so much when they were first married. To prevent this happening keep a constant stream of fashion magazines flowing into the house and leave them around carelessly open at styles for the over forties, give her some fabulous book on oriental cookery as a Christmas or birthday present and arrange jaunts together. If you have decided on a new hair style persuade her to have one too. Make a day of it, hairdresser in the afternoon, and a theatre in the evening. Treat her as your contemporary and equal. Some say that an evening out is only fun with a dashing escort. But I find it a welcome relief to have a companion I know so well that I can relax and enjoy the show, without bothering about whether I am making the right remarks, and if I am being sufficiently dazzling and witty.

A mother is an invaluable asset. Heredity means that she sympathises a good deal with your faults and failings. Her advice is free and help is always forthcoming in times of crisis. She has after all, had quite a bit of experience with your father and the disappointed ones who came before him, so should be able to guide you through your entangled emotional affairs if you are honest enough to tell her all. If possible avoid quarrelling with the son of one of her best friends; as though she will almost certainly be on your side and back you through thick and thin, she may not thank you for the temporary loss of her friend.

> One final reminder, instructions on 'How to get the best wear out of these stockings' always end on a reminder, if you ladder your nylons there is always another pair to be had. But you have only one mother so please handle with care!

The Royal Institution

At the time of my father's appointment, it was assumed that the family would live in the elegant Georgian flat above the shop. The drawing room, study and dining room were beautifully proportioned rooms and suitable for entertainment, as well as for family life. On the floor above it was a higgledy-piggledy arrangement of interconnecting rooms with space for all us grown-up children. We had table tennis on the roof. Every now and then, water leaked down from the third-floor labs into my mother's bedroom.

It was at this juncture that my mother's loyal support and skilled handling of people came to the fore. She threw herself wholeheartedly into the task of restoring the Royal Institution's status and reputation. She was a dedicated hostess, did her homework seriously and entertained a wide range of guests from both the world of science and that of the arts. She had style and she had flair. Mother understood not a word of science, talking about NDA (DNA) and unable to master any of the fascinating challenges in my father's research. It did not matter; she acted the fascinated listener, nodded intelligently and got away with it. My father would explain in careful, simple language the main theme of the Friday evening discourses that were part of her new world. For a brief few minutes my mother and I would understand what was being said. But then the images faded inexorably and we sank back into ignorance. My father was rueful about this but never angry or reproachful.

My mother loved entertaining and of course the guests were a very distinguished crowd. After the children's lectures, a select few would be invited to tea in our flat upstairs. For many of those young people, it was a memorable occasion which influenced their future career. We often had guests in the house with very famous names. A stock of black ties, white ties, socks, cufflinks, etc. was kept on hand; the guests almost always forgot something. We would be sent out to buy vitamin pills for Linus Pauling and knew from the maid that Bernal never brought pyjamas.

My father had always encouraged my mother to write.

24 April
When this bloody war is over you must write, I think you could do something in the Delafield style, which would be fascinating and I laughed till I was weak at the description of David studying the hens and even sitting in their house and then coming back to make all their noises.

It was during her time at the Royal Institution that my mother took to writing articles for newspapers such as *The Manchester Guardian* and *The Times*. I found a pile of them in an old white attaché case that had washed up in our attic. Her reaction to the move to London was expressed in an article written for *The Manchester Guardian* in 1955:

March 16th 1955

We moved to the heart of London a year ago from a country town, and our friends ask us how we stand up to it. Are we exhausted? Is it a strain on our nerves? The answer is that we find it different but peaceful. We are, in fact, relieved of certain strains in our present home. It is in the middle of London in a street off Piccadilly, where as far as I know no one actually lives except us.

Most households, unless buried in the country, have neighbours. We have always had friendly people all around, to whom I could run to borrow anything from an egg to a camp bed, or just for a bit of moral support and a bit of a chat. In our street there are mostly Court hairdressers and jewellers, or offices. At the weekend if we run out of anything we have had it, though I have my eye on the hotel opposite for the odd loaf in a crisis. But this is peaceful; the jewellers and hairdressers do not pop in; one's friends are scattered all over London, and telephone first to see if we are in.

When I went out at home I met friends all the time; it was convenient to catch me—did I know who would take a Swedish boy for a few months, would I jot down his name and address, would I remember the jumble sale and have something ready, would I think of a speaker for this and that? I had to keep alert, and I always seemed to go home with a list of jobs and my memory doing overtime. Now when I go out there is no one I know. I can dream, I can watch people and drink in my surroundings undisturbed.

I was told that when I left my front door in Mayfair I should have to be in black and pearls, but when I do not want to do this I don't. I am anonymous and if I emerge in golfing shoes and no hat, who cares? This is very restful.

At home I bicycled everywhere or I drove the car. I have my bicycle here, it's true, but I am allowed to ride it only on Saturday afternoons and Sundays, because of the traffic. I ride down Piccadilly on Sunday early in almost a solitary state with my head in the air, gazing around. Incidentally I saw a hawk hovering above me last week. Also I bicycle all round Hyde Park, although I have been led astray there through following Teddy boys who suddenly shout, 'Put your feet down miss, Miss, there's a copper coming!' making me realise that I am on a forbidden path where only postmen are allowed to ride. I haven't the car because there is nowhere to keep it. I miss it, but it's a saving of nervous energy; no negotiating traffic and going round and round for parking spaces. I walk or go by bus or taxi, not underground—everyone is in a rush down there. Even the dust rushes up at one, and although people are scrambling into the trains someone is always yelling 'Hurry along'. A more soothing exhortation would be 'Quietly, please—relax along there.'

We are asked if we can go to sleep peacefully. There is no trouble about this, for the street is quiet, and after the dinner and theatre rush of traffic it is almost deserted. Not quite, for if I stroll out later to the post office box at the end of the street there are a few ladies standing about in furs with very green eyelids, and red or yellow hair, and though I know them by sight they do not encourage even 'Good night' from me. Once in bed I could sleep for hours; the only sounds that have ever woken me came from two early office cleaners who were both locked out, screaming cockney witticisms to each other from opposite ends of the street; and once from a pigeon on my windowsill cooing.

I mentioned a post office in our street, but it is not the sort that one can do much in, in a small way. The people who go there buy insurance stamps in sheets, and post parcels in armfuls. Our post office is really not for the private individual, so one hardly ever goes there—another effort saved. At home the post came between 7 and 7:30 a.m. and if you were not about by then your registered letter went off with the postman. Here, in the heart of London, it's impossible to have a worrying letter at breakfast because the post does not come till long after that, sometimes not till we've gone out. Strange, but rather soothing.

People from home ask us if we keep sane without any exercise and still more without any garden. Well, we found a tennis court about five minutes away, and we found a garden and what could be more comforting than carrying across Piccadilly the fork and clippers one thought stored forever, or what more soothing than raking the dead

> leaves toward the bonfire with the hum of traffic so close, and yet so shut away?
>
> We are homesick, of course; one cannot leave the old life and the old friends without feeling a blank, and a view of the windows is a poor exchange for trees and flower beds. But don't tell me life in the very heart of London is all noise and nerve strain—unless one wants to make it that way.

There was another side, however, to this new job. The feuding, the wrangling both internal and external about Dad's controversial appointment there, and the weight of disapproval by the Royal Society cast a heavy cloud over Dad's move and were all an integral part of our family life when we moved there. My father was angry, and my mother did not always calm him down. In fact, she was quite capable of getting him worked up.

WLB felt that he had the qualities and drive to serve the Royal Institution and to restore its status in the scientific world. He was confident that this was the priority, and he was prepared to take the consequence that his reputation might suffer. That was at one level. At quite another, he was mortified on behalf of my mother by the snubs and hostility that they had to endure. Old friends at the Royal Society sat on the fence, though George Thomson was a loyal supporter. People like Robinson at the Royal Society moved away ostentatiously when my father approached and refused to speak with him.[27] He was denied a role when royalty visited. His status as one of the great scientists of the century was ignored and he was passed over when they chose a new President, although he was the obvious man for the job. My father was hopeless at committee politics; he wrote to my mother that he wished he had her tact. He could not read people's faces or gauge the strength of their feelings. This had applied in the past too, when he had felt odd man out as a Fellow of Trinity College, Cambridge. People did not react as he expected and this perplexed him. He would wake at night and bang his head against the headboard of the bed in frustration, as he sank deeper into the maelstrom of identifying and reconciling the two camps.

When my parents left the Royal Institution, they rented a flat in the Boltons. This was a perch, not a nest. Waldringfield was home, and the

[27] Robert Robinson (1886–1975) was an English organic chemist, awarded the Nobel Prize in 1947 for his research on plant dyestuffs and alkaloids.

flat was merely convenient. My mother stayed on at Quietways for a while after my father died, but not for long. Waldringfield was too far from the main stream and, frankly, too dull for her. She sold up and moved permanently to Cambridge, where she had spent many of the happiest and most fulfilled years of her life.

Gilmerton Court, Cambridge

Gilmerton Court was a rather dull modern flat outside Cambridge but highly fit for purpose. Mother had sensibly realised that there was too much for her to cope with in the Waldringfield house and garden. To my amusement, she decorated it to her own taste—previous homes had been done up for public show or to please my father. Here she entertained undergraduates to Sunday lunch.

After her death, I met a former medical student who had been asked by my mother to bring some pot with her to lunch. Since she had never smoked it, Mum felt she had missed out. The student obliged and I only regretted that I did not hear about this adventure until after she had died, so I could not quiz her about it.

She visited the Guildhall to make sure that plans were properly afoot to fly the flag at half mast when she died. On her eightieth birthday, she went to a large department store and asked at the cosmetic counter for a 'lipstick suitable for a grandmother'. 'So how old is your grandmother?' asked the dozy Saturday girl.

Mother enjoyed driving into her eighties. When she finally gave up her car, she discovered her ideal taxi driver. He used to come up to her flat and do up the back zip on her dress before she went out in the evening. In return, my mother would bake him his favourite seed cake.

When my mother was eighty, she decided that she was getting too old for foreign travel. But five years later, she changed her mind. 'Where would you like to go?' I asked.

'Concarneau again,' she said, without a moment's hesitation.

'And when were you last there?' I asked.

'Just after the First World War,' she said. 'The whole family went, including nanny. We stayed in a boarding house, I'm sure I could find it again.'

'Concarneau has changed,' I told her. 'I think you'd find it unrecognisable.'

'I'll go anyway,' she said, and we went.

Mother's mastering of foreign languages was like her mastery of cookery. Just as she had learnt to cook a few impressive dishes, so she learnt some neat phrases. 'Est-ce que la fumée vous gêne?' she would ask the neighbour at the next table as she lit up. Her accent was impeccable. My sister reported from Rome that she had not only mastered 'it's raining cats and dogs' in Italian but also 'it's *not* raining cats and dogs'.

My mother was healthy throughout her life, apart from, in her seventies, breast cancer, which was successfully treated. In her eighties, she had to go into hospital for an eye operation. I rang that evening to find out how she was feeling. 'Not at all well,' she said crossly. 'I rang for whisky at 6 p.m. and the NHS couldn't produce any'.

My mother lived in Gilmerton Court until her death from pneumonia in 1989, in her ninetieth year. It was here that my mother wrote her autobiography, which was entitled *The Half Was Not Told*. I hope I have been able to supply something of that other half.

She died peacefully, with her family around her, having caught up with the omnibus edition of *The Archers* the day before. The flag on the Guildhall was duly hung at half mast.

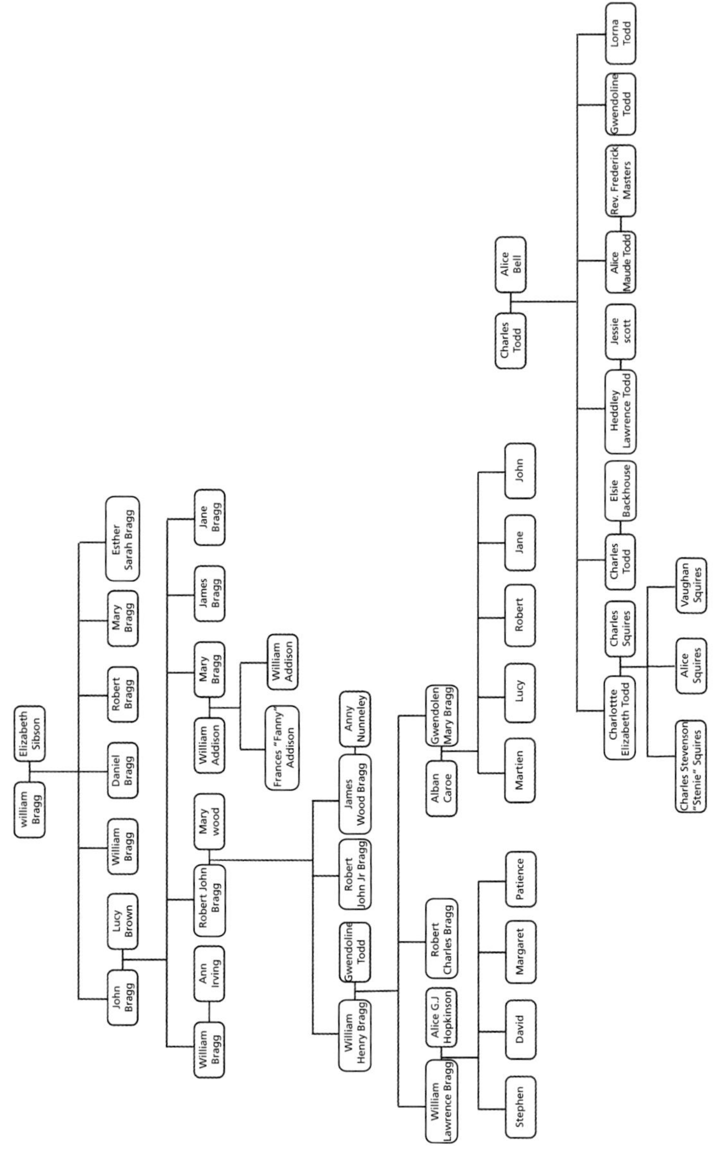

Adapted from Jenkin "William and Lawence Bragg. Father and Son "(OUP)

2

William Lawrence Bragg (In his own Words)

2.1 Growing Up in Australia

At the time of my birth, on 31 March 1890, our family lived in a semi-detached house in Lefevre Terrace in North Adelaide, and we continued to live there till the visit to England in 1897. I can remember the house and the garden vividly, though when I saw it again in 1960 I realised how much smaller it was than I had imagined it to be. It faced over the 'Park Lands'. When Adelaide was still quite a small settlement, a town plan was laid out by the famous Colonel Light.[1] Adelaide is approximately square, with north, west, south and east terraces, and is surrounded by a strip of Park Lands about half a mile wide on which no building is allowed. North Adelaide is separated from Adelaide by a broad strip of parks, traversed by the river Torrens. A similar strip of Park Lands surrounds North Adelaide. A low fence separated the two gardens in front of the semi-detached houses. There was sometimes friction with the widowed lady who lived next door, as she used to snip the stems of our nasturtiums if they wandered through to her side of the fence, even though they came back to our side; but there must have been amicable arrangements also because I remember my mother telling me she used to share with our neighbour in buying half a lamb, a whole lamb at that time costing seven shillings. Next door again lived the Gills. Harry Gill was the leading artist in Adelaide, and my mother attended his classes.[2] His boy, Eric Gill, was just my age and my great crony.

[1] William Light (1786–1839) was a British military officer and the first Surveyor-General of the Colony of South Australia.

[2] Harry Pelling Gill (1855–1916) was an English-born Australian art curator and painter.

I can remember my brother Bob as a baby in long clothes and in the charge of the monthly nurse. I would be about two and a half years old at the time. My mother had been very ill indeed at the time of his birth and it was some time before I was allowed to see her. It is related in the family that my first words to her were 'Mummy, do you know that I have got a baby brother?' While in bed in her convalescence she used to tell me stories and I can remember my fury and tears when a visitor interrupted one. Bob and I were wheeled out in the pram together. I dreaded these excursions because the larrikins, the rough boys of the neighbourhood, would shout gibes on seeing so large a boy in a pram. The indignity was heightened by my mother's artistic taste, which led to our long hair being done in sausage curls formed around the nurse's wet finger. We later had, for best wear, blue tunics with red belts and broad-brimmed straw hats, when I felt my dignity demanded trousers and coat like other small boys of my age.

The cook and housemaid were called Tilly and Naomi. I had a great affection for them. They afterwards ran a boarding house together and I used to visit them from time to time when I was much older. I cannot recall the early series of nursemaids except one who had to be sent packing because she had nits which she passed on to us. But very soon Charlotte must have arrived on the scene.[3] Charlotte became a family institution, staying with us for nearly thirty years. She came from that part of Denmark acquired by Prussia in the war of 1867, and she remembered making bandages for the wounded as a girl. She afterwards was a nurse to families in Prussia and in Holland. Her sister and brother-in-law had emigrated to Australia, and the brother-in-law had persuaded her to invest her savings in some venture which failed. As some recompense he offered her a home in Australia till she could find work there.

Looking back, I realise that Charlotte was not the right person to be a nurse; she was neurotic and fierce. When she had been a nurse in another family, she had dropped the baby on its head and, if I remember the story rightly, its brain was affected permanently; and this incident may have been responsible for her queerness. I do not think my brother Bob was much affected by her—even as a child he had considerable calm self-confidence—but I was very impressionable and unsure of myself and I am certain that Charlotte was very much the

[3] Charlotte Schlegel.

wrong person for me. She had a passion for our clothes always being smart, and she sternly repressed any game devised by Eric Gill and myself by which we could conceivably get dirty, a severe restriction for small boys and very daunting to initiative. Our playground was the gravelled 'backyard' so characteristic of Australian houses, with outhouses along one side and a huge woodpile in one corner. Domestic fires were all fed with wooden logs, and the quantity required for a household was immense. Flowers and shrubs were reserved for the front garden, which had to be watered every evening in summer. Our house was a corner house with an extension of the backyard running between the house and the side street. There was a tree in which some previous tenant had built a platform reached by steps, which overlooked the street and which was a great joy to us. My playtime was divided between our backyard and that of the Gills.

I was sent, I suppose when five, to a convent school on the other side of North Adelaide; I used to walk both ways. I must have been a very conventional and timid small boy, because I remember once the butcher, dashing by the school in his smart cart as I came out from the morning session, offered me a lift home. I was too dumb to accept so great a break with custom, to the astonishment of my companions. Again, the same butcher once chanced to pass when I was being subjected to some mild bullying by the larger boys and girls. Next day the whole school chanted, 'Tell tale tit' at me; I had not breathed a word to anyone about the bullying—it was of course the butcher who had reported it to my parents, who had taken the matter up with the headmistress. How powerless the young are when dealt injustice! I could not possibly have explained. I remember, too, a fierce argument with one of the nuns about the way a mirror worked. She was of course right but it perhaps showed an early interest in science.

On Sundays we traditionally paid a visit to our grandparents at the observatory. We went in a four-wheeler with two horses which had a rich smell, partly horse, partly cab, and I suspect partly driver. My grandfather combined the posts of Astronomer Royal and Postmaster General, which I thought inseparable in my young days.[4] The

[4] Sir Charles Todd (1826–1910) was employed by the government of South Australia as an astronomical and meteorological observer, and head of the Electric Telegraph Department. He set up the first trans-Australia telegraph line. His wife, Alice Gillam Bell, gave her name to Alice Springs. They had the following children: Elizabeth, Charles, Hedley, Gwendoline, Maude and Lorna.

observatory was a wonderful place for small boys. It was a rambling two-storey house on West Terrace with deep latticed verandahs and balconies in front. There was a circular drive surrounding a garden in which the central feature was an enormous Norfolk Island pine. To the right, in a lawn of buffalo grass planted with almond trees, was a cluster of buildings which housed the offices, the transit telescope, and other astronomical and telegraphic equipment. At the back of the house was another wide verandah and at one end of this was the bathroom. Either this room was open to the outer air, or the window was always open, because swallows nested inside each year. The bath was made of large slate slabs fastened together with angle irons and had only a cold tap. If a 'hot bath' was required, the hip bath was trundled out and filled by kettles. When we stayed at the observatory we had a 'hot bath' once a week, superintended by mother or an aunt to see that all corners were attended to. The bath on the verandah was the centre of endless games. Stones and bricks made islands and harbours for our boats.

At the back was the usual backyard, with a row of outbuildings, stables and storerooms on the left-hand side. The outbuildings were fascinating because all sorts of junk which could be incorporated into our games had accumulated there. Uncle Hedley had kept his horses in the stables, and the bins still contained chaff and corn.[5] Another outhouse contained souvenirs our grandfather had brought back from the interior when he put up the overland line, in particular gorgeous shells from the Tasman Sea, and there were boxes of old letters which we ransacked for stamps. The carriage gate at the bottom of the backyard, overhung by a large fig tree which had especially luscious figs, led on to the back drive and beyond that was the large equatorial telescope in its dome, surrounded by various smaller erections housing the meteorological instruments. We used to accompany Grandfather when he went the round and made the readings on Sunday. The evaporation tank was part of the programme, but its readings must have been rather untrustworthy when we were there because we fished for the tadpoles it contained. The cellars under the main telescope building also contained a glorious collection of junk, insulated wire, battery elements and chemicals which, when we were older, we used for the electrical gadgets we made. To the left of the telescope

[5] Uncle Hedley was Hedley Lawrence Todd; one of the children of Sir Charles Todd.

paddock was a truly enormous paddock which in my memory seems to have been about half a mile each way but must really have been smaller. It was surrounded by a belt of pine trees and, in the mess at their feet, we hunted for the cunningly concealed hinged flaps of the trapdoor spiders. Some years a riotous growth of mallow spread over this paddock, and a flock of sheep was then imported to eat it.

The Sunday lunch was presided over by our dear, placid, vague grandmother in her old lady's cap with lace frills. I cannot remember much about her. She died when I was seven years old when we were in England. Knowing my love of custard, she always provided for me a custard in a stemmed glass. After lunch my father and grandfather smoked their cigar of the week, and we were despatched to play in the grounds.[6] All has gone now; South Australia has ceased to have an Astronomer Royal of its own, and the lovely old house, as well as the astronomical buildings, has been demolished.

It is sometimes very difficult for children to understand the passions and motivations of grown-ups. The lawn in front of the transit

Figure 4 The observatory, from an air photograph. The house is in the centre distance flanked by the offices and anemometer tower on the left. The telescope dome is in the foreground. The radio mast and transmitting hut can be seen in the large paddock on the right. The evaporation tank enclosure is in the centre foreground. (Sketch by WLB; Courtesy: the Royal Institution London)

[6] The father is William Henry Bragg (1862–1942); in 1915 he shared the Nobel Prize in Physics with his son William Lawrence Bragg (the author of this manuscript) for the discovery of X-ray crystallography.

Figure 5 Grandfather reading the rain gauge. The evaporation tank is on the left. (Sketch by WLB; Courtesy: the Royal Institution London)

buildings led, by a wicker gate, to the large paddock I have mentioned above, and in the nearest corner of the paddock was the fowl run. The technical assistants would often have their midday snack on this lawn, and I once mentioned as an interesting item of current news that they were supplementing their lunch by sucking eggs. I was completely at a loss to understand why the assistants, hitherto good friends, reproached me so bitterly next day and why relations between us remained strained for so long.

Around the time of the Queen's birthday, any aborigines within reach of Adelaide would arrive and camp in the Park Lands around the observatory. In the nineties there were still quite a number who came each year. A man received a blanket from the government, and a woman a pair of stockings. They could also get medical treatment; my grandfather used to say that, if the men were given medicine, they made their lubras drink it.[7] I remember a party of them presenting

[7] Aboriginal girls or women.

Bob and myself with an emu egg, which we naturally accepted joyfully as a present; but the grown-ups were much embarrassed by the blacks of course appearing later on the verandah to claim payment. The aboriginal women and men had strange voices, deep and guttural and yet liquid. These aborigines were the last relics of the tribes which had lived in that area, and a few years later there were no more of them to claim the bounty.

Most of our holidays were spent at the sea, some seven miles away from Adelaide. There were several seaside settlements: Henley Beach, Grange and Semaphore, each with a long wooden jetty. At one of them, I think it was at Largs, the train actually ran along the jetty to a covered station at the end for the convenience of liner passengers.[8] The liners had to anchor a mile or two out to sea because it was so shallow inshore, and they were served by small steamboats which plied between the jetty and the liner. Also anchored out to sea in the St Vincent Gulf were the grain ships, mostly four-masted, sometimes as many as twenty or more waiting for a good price for their freight. They would be towed by a tug up the Port River to Port Adelaide for loading. They had a painted chequer pattern along their sides to simulate gun ports, I suppose a relic of the old days when they feared privateers. There was generally a mirage over the calm sea in the early mornings, and the hulls looked like a picket fence as high as they were long.

The journey to the sea was made in a horse tram from Adelaide to Henley Beach, and there were two great excitements on the way. At one point the road crossed a bridge over a stream bed, and how we craned out of the tram to look down, because sometimes one could glimpse WATER! It is hard to convey what a thrill this was; there were many stream beds in the country round Adelaide worn deep into the ground but they were quite dry except when there was torrential rain. I do not think my brother and I ever saw a running stream till we came to England. The other thrill came on the return journey. As we approached the plateau on which Adelaide is built there was a fairly steep hill leading up to West Terrace. At the bottom of this hill was a tin shed, the habitat of a very large horse and a very small boy. When the tram hove in sight the boy trotted out on the horse, flung a cable to the driver which he cleverly hooked on to the moving tram, and

[8] Largs is a suburb of Adelaide.

our steeds now augmented to three were lashed into a hand-gallop which bore us triumphantly to the crest, where the extra horse and its postilion were released and returned to their post. There was a dash about the whole affair which appealed to us greatly.

When I was six I had an accident which might have had worse results if it had not been for the skill of my doctor uncle (Uncle Charlie).[9] We used to play in the afternoon in a square at the centre of North Adelaide, and once when I was riding my tricycle Bob jumped in behind and upset me. The weight of both of us fell on my left elbow which was smashed into numerous pieces. I remember well the walk home with my arm feeling strangely stiff and the consultation round my bedside. The family doctor, Dr Lendon, thought the smash to be beyond repair and could only advise that the arm be allowed to set stiff in the most useful position. Uncle Charlie, however, determined to do better. Every few days I was put under with ether, and the doctors flexed the arm backwards and forwards so as to coax a new joint to form. How I hated these occasions! I would be quietly playing and then hear the dreaded voices in the hall which announced their arrival and start yelling at the top of my voice. The treatment was successful and left me with a very useful left arm, though it is out of the straight and shortened. It had a curious sequel. Forty years later I noticed that there was something queer about my left arm and that the hand was becoming paralysed. Platt, the Manchester surgeon, diagnosed that the bone was pinching the nerve at the back of the elbow and operated on me to relay the nerves down the inside of the arm.[10] This arrested the paralysis, though I have never regained the full use of the hand.

Incidentally I must have been one of the first patients to be X-rayed in South Australia. Very soon after Roentgen's announcement of the new radiation, my father set up a tube worked by an induction coil, and he took radiographs of the broken elbow. I was scared stiff by the fizzing sparks and smell of ozone and could only be persuaded to submit to the exposure after my much calmer small brother Bob had his radiograph taken to set me an example.

[9] This is Charles Todd, son of Sir Charles Todd.

[10] Harry Platt, First Baronet of Rusholme (1886–1986), was an English orthopaedic surgeon.

Sometimes, instead of having a holiday on the nearby shore of St Vincent Gulf, we went to Port Elliott on the ocean coast. I suppose the train journey was some sixty miles, but it seemed to take the best part of the day. I think we changed at Mount Barker on to a local line which ran to Port Elliot and Victor Harbour. A landmark on the way was a huge gum tree, from which a rectangular section of bark had been cut out by the aborigines to make a canoe. At some time, I believe, there had been an idea of making the capital of the state at Victor Harbour or, at any rate, of making it an important port. Breakwaters, quays and spurs of the railway had been constructed, and cranes and warehouses erected, but even in my boyhood these were falling into decay. The Bragg and Todd families combined to take a large part of a boarding house at Port Elliott. There is a watercolour, painted in 1896 by my mother,[11] that shows Aunt Maude,[12] Aunt Lorna,[13] Grannie and my father grouped around the table in the sitting room of this boarding house. Port Elliott was a tiny township, with just one main street with the station at one end and our boarding house at the other seaward end.

Figure 6 William Henry Bragg and his wife Gwendoline Bragg (née Todd); the photographs are believed to have been taken in December 1908. (Courtesy: the Lady Adrian)

[11] Gwendoline Todd married William Henry Bragg in 1889.

[12] Alice Maude Todd.

[13] Lorna Todd.

My father joined a few friends for a before-breakfast bathe each day and took me with him. Once he had me on his back in deeper water when a wave bowled us completely over, but he managed to retain a grip of my ankle. I am told the first thing I said, when able to speak again, was 'Daddy, you shouldn't have done it'! One member of the party, as we returned home, used to regale us with delicious pears. My mother spent much time sketching, and my father was prompted to try his hand at it. What must have been almost his first effort is the sketch we have of a group of aloes and a broken bridge, actually the bridge over the railway spur which led to the abandoned port. My father also joined in making some sort of golf course in the rough ground in front of the hotel. He had a great interest in games of all sorts. He was in the first eleven at King William's.[14] He was one of the pioneers in starting hockey at Cambridge. He introduced lacrosse into South Australia, where it became a very popular game. He was a first-class tennis player and I believe he played for South Australia. He was also a fine golfer, playing first at Glenelg and later being a leader in planning a new course at Heaton, which has since become famous.

My great friend and confidante on these holidays was my dear Aunt Lorna, the youngest of the aunts, who was then still in her teens. She looked after me, invented games for me, read Grimm's Fairy Tales to me and altogether constituted herself my guardian. We used to make excursions to Victor Harbour, a romantic place where Granite Island was connected to the shore by a long wooden jetty, along which plied a small horse tram. The ocean rollers, falling on the granite cliffs, were always a thrilling sight. In the other direction we went to Middleton, a long sandy beach where the same rollers broke in a terrace of foam lines seven or more behind each other, and all the shells were stained blue by some chemical coming from the seaweed. I can remember well the deafening roar of that long series of rollers thundering on the beach.

In 1897, when I was seven years old and Bob four, our parents made the momentous decision that my father should take a year off and the whole family should spend it in England. I think the main object was that Uncle William, who lived in Market Harborough, had brought up my father as a boy.[15] My father's success in school and at

[14] King William's College, Isle of Mann, where WHB was educated.

[15] Uncle William is William Bragg (b.1828).

Figure 7 Grandfather Todd, 1906. (Sketch by WLB; Courtesy: the Royal Institution London)

Figure 8 My mother sewing, March 1908. (Sketch by WLB; Courtesy: the Royal Institution London)

Cambridge had been a source of pride and joy to Uncle William, and it was a great blow to him when my father left England for the post at Adelaide University. My father felt he owed everything to Uncle William and that he had to see him again as his uncle was getting to be an old man.

My Aunt Lizzie was then on a visit to Adelaide, and the plan was that my father and mother should start first, making a tour of Egypt and Italy.[16] Bob and I were to come later with Aunt Lizzie, with Charlotte in charge of us, and join the parents at Marseilles. My father kept a very detailed diary in triplicate of their tour, one copy being sent on to 'The Uncles' at Market Harborough and one to the observatory; we have this diary and also some of the sketches done by my mother in Egypt, with a dragoman holding a parasol over her. We were parked at the observatory until we sailed for England. I can remember the departure of the parents from the observatory, a very tearful farewell on my mother's part and a completely unemotional one on ours; the difference between an absence of a day or of some months meant nothing to us. I can also remember the arrival at the Marseilles quay, when I was far more interested in watching intrepid Frenchmen sail canoes around the liner than in the parents we were joining. Our P & O liner was called the *Oceania*. She had four tall masts with yards and we used to set some sail, though more, I think, to steady the boat than to assist our progress. Colombo produced a great impression on me. The gentlemen with complete European upper garb, but with only a kind of dishcloth from the waist down, the punkahs and the punkah boys in the shops, the naked children who ran after our carriage begging, 'me no father, me no mother, me no grandmother, me no. . .', running despairingly down the scale of relations as they were outdistanced. The outrigger canoes and the lighters rowed across the harbour by men with very few ineffective-looking paddles and keeping time to a chant. These, and the heat, and the crowd of bargaining pedlars on board, were a great thrill. I also remember marvelling at the horny feet of the lascars, who worked the cocks of the hydraulic cranes on board with their toes.

We first made a long stay with Uncle William at Market Harborough. Uncle William had created for himself a leading position in town. He

[16] Aunt Lizzie is Charlotte Elizabeth Squires, the sister of Gwendoline Todd, i.e. daughter of Charles Todd and Alice Bell.

had built a house in the market square, which is now I believe a bank. It is a very ugly Victorian semi-detached house, but the view over the market square was very attractive. On market day a bullock got loose and ran into our hall, a great excitement! Uncle William and Uncle James kept a bachelor establishment together; Uncle William had run a chemist's shop, and Uncle James a grocer's business, but I do not think they were still actively engaged in trade when we were there.[17] Uncle William also owned a good deal of property in the town, including brickworks, which we enjoyed visiting. I loved seeing the clay oozing out as a long rod and being cut into bricks by wires. Behind the house, there was a garden, which seemed enormous to us and, at the far end, stables, and Uncle James's workshop. Uncle James was very gentle and kind and made toys, whistles, kites and endless other things for us; we were very attached to him. I think he was rather a simple character, quite under the domination of Uncle William, but it is of course just such simple people who so endear themselves to children. Uncle William was the complete bachelor, into whose life women and children had hardly come at all. I remember that Bob and I demolished most of his rockery to make a fort in one corner of the garden under a large elder tree, and any annoyance felt by Uncle William was mastered by his amazement at our imagination and at the ant-like perseverance with which we had moved, inch by inch, stones almost our own weight. He did, however, take the precaution of keeping his beehives in the strawberry bed. He was rather a vulgar old man—I remember his pinching us behind and saying our bums were like little cheeses. But even we realised his powerful personality and stood in great awe of him.

During our stay our parents and Uncle James went off for a bicycle tour in Wales while we remained at Market Harborough in the care of Charlotte. My father had bought two magnificent Humber bicycles for my mother and himself, which were afterwards taken to Australia. Bicycles were still a comparative novelty. When my parents were married, bicycling parties were the order of the day, when people met to go for rides, have bicycle gymkhanas and compare notes about their machines. This cannot have been my father's first machine; he must have had one when we were living on Lefevre Terrace, because I remember his indignation when he was summoned and fined by the

[17] Uncle James is James Bragg (b. 1839).

police for dismounting in the road and wheeling his bicycle across the footpath. He should, I suppose, have carried it. In his younger days he rode a penny-farthing bicycle, and he had a scar on his forehead due to pitching over the handlebars. I remember my father's story of a Welsh ferry where bicycles were so unknown that no provision was made for their tariff. The girl in charge deliberated and then said, 'I'll charge you as pigs.'

I am confused about the timetable of our round of visits in England. I know we stayed for quite a time in Cambridge, boarding in a house on the corner where Station Road diverges from Hills Road. Our living room had two pictures illustrating the broad road which leads to Hell, and the narrow path which leads to Salvation. I found these very edifying; I can see them with my mind's eye now, with a lady encouraging a child to enter the wicket gate which led to the narrow path. I played with Stenie Squires in Uncle Charlie's garden, which was in another street some way from Vale House.[18] Once a circus came to Cambridge and we were all to go to it. On the previous day I had some minor ailment, and it was judged that it would not be advisable for me to go to the circus; I was taken instead to a flower show on the same afternoon. This fine distinction caused me much grief. We had a holiday with the Squires at Hunstanton, and it was then that we heard of my grandmother's death. She was a dear old lady—I wish I had been older during her lifetime and known her better. I remember astonishing the family there once; I had got hold of my cousin Stenie's French primer and determined to learn French. I announced solemnly one evening, 'Donnez-moy dez belly bills', pronounced the English way.

Then we paid visits to Dad's cousins. Uncle William, as well as having my father to live with him as a boy, also persuaded his widowed sister to let him have her daughter, Fanny Addison,[19] and later when the mother came to live with Uncle William her son Will Addison came too. He was about my father's age and apparently was a truculent boy who caused my father much distress. Dad was very fond of Fanny, who married a farmer called Kemp-Smith. We stayed with the Addisons in Croydon, and history repeated itself. Will Addison, now a doctor, had taught his boys to box and set them on to us, when I am afraid the Braggs put up a poor resistance. We did not enjoy our

[18] Uncle Charlie is presumably Charles Squires.

[19] Fanny Addison (b. 1862).

stay. On the other hand I have pleasant memories of our stay in the country with the Kemp-Smiths. The daughter, Freda, was about my age. She had the finest set of properties for pretence meals that we had seen, all sorts of china joints and vegetables on plates. During this stay my father started to tell me bedtime stories and they were always the same—about the properties of atoms. We started with hydrogen and ran through a good part of the periodic table. Also, I remember, I had been given a large volume about Captain Cook's voyages; it was really a grown-up book. How little it matters to a child whether he understands all of a book; it is the atmosphere he likes. I ploughed through it from end to end. We did the sights of London, I expect, when staying in Croydon. I am glad that I remember the hansom cabs, with the clip-clop of all the horses on the wooden sets, and the adventure of riding behind the apron with the cabbie perched up above. I have little recollection of our return voyage when I was eight, except for staying in some hotel in Marseilles where the chambermaid recognised Bob and me and said, 'Bébés upgrown', which we thought very insulting.

When we returned to Adelaide we first stayed at the observatory. My father at once set about building a house. It was impossible to get my mother to understand any plans, so my father made a model of stiff paper fastened by a clipping machine. I remember Grandfather laying the foundation stone and naming the house 'Catherwood House' after Uncle William's house in Market Harborough.[20] Alice and I visited the house in Adelaide in 1960;[21] we were very pleased to find that it had been converted into a club for old boys from the larger Adelaide schools, so that it will be preserved much in its old form. It is an attractive house, of fair size, looking out over a strip of the Park Lands, with the racecourse beyond it.

I was sent to a preparatory school called Queen's on the far side of North Adelaide. This meant a walk to catch the tram, quite a long journey, and then a walk at the far end. The trams were double-deckers, with an awning top and canvas sheets which could be lashed along the sides of the upper story in wet weather. There were no points at a junction of routes; the driver skilfully swung his pair of horses to one side so that the wheels should follow the right grooves. Sometimes they did not and then we all had to help manhandle the

[20] The house is now a Caffè Nero and was formerly a branch of Lloyds Bank.

[21] WLB married Alice Grace Jenny Hopkinson in 1921.

tram on to the rails again. Both the walking sections of the route were fraught with danger. Returning from school in North Adelaide we were from time to time set upon by 'larrikins' so we avoided returning singly. Our best weapons of defence were our satchels, which we swung round our heads by the straps. On the spot I bore up bravely and then talked of our prowess with my companions the rest of the way to the tram terminus, but when I got home and told my mother about it I was apt to burst into tears.

At the other end, between tram and home, there was a boy with whom I had a vendetta. He got home from school at about the same time as I passed his house, so we often met. I could have returned by another street, but somehow honour forbade that easy way out. I think our feud started because I came on him once beating up his younger sister. Knight errantry prompted me to interfere, whereupon both brother and sister attacked me, a lesson in human reactions. Our scraps generally ended by his mother dashing out of the house and driving me away with a broom.

Our headmaster was called Hood.[22] He was a believer in corporal punishment. In the main classroom there was a cupboard full of the textbooks which were issued to us, and thrust between this cupboard and the wall was about twenty or thirty feet, doubled up into loops, of

Figure 9 Catherwood House, East Terrace, Adelaide. 'My father built this house shortly after our return to Adelaide in 1897 after our visit to England.' (Sketch by WLB; Courtesy: the Royal Institution London)

[22] Robert Gordon Jacomb Hood (1865–1926).

Figure 10 Catherwood House, Adelaide, today. (Courtesy: Professor Chris Sumby)

some species of cane. Hood cut a new length off this when the old one got worn out. We used to be given sets of ten words to spell. We were allowed, if I remember rightly, to get two wrong but had a cut of the cane on the palm of the hand for each mistake beyond that number. I was fortunately a good speller and only once had a cut; I can still sense the numbness which lasted for the rest of the morning.

There was a fair amount of bullying. The elder boys had shang-eyes (catapults) which they charged with lead pellets. The smaller boys were made to touch their toes, their backsides being the targets, and a hit registered by an obvious reaction. After the exercise we were made to search for the expended pellets. I was a misfit at school, being so very immature in some ways and so precocious in others. We had lunch in the boarding house attached to the school, and after lunch a scratch game of hockey was organised by the masters. Boys were asked if they wished to play and then supplied with sticks. Now it would have done me the greatest good to have played in these games, but I could not do anything so decisive as announcing I wanted to play and so moped rather miserably in the lunch hours. I think that in these days such a trouble would have been diagnosed and that I would have been better looked after at school. On the other hand I was precocious in lessons. We generally had more than one class in each room, and I remember being in the same room as a very senior class doing Euclid. From what I overheard I realised what it was all about. Somehow Hood must have caught on to what was happening, for he pulled me, a very small boy, out of my class and made me explain the theorems to the large boys while he crowed with delight.

While I was still at Queen's we moved into our new home. I started there the love of gardening which has always stayed with me. Bob and I were given two small plots at the back of the house and I remember vividly the thrill of seeing a green tip appear from a daffodil bulb I had planted, and eventually the formation of the flower. I had a prolific peach tree in my garden, but the pride and glory of the garden was an immense vine. Someone had given me a foot-long twig of vine, which I planted, and it grew until it covered the trellis along the whole back of the house, with gigantic bunches of black grapes, though they were rather tasteless. We had so many that we used to put them in a clothes basket outside the back door for neighbours to help themselves, and we pressed the tradesmen to take all they wanted. I experienced a curious coincidence in connection with this vine. Once when waiting at London Airport for the departure of our plane to America, I started chatting to my neighbour, found he was Australian from Adelaide over for a short visit and asked him if he had done all the missions on his programme. 'All except one,' he said. 'I have a friend who lives in a house on East Terrace which has an enormous vine growing over the back. There is a story current that the vine grew from a cutting of the famous vine at Hampton Court. My friend knows that a man called Bragg once lived in that house and asked me to track him down in England and find out if the story is true, but I have not been able to do this.' I of course was able to say, 'I planted that vine'. I saw it during our stay in Adelaide in 1960 and in sixty years it has grown into a noble vine, with a trunk as thick as a man's leg.

At about eleven years of age, I was sent to St Peter's College, the premier Church of England school in South Australia. It must have had between three and four hundred boys in my time, with about seventy boarders. It was a good school. The headmaster was Girdlestone, a vast and impressive man with a china-blue eye and a small yellow beard.[23] He carried in summer a small baton with a switch of hair at the end of it. Fresh from Queen's, I supposed it to be some instrument of punishment, but actually it was for attacking flies and not small boys, though I am not sure the headmaster thought them very different. He had a way of saying 'Boy, you are a humbug' which shrivelled one up. He had a passion for good English and spared no pains in correcting the essays of the boys in his sixth form. I have always felt very grateful for this training.

We did not specialise so much in those days, as boys have to do now. I took English language, English literature, French, Latin, Greek, Scripture, Mathematics and Chemistry, all to an equal level. The only subjects on the school curriculum which I did not take were German and physics, and I have always been sorry I learnt no German. My great friend at St Peter's was Bob Chapman, the son of Robert William Chapman, Professor of Applied Mathematics at Adelaide University.[24] The master of our maths class was rather feeble. Bob Chapman and I were rather hot on mathematics and we had the advantage of two fathers ideally equipped to help us with our homework problems. Often the maths master could not himself understand the answer to the problem he had set us, and he got out of his difficulty by making Bob Chapman or myself run over it to the class.

I had the same handicaps at St Peter's due to my being so immature and bad at games on the one hand and, on the other, being so precocious in my lessons, the latter condition leading to my being in a class of much older boys. I got into the sixth form when I was fourteen, and at fifteen my father decided that a further stay at school would not be profitable and I entered Adelaide University. The games at St Peter's were played in the lunch break, and the boys were divided into groups of roughly the same ability who played together. My prowess, or rather the lack of it, indicated my being placed in one of the lowest sets,

[23] Henry Girdlestone (1863–1926).

[24] Robert William Chapman (1866–1942) replaced WHB from 1910–1919 as Professor of Mathematics and Mechanics. He was a keen astronomer.

while my fellow sixth-formers were the prefects and glorious heroes of the school teams. I just could not disgrace myself playing games with the little boys in the lowest forms and so had a lonely and aimless time in the lunch break. Boys who do not fit into the normal pattern do not have an easy time at school, though it must be said that there was a fine spirit of tolerance at St Peter's. I was regarded as an amusing freak and interesting specimen and had very little teasing or persecution.

I found my physical outlet in rowing. We rowed on the river Torrens, which was really a dammed-back lake in the old river bed running through the parks between Adelaide and North Adelaide. The headmaster was an Oxford Blue and had a beautiful style. He used to stroke us in a four and, as the seats were offset and he weighed as much as the rest of us put together, we used to fill a large can with water and put it well to starboard to balance him. The headmaster was clever with his hands, and we did much of the repairing of the boats under his direction. The great race of the year, between St Peter's and Geelong Grammar School in Victoria, was rowed in the Port River.

We had wonderful holidays, sometimes in the hills of the Mount Lofty range, but more often at the seaside in St Vincent's Gulf. Adelaide was a very hot place in summer. There would be spells of days when the temperature would not drop below 100°F at night and it had been known to go up to 115°F in the shade during the daytime. It was a great relief to escape to the relative coolness of the hills or the sea around Christmas time. Places we stayed at, in addition to the nearby ones which were practically outlying suburbs of Adelaide, were Port Noarlunga, Aldinga and Yankalilla. Port Noarlunga at that time had only one farm, the Pocock's,[25] where we stayed, and a fisherman's hut. The river Onkaparinga ran out to sea at Port Noarlunga and was tidal where the bridge crossed it. We had a boat on the river and had happy days fishing and exploring the reefs at low tide. It was so deserted that Bob and I often wore no clothes except our hats with the essential fly-veils round them. The flies were a curse, but they could be kept at bay by the veil which was a very open string net with tassels hanging from it; a jerk of the head whisked the flies away.

[25] Probably refers to Frederick Charles Pocock (1851–1925), a well-known farmer in Port Noarlunga.

Port Noarlunga is about twenty-five miles from Adelaide and is now quite a seaside resort reached of course by car in less than an hour. In those times it took a good part of a day to get there. We started out from Hill's Coaching Yard in a coach and five. Bob and I together with our fox terrier Tommy generally managed to secure a place on the driver's seat. Below this was the boot, a large receptacle where the luggage was stowed, and this was a very convenient place for Tommy to retire to when the motion made him sick. The driver had a pile of mail for the houses along the road beneath his legs, and we were impressed by the way he could flip a packet of letters so that it sailed over the fence and fell on the mat before the front door. We changed horses after twelve miles or so, and when we came to a steep hill the passengers got out and walked. At Willunga we were met by Mr Pocock in his buggy and were driven to the farm. There was no made road. The land had been parcelled out into rectangles irrespective of contours, with strips about fifty yards wide left between the lots. A zigzag path along these strips had evolved by usage, covered with ruts as the traffic moved from those which were too deeply worn to a less-used part. One summer my mother was expecting a baby (my sister Gwendy[26]), and my father hired a buggy with two horses and a driver from one of the livery stables so that my mother could go for excursions. Before breakfast Bob and I rode the ponies bareback into the sea until they swam and we were towed by their manes; afterwards, we galloped on the sands till we were dry. I remember on one picnic we went to a neighbouring beach where there was a very steep approach. Jack (the driver) took an axe with him and cut down a small tree at the crest, which we towed behind the buggy, branches first, so as to break our descent. The bottom of the hill was littered with discarded trees used for a similar purpose.

This summer was also memorable because two girls of about our age, Hilda Fisher and Frances Hawker, were invited to stay with us on the Pocock farm. It was the first time Bob and I had really met girls, other than at formal occasions like dancing classes, and what a time we had showing off all the features of the place to them.

In the backyard at Catherwood House, there was a shed of galvanised iron which was allotted to us as a workshop. Dad arranged for one of the assistants in the laboratory to show us how to carpenter. We made endless gadgets, particularly electrical ones. I made a simple

[26] Gwendolen Mary Caroe, née Bragg (1907–1982). She married Alban Caroe.

form of motor, the armature being a toothed wheel. Power was supplied by a bichromate battery which we rigged up. I remember well how astonished I was when it worked (I had only read about it in a book) and, the crowning touch, that when I left it alone and went a walk round the garden, it was still running when I came back. Then Bob and I rigged up an electric bell in the workshop and connected it to a push-button in the nursery, so that Charlotte could summon us when tea was ready. It was too wasteful, however, to leave the battery connected up, so Charlotte had to yell to us to put the battery on before pressing the push-button. We made a telephone of the original Bell type, a clock and, inspired by a new instrument just then installed at the observatory, a beam seismograph. This last did not record earthquakes but was very sensitive to small tilts produced by our walking about the room.

While we were both still at school, Dad and Grandfather set up the first Marconi wireless link in South Australia. The messages were sent from a tall mast erected in the large paddock at the observatory, with a small tin shed at the bottom for the instruments. The oscillating system consisted of two brass spheres, connected to an induction coil. Oscillations occurred when a spark passed across the small gap separating them. The receiving aerial was set up on the sand hills at Henley Beach, seven miles away, and a Bradley coherer picked up the signals. This was a small evacuated tube containing nickel filings, connected between aerial and earth. The signal caused small sparks between the filings which welded them together and increased the conductivity, so that a current passed which worked the Morse receiver. At the same time the current rang a bell which rattled the coherer, so that it ceased to work when the signal ended. All worked according to plan. Bob and I took a great interest in these experiments, especially because it meant a picnic on Sunday afternoons when my grandfather and father drove to Henley Beach with us in the official Post Office wagonette to see the signals coming in.

2.2 The Journey to England

The minimum age for entry at Adelaide University was sixteen, but somehow my father managed to wangle me in while I was still only fifteen. My course was Honours Mathematics, in which I got a first class in 1908. I had my first introduction to physics at the university, physics and chemistry being the two required subsidiary subjects for an Honours Mathematics degree. My father was Professor of both Mathematics and Physics, so most of my instruction came from him.

Although I was fifteen when I entered Adelaide University I think my emotional age was about twelve or less, and my fellow students were mature young men and women. Such a disparity has a cumulative effect. Anyone handicapped in this way is debarred from taking part in the normal activities of his age group and the very fact that he cannot enter into their plans, schemes, differences of opinion, exercise of authority and so forth means that he loses the earlier experience which would teach him how to take his place later in life in the world of affairs. He loses touch with what is going on round about him and he thinks of the people who guide the course of events as 'they' and not as 'we'. He develops a defence mechanism to hide his inexperience from those he meets and this again makes him shy of asking the questions, the answers to which would keep him in touch. He is like a hermit crab with a soft tail which he tries to conceal in a protective shell.

How is such a one to face in later life the demands made upon him when his position is one of authority? He is greatly helped and fortunate if he can find a colleague who will act as a 'blind man's dog', who provides a pair of eyes to guide him through the jostling contacts with other people and across the roaring traffic of current events. He cannot fully see the obstacles and dangers himself. His disability has a compensation in that his very lack of appreciation of what is going on around him gives him a power of intense concentration and enables him to exceed in the subjects of his interest. His colleagues, however, seeing his competence in these special lines expect him to have at best an ordinary competence in the affairs of life and their disappointment when they realise his failures can be very mortifying to him. I can never be sufficiently grateful to those colleagues who have so often helped and guided me.

My main hobby at that time was collecting shells, and I built up quite an interesting collection of about 500 species. The shells were not as large and magnificent as those in tropical waters, but there were many interesting and striking ones. I found a new species of cuttlefish which was named *Sepia braggi* by Dr Verco in Adelaide.[27] He had the finest collection of shells in South Australia and was very kind to me

[27] Joseph Cooke Verco (1851–1933) was an Australian physicist and a collector of shells.

as a young enthusiast. I remember when I took some specimens of the 'bones' of these little cuttlefish to him, and he had verified that there was no previous description of the species, he said, 'I will call it *Sepia gondola* because of the shape'—then, seeing my face fall, he rapidly altered the name to *Sepia braggi*. I also found a new species of ischnochiton,[28] and several species which were new records for South Australia. St Vincent's Gulf was protected from the ocean rollers, so the shells were washed up on the beach in good condition. Bob and I used also to explore the reefs at low tide, turning over the boulders for chitons and other shells beneath them. I particularly remember one incident. We saw an octopus in one corner of a pool waving its tentacles, and there was a semicircle of crayfish in front of it, slowly advancing towards it step by step. Suddenly there was a great flurry of sand, and when it cleared octopus and crayfish had disappeared. I am told that this fascination of the crayfish by an octopus is a known phenomenon but has very rarely been observed. I ultimately gave my collection of shells to the Manchester Museum.

In January 1904 the Australian Association for the Advancement of Science met in New Zealand, and my father gave his presidential address in the mathematics–physics section on the electron and on radioactivity. Up till then he had done no original research of any kind, but the preparation of his address stimulated him to think about, and experiment with, the passage of particles through matter. This led to his famous work on the penetration of matter by α-rays and on the secondary β-rays. He had predicted in his New Zealand address that the heavy α-particles, suffering only a slight loss of energy in each atom they traversed, should pass almost undeviated through matter with a definite 'range'. Experiment brilliantly verified his ideas. I remember one awful time when he had heated the radium preparation on a plate, to drive off the products, and then replaced the plate in his chamber. As he decreased the range, he found no effect and thought he had lost his radium. To his great relief it suddenly came in at the short-range characteristic of the α-rays from radium itself.[29]

[28] A genus of mollusc.

[29] WHB plotted the energy loss of ionising radiation through matter. For protons, α-rays and other ion rays a pronounced peak occurs before the particles are entirely absorbed. This peak is known as the Bragg peak. This effect is exploited in the particle treatment of cancers, since it enables one to determine where to concentrate the radiation on a tumour.

He tried out his ideas and his papers on me. I remember his telling me of a great new idea which had just come to him, that the γ-rays were neutral particles, as we were mounting into the horse tram to take us to the observatory. His experiments were carried out in the basement of the main university block. His instrument maker Rogers was a real genius, and the α-ray apparatus was a gem. My father aimed at a high standard of perfection in design and construction. His measurements when plotted always lay on a smooth curve, and his experimental notebooks were a model of neatness and order. He even tamed that intractable instrument, the quadrant electrometer, which he used to measure ionisation. His experiments between 1904 and 1908 made him famous in the world of physics, and in 1908 he was invited to take the chair of Physics at Leeds University.

We came to England in the *Waratah*, of the Blue Anchor Line, which later met with such a disastrous fate.[30] She made her maiden voyage to Australia in 1909, and we joined the ship at Adelaide on the return journey from Sydney. My mother was very exercised about a supply of food for our baby sister, Gwendy, who was then only a year old. She wished to give instructions to the head steward but missed the chance of doing so while the ship was in port on her way to Sydney. But the ship remained for a short time anchored two miles offshore, and my mother persuaded my poor father to charter a small steam tender and take us all out to the *Waratah*, so that the steward might be briefed about the milk supply. Halfway out to the ship in a very choppy sea, my father's heart failed him. He could not face what he felt to be the ridiculous situation of boarding the ship on such an errand and he ordered the tug to be put back. The milk question was, I think, solved by correspondence.

There was something very wrong about the construction of the *Waratah*. She was a boat of 10,000 tons, and the shipyard where she was built had never before produced anything much more than half this size. Captain Josiah Ilberry was very worried about her behaviour at sea and consulted my father as a man of science.[31] She could not be brought to equilibrium on an even keel; however he trimmed

[30] The SS *Waratah* has been called Australia's *Titanic*. It disappeared with all hands in 1909.

[31] Captain Josiah Edward Ilberry (1840–1909) was Commodore of the Blue Anchor Line and was lost at sea when his ship the *Waratah* sank without trace.

his water tanks she listed over to port or to starboard. She was very sluggish in rolling; it seemed as if she would never right herself when thrown over to one side. I remember once, when having my bath, seeing the shower slowly cant over the side of the bath and fall on the floor for what seemed minutes, before it returned with the righting of the boat. She had a great castle of decks, and a large extra coal bunker on the top one.

Anyhow, she went out to Australia, we came home in her, she went out again, and then on the return voyage she disappeared completely between Durban and Cape Town. No trace of her was ever found. My father gave evidence at the enquiry and it was presumed she was top-heavy and must have rolled right over in rough weather. She had taken on much extra coal at Durban in her deck bunker.

We landed in Plymouth early in 1909, and Gwendy, Bob and I stayed there while our parents went to Leeds to find a house. They found a temporary home in a house near Shire Oak belonging to Miss Baines and later rented a fine house called Rosehurst, in Grosvenor Road.[32] At first mother was horrified at the grime of Leeds, and my poor father felt very remorseful at having brought her there from Australia. Actually, looking back, I think the years my mother spent in Leeds were the happiest of her life. She made friends easily and found some kindred spirits whom she loved dearly. Although our stone house was black with grime outside, it had fine rooms and was very attractive inside. And then, fairly soon after we had settled down, we acquired a cottage at Deerstones near Bolton Abbey, where we went for the holidays. There we were in wild country, surrounded by the moors, right away from the smoky city.

We had landed in Plymouth early in the year, and I was not due to go to Trinity till October. I always have felt sorry that this interlude was frittered away. It would have been a grand time to go abroad and learn French or German or to go to art classes in Leeds, which would have added greatly to the pleasure I have always had in drawing and sketching. Instead, I went to Cambridge for the Long Vacation term, when I was in an anomalous category and did little useful work. I expect it was my own fault; if I had been more mature and clear about my ambitions my parents would have agreed to any plans I made. My tutor at Cambridge was Barnes, who was later Bishop

[32] The house at 8 Grosvenor Road, Headingley, is still a guest house today.

of Birmingham.[33] We had mathematics lectures from Whitehead,[34] Hardy and Forsyth,[35, 36] and I was coached by Herman.[37] I liked Whitehead's lectures on mechanics, and Forsyth's on differential equations. I never really felt sympathetic with Hardy's logical treatment of infinite series and their convergence or otherwise. I played hockey and lacrosse extremely badly in the winter, and tennis in the summer. My lack of success at games was, I think, due to faults of temperament rather than unskilfulness. I also ran at Fenners, short-distance sprinting being my forte.

In the spring I tried for a scholarship. A short time before we were due to sit for the papers, I developed a violent cough which was diagnosed as bronchitis by Dr Cook and I was told to stay in bed for a few days in my lodgings. He did not come to see me himself but sent his understudy, who must, I think, have thought I was malingering, for he told me to go out in very cold weather for a brisk walk. As a consequence I went down with really bad pleurisy and pneumonia and was very ill. Mother came down from Leeds to nurse me. I still had quite a high temperature when the time for the exams came but was allowed to take them in bed. I think my brain was stimulated by the temperature. The essays were read by the Master, Butler, and he commented on the brilliant imagination shown in mine![38] Anyhow, I got a major scholarship in mathematics.

I concentrated entirely on mathematics in my first year and took Part I at the end of it. My tutor expected me to continue in the same line, but my father strongly urged me to switch over to physics. I therefore spent two years on Part II Honours Physics and got my degree in 1912. C. T. R. Wilson ran the Part II practical class and lectured

[33] Ernest William Barnes (1874–1953) was an English mathematician and scientist and became Bishop of Birmingham in 1924.

[34] Alfred North Whitehead (1861–1947) was an English mathematician and philosopher.

[35] Godfrey Harold Hardy (1877–1947) was an English mathematician famous for his work on number theory and mathematical analysis. In 1914, he became the mentor of the celebrated Indian mathematician Srinivasa Ramanujan.

[36] Andrew Russell Forsyth (1858–1942) was a Scottish mathematician.

[37] Robert Alfred Herman (1861–1927) was a Fellow of Trinity College, Cambridge, and coached many students to a high wrangler rank in the Cambridge Mathematical Tripos.

[38] Henry Montagu Butler (1833–1918) was Headmaster at Harrow School before becoming Master of Trinity College Cambridge.

on optics.[39] He taught me most of the physics I learnt. His delivery of his lectures was appalling, but the matter was marvellous. He must have been the main influence in all the subsequent teaching of optics in this country, through his pupils and in turn their pupils. He was also excellent in the practical class. He would not let us rush through experiments; he made each of them into a little research project for us. Searle gave deadly dull lectures on the topic of heat.[40] We went, for instance, at great length into the creep of the zero in thermometers made of different kinds of glass. J. J. gave us stimulating fireworks.[41] I also got very excited over some lectures of Jeans, because they opened up a new world of statistical mechanics.[42] After them a strange young man used to draw me aside and explain at enormous length just where Jeans was wrong. This was Bohr![43] He came first to Cambridge but soon realised that Manchester was the great centre for physics in this country and went there to sit at Rutherford's feet.

Two things of the greatest importance happened to me at Cambridge. In the first place, I became a member of a small group of close friends. Townshend was a mathematician,[44] Higham a historian,[45] Tisdall a classicist,[46] and Gossling and I were physicists.[47] The bond was, I think, our interest in intellectual exploration. We read papers

[39] Charles Thomas Rees Wilson (1869–1959); inventor of the cloud chamber and tilting electroscope. Awarded the Nobel Prize in Physics in 1927.

[40] George Frederick Charles Searle (1864–1954) was a British physicist and teacher. He was known for his work on the velocity dependence of the electromagnetic mass.

[41] This refers to John Joseph Thomson.

[42] James Hopwood Jeans (1877–1946) made important contributions in many areas of physics, including quantum theory, the theory of radiation and stellar evolution.

[43] Niels Henrik David Bohr (1885–1962); Danish physicist, famous for his understanding of atomic theory and quantum physics. Awarded the Nobel Prize in Physics in 1922.

[44] Hugh Townshend (1890–1974) graduated with first class honours in the Mathematical Tripos and thus became a Wrangler. He served with the Royal Engineers in the First World War, after which he was appointed Director of Overseas Telecommunications with the General Post Office.

[45] Charles S. S. Higham (b. 1890).

[46] Arthur Walderne St Clair Tisdall (1890–1915) was a British recipient of the Victoria Cross, the highest and most prestigious award for gallantry in the face of the enemy that can be awarded to British and Commonwealth forces. He was killed at Gallipoli in 1915.

[47] Brian Stephen Gossling (1890–1963) later worked on thermionic valves for naval use.

to each other and sat up to the small hours discussing the nature of the universe. I remember Townshend and me giving a joint paper on Minkowski's interpretation of relativity.[48] At my suggestion, I believe, the paper was entitled 'Some hitherto little-understood ideas'. Townshend started and gave all the meat we had managed jointly to extract from the paper and then turned to me and said, 'Now you carry on'. This was the first time in my life that I had a simple intimate relationship with a group of kindred spirits, and I revelled in it. I was still a queer fish at whom they often laughed, but our relations were quite easy.

The other formative influence in my life was my friendship with Cecil Hopkinson. He came from a famous family of engineers. His father, John Hopkinson (of the Edison–Hopkinson dynamo, the Hopkinson test and the Hopkinson theory of the magnetic circuit), had been a leader in the development of electric power and lighting.[49] He, together with a son and two daughters, had been killed in a tragic Alpine accident.[50] The eldest of the family, Bertie Hopkinson, was at that time Professor of Engineering at Cambridge.[51] A sister, Nellie, was married to Alfred Ewing, Principal of Edinburgh University.[52] Cecil was very much the youngest of the family; he was about my age and studying engineering. It is hard to understand why we were so drawn together; it was the attraction of opposites. I had grown up with no experience of physical adventure. There was no tradition of it in the Todd or Bragg family. This statement must be qualified because Grandfather Todd had gone through great adventures and hardships in the erection of the overland telegraph line, but he had not passed

[48] Hermann Minkowski (1864–1909); German mathematician best known for his work on relativity.

[49] John Hopkinson (1849–1898) was a British physicist and electrical engineer who invented the three-phase system for distributing electrical power.

[50] This happened on 27 August 1898 in the Alps. John Hopkinson set out from the Swiss village of Arolla with three of his five children to climb the Petite Dent de Veisivi in the Alps. The daughters were Alice (aged 23) and Lina (aged 19), and the son was John (Jack) Gustave Hopkinson (aged 18). They were found roped together just below the summit.

[51] Bertram Hopkinson (1874–1918) was a patent lawyer and Professor of Mechanism and Applied Mechanics at Cambridge University. He researched the properties of flames and was a pioneer designer of the internal combustion engine.

[52] James Alfred Ewing (1855–1935) was a Scottish physicist and engineer, best known for his work on the magnetic properties of metals.

the urge on to any of his children. Our early upbringing under Charlotte, when we were made to feel wicked if we had any exploits which might possibly lead to damage to body or clothes, had perhaps reinforced this trend.

Cecil, like all the Hopkinsons, loved adventure and hardship spiced with danger. He differed from them in that, while many of the Hopkinsons tended to be foolhardy, Cecil felt a kind of artistic shame about taking unnecessary risk, though fully prepared to take a necessary one. He introduced me to skiing, sailing, shooting and climbing. I well remember how it started. Bob, Cecil and I were walking along Trinity Street together when Cecil, whom I barely knew as yet, diffidently asked if I would join a skiing party at Vermala in Switzerland. I had my usual hesitation in letting myself in for any experience of an unknown nature, but dear Bob leapt in and insisted that I should accept. The party was presided over by Cecil's wonderful gallant mother, with her indomitable spirit and vitality. At that time I called her Mrs Hopkinson, but she became 'Aunt Evelyn' when I became engaged to and married Cecil's cousin, Alice Hopkinson.

Skiing was a very different matter in those days. There were no lifts; running downhill was a brief interlude at the end of a long day, and our main object was an expedition into places which could not be reached except on skis. Cecil gave me two or three days to find my feet, and we then went on an expedition to the top of the Wildstrubel, about 10,000 feet high. He was very proud that my legs lasted out, although I was such a tyro and the other members of the expedition were experienced skiers. Cecil himself was brilliant; he had won the Kandahar Cup in a precious year.

Then I was invited by Cecil's mother to their summer place on Loch Spelvie in Mull opposite Oban. It was a large farm belonging to a Livingstone who claimed to be a direct descendant of the great explorer. He was an old tyrant, who had not let any of his sons and daughters marry because he wanted them to work for him. The eldest boy, then almost fifty, had been allowed to get engaged. The family gave up the main part of the farm building to us and lived in the back part during our stay. When Aunt Evelyn had a bone to pick with the old man, as happened frequently, he would retire behind a smokescreen of Gaelic, saying he had no English. The Bertie Hopkinsons, with their large family of small girls, took a place about five miles away, on the other side of the neck of a promontory on which our parties had the shooting rights. It was just rough shooting and the bags were very mixed—grouse, duck, blackcock, snipe and hares—which we walked up. The gamekeeper, McPhail, unfortunately belonged to a clan which had a traditional feud with the Livingstones, and a ragged bevy of young Livingstones would sometimes scour the country a mile or so ahead of us and drive the game away, so as to disgrace McPhail. I well remember the thrill of my first right and left when two brown objects, flying considerately slowly, burst up in front of me. Unfortunately they turned out to be young pheasants, then quite out of season. Aunt Evelyn kindly made them into a pie. One morning when Cecil and I were out with McPhail we put up endless blackcock, all of which we missed. McPhail was thoroughly disgusted. He said if we saw the bird just above the sight when we fired it would drop to the ground 'chust like that' and to prove his point made a rough blackcock of heather and threw it violently to the ground.

The Bertie Hopkinsons had a seventeen-ton steamboat, looked after by one of the young engineering assistants called Impie, who lived up to his name by being always black and oily. The boat kept up connections with Oban and also brought all the little girls for the day to Loch Spelvie. Once Cecil and I set out in the boat to Oban to collect his Aunt Mary Hopkinson and the lady head of an agricultural college, both of whom were to stay with us. Unfortunately, a condenser pipe burst, and we could not bring them back that evening. With some difficulty, Cecil found a room for them in crowded Oban, but the ladies dug in their toes and said that the idea of their both sleeping in the same room was preposterous. So Cecil managed to hire the harbour master's launch by promising him that there was a good landing place at Loch Spelvie, as the harbour master was scared of his boat being holed by a rock. There was of course no harbour, but a stone wall ran out some way into the sea to prevent cattle getting past at low tide. We landed the ladies on this very loose rattling wall, supporting them by wading waist deep in the sea. If they had known what peril of a ducking they were in, they probably would have preferred to share a room.

Aunt Evelyn hired for us a large open fishing boat, and we went for expeditions in it, sleeping in the bottom of the boat at night. Cecil taught me to sail. He never said anything directly; his instruction was most efficiently conveyed by his expression and by approving or disapproving grunts. I was generally cook on these expeditions and got quite a reputation for camp cooking.

One summer Bertie Hopkinson hired a large sailing boat at Falmouth, the crew being Cecil, Russell Clarke, myself and a very small boy who acted as cabin boy.[53] They planned to sail across to Cork where I was to join them. I duly went to Cork and stayed in a hotel awaiting their arrival, and after two or three days I had a telegram to say they had landed up well to the East of Cork, at Youghal. On joining them I heard that the voyage had been somewhat adventurous. First, the propeller (the boat had a small engine) had got tied up into a fishing net and Bertie had to dive overboard to free it. Then they had a storm and, while they were trying to heave to with a lashed helm, the tiller was broken. I think they also lost their bearings and it was more or less by luck that they hit the Irish coast at Youghal. There was only one man in Youghal who could mend the tiller, and he had to be handled carefully. He could do no work if sober, and none if too drunk. Bertie made the mistake of giving him too much in advance, with the result that he was out of action for three days. He had to be primed with just enough to keep him going. When we finally got away, we sailed first to Cork and then on to Castle Townshend. I was not feeling too good—Russell Clarke took my temperature and found it to be over 104°F. It was a problem what to do with me, and the only solution was to put me in the Skibereen workhouse, which had an infirmary. So I was dumped there in a ward with half a dozen very old men, and we were looked after by nuns. I was having quite a severe attack of pneumonia. It was a very interesting experience. The C. of E. vicar visited me regularly, and I made great friends with a local schoolmaster, Jerome O'Regan, who held long talks with me. He told me at great length about his garden and it grew and grew in my imagination till it seemed quite an estate. When I was able to stagger about again and visited his house, it turned out to be a strip about ten feet by three running alongside the path from the gate. I remember also he had a reproduction of Mona Lisa of which he was very proud, but he thought that there was rather too great an exposure of her bust, so he had painted a lace frill over it. Last year I had a letter from him in which he said that he had 'moved with the times' and taken the little frill away again.[54]

[53] Edward Russell Clarke was an expert adviser to the Naval Staff on Wireless Telegraphy during the First World War.

[54] After fifty years.

One morning, the doctor said, 'What have you been doing? The nuns have had a special service on your behalf in the chapel this morning.' It turned out that the schoolmaster and I had discussed books of travel and I had tactlessly asked him if he had read *The Voyage of the Beagle*. As this was by Darwin it was of course on the list banned by the Roman Catholic Church; the dear nuns thought how sad it was that so young a man was going straight to Hell and had prayed for me.

We had many diversions. There was a young man always referred to as having 'a nice nature and a gramophone' who used to bring it out and place it on a soap box in the courtyard. We would hobble or be carried out to hear it. Then when I would be walking in the grounds with some of the inmates, one would say, 'Hist, do you hear that now? 'Tis the steam roller.' And we would make up a party to see the famous machine working; it was the only one in the south-west of Ireland. We had to sneak past the porter at the gates, an activity which added spice to the adventure.

When I was convalescent, the kind Townshends of Castle Townshend had me as a guest for three weeks. I met there Miss Somerville, of 'The Irish R. M.' (Somerville and Ross[55]) fame. It was quite a summer colony of hunting people.

When I returned to England my mother met the boat at Liverpool. This was a tremendous event to her; I think it was the first time in her life that she had travelled alone and she took the precaution of getting from Leeds friends the addresses of their friends in Liverpool in case she needed help. My mother could not bear ever to be alone. If my father was away for the evening she always arranged for someone to spend it with her. It was really counter to her nature to try to think over anything quietly by herself; she had to talk about it with a relation or friend. This went with her being so gregarious and so very clever at making people enjoy themselves and have a good time.

Why Cecil should have selected me as a friend is hard to understand. I was utterly unadventurous where he was so adventurous, and sometimes I tried his patience sorely. I think he may have found in me something he could not get from his own family. They were inartistic, practical, bold, realistic and unimaginative. Cecil was a breakaway—he longed to understand art, he was fascinated by new ideas, he was

[55] Edith Somerville (1858–1949) and Violet Florence Martin (1862–1915) (pseudonym Martin Ross) wrote books that were later made into a TV series.

delightedly amazed by anything quaint or bizarre in points of view and you could see him muttering and chuckling about such points as he mulled them over. What he gave me was like water in a thirsty land. He dragged me into adventures which I thoroughly enjoyed once I was launched into them and remembered afterwards in a way which bolstered up the self-confidence in which I was so sadly deficient. Cecil was severely wounded in the head in France and though he lingered for a long time he eventually died. Bob went out to Gallipoli with his battery and was killed there towards the end of 1915, so the war robbed me of the two people who meant most to me. Bob was sent first to school at Oundle when he came to England, and when he came up to Cambridge he read engineering. He was in many ways my opposite. He was sturdy, competent, practical and neat. He was always smartly dressed, an excellent dancer and participant in all the leading social events in Trinity. I have some of his notebooks, which are models of accuracy and neatness; I think he would have been a fine engineer.

I took my finals in 1911. I had not organised my time well. Part II is a one-year course, and I had two years in which to take it. I branched off into interesting side issues and did not allow enough time to refresh my memory of the work done in the first year. I was in rather a state and could get little sleep while the exam was on; I was lucky to get a first.

2.3 The Cavendish Laboratory

Then came a time of research in the Cavendish.[56] It was a sad place at that time. There were too many young researchers (about forty) attracted by its reputation, too few ideas for them to work on, too little money and too little apparatus. We had to make practically everything for ourselves, and even for that the means were meagre. For instance, we had to do our own glass-blowing and there was only one foot-pump for the blowpipe. One poor lady student had managed to get hold of it after waiting for weeks; I passed her room shortly afterwards, saw through the open door that she was not there and pinched the pump. Later my conscience was smitten to see her bowed over her table in tears—but not sufficiently smitten to make me return the pump. J. J. set me on some problem on the variation of ionic mobility

[56] In 1912 WLB obtained an 1851 Exhibition Scholarship at Cambridge.

Figure 11 Members and research students of the Cavendish Laboratory in 1913. Back row: W. D. Rudge, R. W. James, W. A. Jenkins, J. K. Robertson. Middle row: WLB, V. J. Pavlov, S. Kalandyk, F. W. Aston, H. A. McTaggart, H. Smith, F. Kerschbaum, A. Norman-Shae, Front row: K. D. Kleeman, A. L. L. Hughes, R. Whiddington, C. T. R. Wilson, Prof. Sir J. J. Thomson, F. Horton, R. T. Beatty, A. E. Oxley, G. Stead. (Courtesy: Cavendish Laboratory, Cambridge)

with the saturation of water vapour. If a first-rate workshop had constructed an apparatus for me, the results might have been of value but, with my self-made crude set-up, they were meaningless. There were a few senior people who had built little kingdoms for themselves with good equipment but most of us were breaking our hearts trying to make bricks without straw. J. J. did his best to think of ideas for us all and guide us; but there were too many of us and he was the only leader of research. C. T. R. Wilson liked doing everything on his own, and no other member of the staff was interested in research.

After a year of this, however, my golden opportunity came. Von Laue published his paper on the diffraction of X-rays by zinc blende and other crystals, and my father discussed it with me when we were on holiday at Cloughton on the Yorkshire coast, staying with Leeds

Figure 12 William Henry Bragg. (Courtesy: the Lady Adrian)

friends, the Barrans.[57, 58] My father at that time believed X-rays to be neutral particles and was of course intensely interested in results which ran counter to his hypothesis. On our return to Leeds, I set up an experiment in my father's laboratory to test whether the spots seen in von Laue's photographs and which he ascribed to diffraction might really be due to neutral particles shooting down avenues in the crystal pattern. Returning to Cambridge, I continued to mull over von Laue's results and soon convinced myself that they must be due to diffraction. My next advance was an interesting example of the way in which apparently unrelated bits of knowledge click together to suggest something new. J. J. had lectured to us on the pulse theory

[57] Max Theodor Felix von Laue (1879–1960) acquired the 'von' in 1913 when his father was ennobled. Together with Walter Friedrich (1883–1968) and Paul Knipping (1883–1935) he obtained the first evidence of diffraction of X-rays by crystals, thus proving that X-rays consisted of waves. Awarded the 1914 Nobel Prize in Physics.

[58] John Jenkin discovered a letter from the Norwegian Lars Vegard sent to WHB on 26 June 1912 informing him about von Laue's discovery, and it was this that started WHB and his son to look into the phenomenon of crystal diffraction. Lars Vegard (1880–1963) was a Norwegian physicist, especially known for his work on the aurora borealis.

of X-rays, a theory which explained them as being electromagnetic pulses created by the sudden stopping of electrons. C. T. R. Wilson, in his brilliant way, had talked about the equivalence of a formless pulse and a continuous range of 'white' radiation. Pope and Barlow had a theory of crystal structure, and our little group had an evening meeting when Gossling read a paper on this theory.[59, 60] It was the first time that the idea of a crystal as a regular pattern was brought to my notice.

I can remember the exact spot on the Backs where the idea suddenly leapt into my mind that von Laue's spots were due to the reflection of X-ray pulses by sheets of atoms in the crystal. A re-examination of von Laue's photograph from this point of view checked that all agreed. There was no fundamental difference between von Laue's analysis and mine as regards optical principles, but we differed in physical interpretation. Von Laue considered the spots as arising from certain definite wavelengths in the X-ray spectrum. I regarded this spectrum as continuous (the pulse concept), and the crystal planes as giving rise to a train of pulses which could be analysed as wavelengths given by $n\lambda = 2d \sin \theta$ (the Bragg equation).

I then tried to explain the intensity of the spots as due to a range of 'white' radiation which was continuous within certain limits—and all was wrong! But here our talk about the Pope and Barlow theory saved the situation. They had made much play with the face-centred cubic lattice of closest packing. I spotted that if one assumed von Laue's zinc blende crystal to be based on a face-centred lattice, instead of a simple cubic lattice, as von Laue had assumed, the intensities of the spots fell into an ordered sequence. This was the first example of X-ray analysis of crystals, being the proof that zinc blende, ZnS, is face-centred cubic. Pope, who was the head of the chemistry laboratory, suggested that I should examine NaCl and other crystals of the series, and I worked out

[59] William Jackson Pope (1870–1939) was Professor of Chemistry at Cambridge. He along with William Barlow suggested models for simple crystal structures before the discovery of X-ray crystallography.

[60] William Barlow (1845–1934) was an interesting character and one of the last of the British amateur scientists. Before the discovery of X-ray crystallography, he suggested many models for simple crystal structures based on the close packing of atoms, and independently he derived the so-called 230 crystallographic space groups that describe the possible symmetries of all normal crystals. There is an excellent article about William Barlow written by Peter Tandy which can be seen at <http://www.sciencedirect.com/science/article/pii/S0016787876800491>.

their structure.[61] At the suggestion of C. T. R. Wilson I tried to get the specular reflection of X-rays from a mica sheet, and this worked. I took the photograph, still wet, to show to J. J. and he was really excited. He pushed his glasses up on his forehead and ruffled his hair in a characteristic way. But oh! If only I had had the facilities available in modern laboratories! A young student in these days, getting a fundamentally new effect like the specular reflection of X-rays, would have a really accurate apparatus rushed through the workshop for him and be provided with the best available source of X-rays and other aids. I had to manage with bits of cardboard, drawing pins and a very poor tube worked by an induction coil. I got so excited with the reflections that I worked this coil too hard and burnt out the platinum contact.[62] Lincoln, the head mechanic who doled out the stores, was very angry.[63] The contact cost ten shillings, and to 'larn' me he made me wait about a month for a replacement. What might I not have discovered if proper gear had been available for rapidly exploring the reflection over a wide range of angles!

My father still wanted to be sure that these diffracted rays were actually X-rays, and he seized on the idea of reflection to build his X-ray spectrometer.[64] With this he not only discovered the X-ray spectra but also provided a far more powerful method of finding out crystal structures. It was at this stage that we joined forces, which was extraordinarily fortunate; if I had struggled on alone at Cambridge I should have got nowhere, but my father had all the resources of a first-rate workshop and of his experience in measuring radiation accurately. He pushed ahead with the investigation of X-ray spectra, while I was able to use the spectrometer results to solve crystal structures. In 1913 and

[61] NaCL is sodium chloride, or common salt. WLB's crystal structure was highly controversial at the time as chemists had expected that the Na and Cl atoms would form molecules. His model instead was like a three-dimensional chessboard of alternating Na and Cl atoms. Published by WLB in *Proc. R. Soc. Lond. A* 1913 **89**, 248–277.

[62] To this day the diffraction spots obtained by shining X-rays on a crystal are known as reflections.

[63] Fred Lincoln was the chief laboratory assistant, a supreme craftsman but a bit of a tyrant who regarded himself as the appointed executive of J. J.'s parsimony. He had entered the service of the laboratory as a young boy in 1892. If someone required a piece of wood, Lincoln would look at him as if he had asked for the moon. He retired in 1944.

[64] This was the ionisation spectrometer, built by WHB's workshop assistant C.H. Jenkinson. This was the forerunner of the modern diffractometer used universally by crystallographers today. Published by WHB in *Proc. R. Soc. Lond. A* 1913 **88**, 428–438.

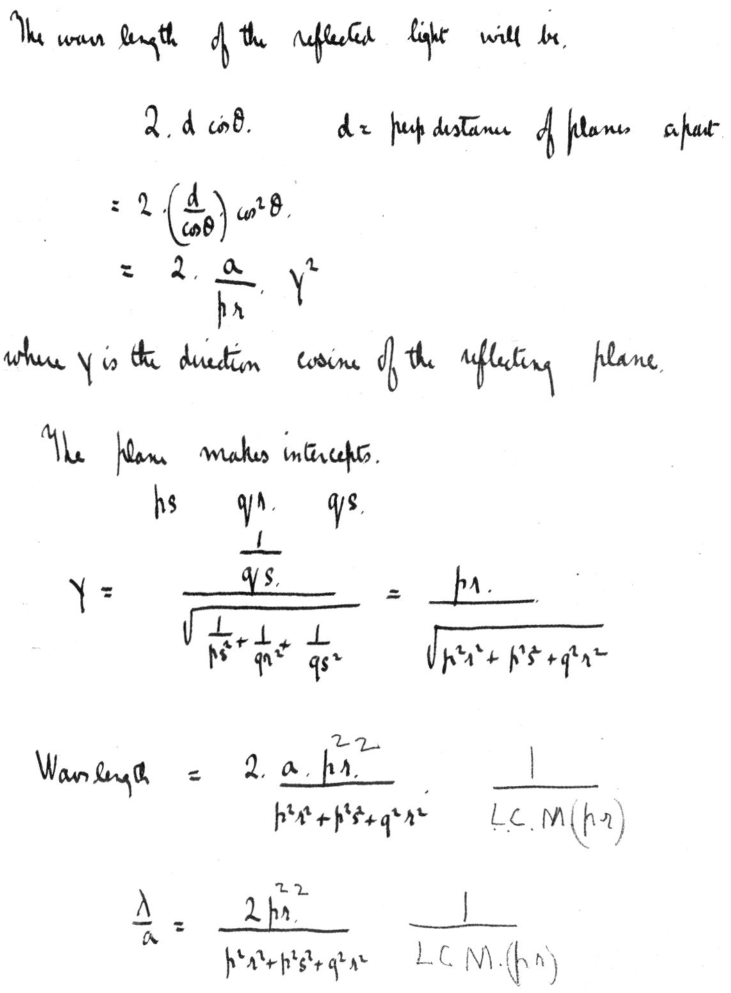

The wave length of the reflected light will be.

$2.\,d\cos\theta.$ $\quad d =$ perp distance of planes apart

$$= 2.\left(\frac{d}{\cos\theta}\right)\cos^2\theta.$$

$$= 2.\,\frac{a}{pr}.\,\gamma^2$$

where γ is the direction cosine of the reflecting plane.

The plane makes intercepts.

$$ps \quad qr. \quad qs.$$

$$\gamma = \frac{\frac{1}{qs.}}{\sqrt{\frac{1}{ps^2}+\frac{1}{qr^2}+\frac{1}{qs^2}}} = \frac{pr.}{\sqrt{p^2r^2+p^2s^2+q^2r^2}}$$

$$\text{Wavelength} = 2.\,\frac{a.\,p^2r^2.}{p^2r^2+p^2s^2+q^2r^2} \quad \frac{1}{L.C.M(pr)}$$

$$\frac{\lambda}{a} = \frac{2p^2r^2.}{p^2r^2+p^2s^2+q^2r^2} \quad \frac{1}{L.C.M.(pr)}$$

Figure 13 Page from WLB's 1912 notebook showing that the X-ray wavelength was given by the formula $2d\cos\theta$, later changed to $2d\sin\theta$. (Courtesy: the Royal Institution London)

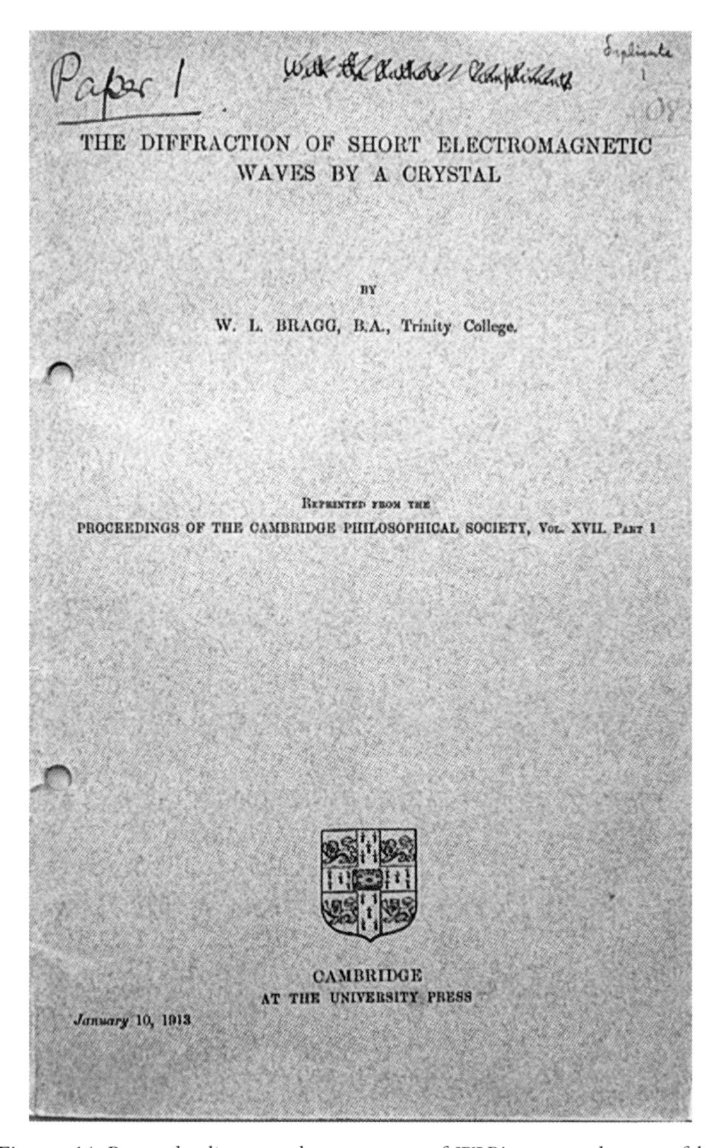

Paper 1

THE DIFFRACTION OF SHORT ELECTROMAGNETIC WAVES BY A CRYSTAL

BY

W. L. BRAGG, B.A., Trinity College.

REPRINTED FROM THE
PROCEEDINGS OF THE CAMBRIDGE PHILOSOPHICAL SOCIETY, VOL. XVII. PART I

CAMBRIDGE
AT THE UNIVERSITY PRESS

January 10, 1913

Figure 14 Recently discovered cover page of WLB's personal copy of his paper to the Cambridge Philosophical Society (read 11 November 1912). The title of the paper was *The Diffraction of Electromagnetic Waves by a Crystal* (note: there is no mention of X-rays, probably in deference to his father's belief, at the time, in particles as opposed to waves). For a discussion of the history of this paper, see A. M. Glazer, *Crystallography Reviews*, 19, 117 (2013). (Courtesy: the Royal Institution London)

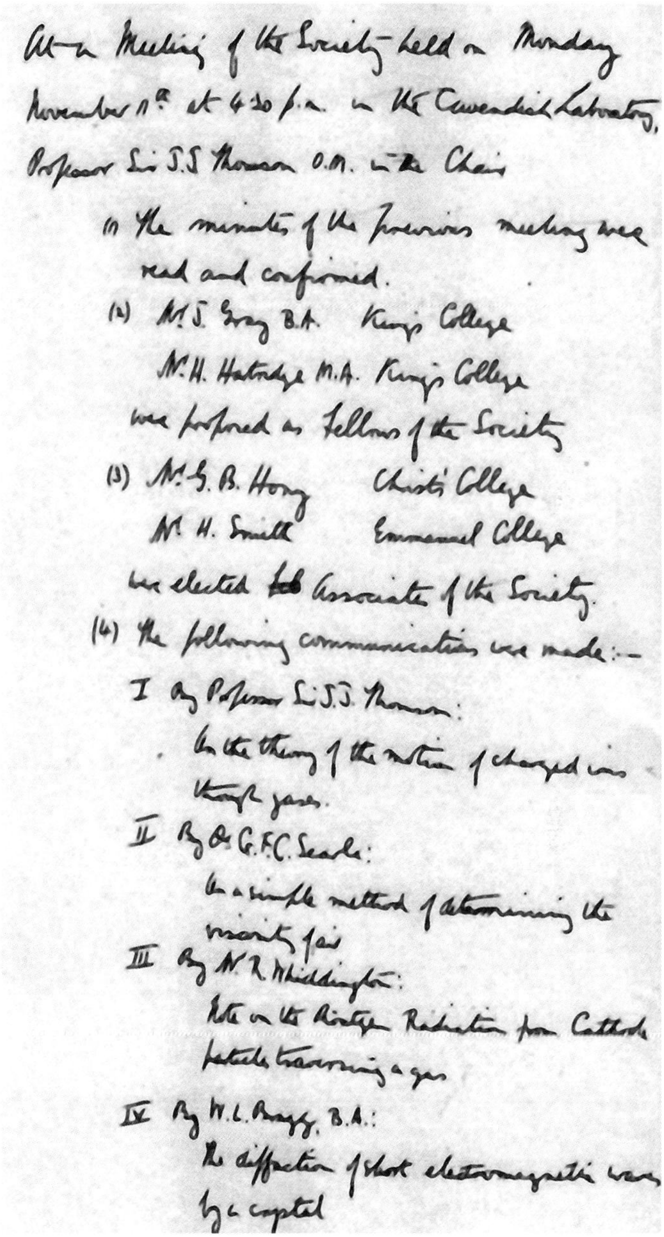

At a Meeting of the Society held on Monday
November 11th at 4.30 p.m. in the Cavendish Laboratory,
Professor Sir J.J. Thomson O.M. in the Chair

(1) The minutes of the previous meeting were
read and confirmed.

(2) Mr J. Gray B.A. King's College
Mr H. Hartridge M.A. King's College
were proposed as Fellows of the Society

(3) Mr G. B. Hong Christ's College
Mr H. Smith Emmanuel College
were elected ~~...~~ Associates of the Society.

(4) The following communications were made:—

I By Professor Sir J.J. Thomson:
On the theory of the motion of charged ions through gases.

II By Dr G.F.C. Searle:
On a simple method of determining the viscosity of air

III By Mr R. Whiddington:
Note on the Röntgen Radiation from Cathode particles traversing a gas

IV By W.L. Bragg, B.A.:
The diffraction of short electromagnetic waves by a crystal

Figure 15 Notes made at the time of the meeting of the Cambridge Philosophical Society 11 November 1912 at 4:30 p.m. in the Cavendish Laboratory. (Courtesy: Cambridge Philosophical Society)

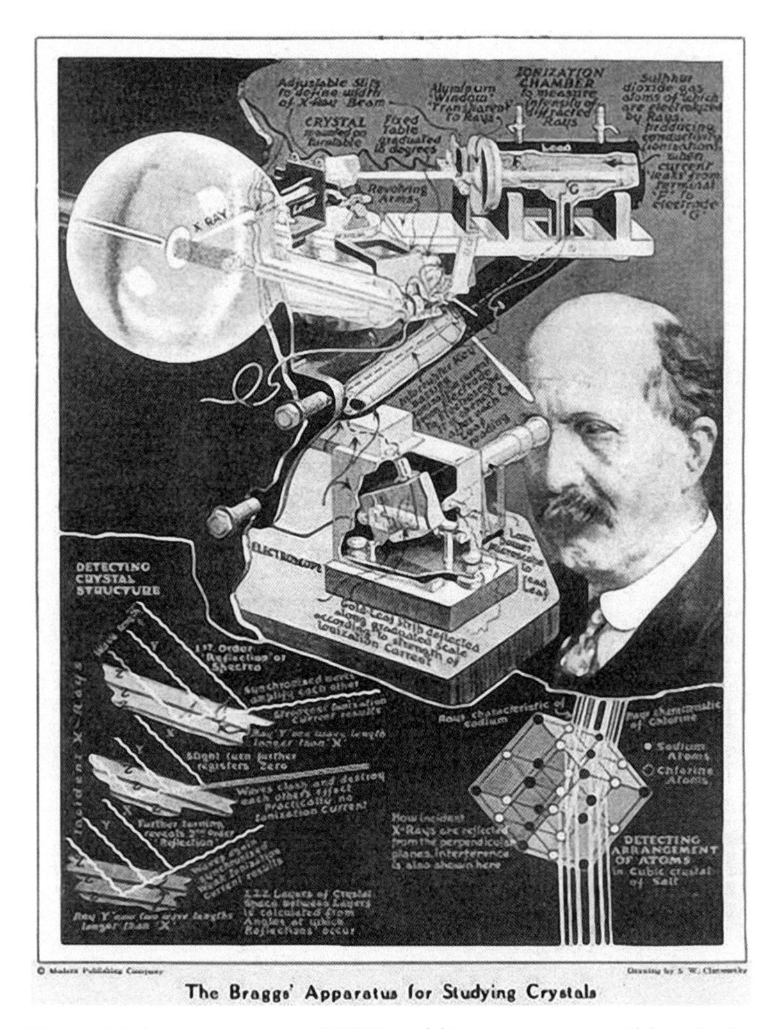

Figure 16 Amusing poster of WHB and his spectrometer. Although the caption refers to the 'Braggs' Apparatus', that is, to both WHB and WLB, only WHB is shown. (Courtesy: the Royal Institution London)

1914, until the war stopped research, we had a thrilling time, with new results tumbling out every week. It was only during vacations that I could work at Leeds, but I had a spectrometer of my father's design at Cambridge and investigated some structures there. Incidentally, the young student who assisted me in these researches was Appleton,

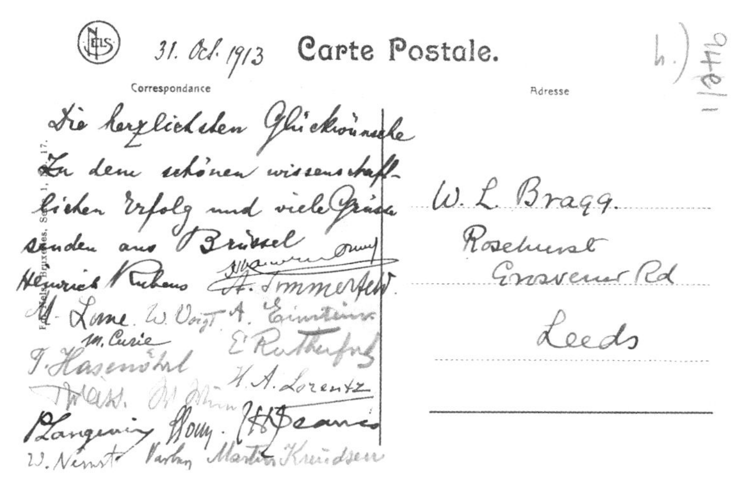
31. Oct. 1913 Carte Postale.

Correspondance

Die herzlichsten Glückwünsche
Zu dem schönen wissenschaftlichen Erfolg und viele Grüsse
senden aus Brüssel
Heinrich Rubens A. Sommerfeld
M. Laue W. Voigt A. Einstein
M. Curie
F. Hasenöhrl E. Rutherford
H. A. Lorentz
P. Langevin J. H. Jeans
W. Nernst Martin Knudsen

Adresse

W. L. Bragg.
Rosehurst
Grosvenor Rd
Leeds

Figure 17 Card sent to WLB, with signatures of members of the Solvay Meeting in 1913. (Courtesy: the Royal Institution London)

afterwards Sir Edward Appleton of DSIR and Edinburgh University.[65, 66] He might have become an X-ray crystallographer had not the work he did in the war, on radio communication with the newly invented valve, diverted his interest in another direction.

It was not altogether an easy time, however. A young researcher is as jealous of his first scientific discovery as a kitten with its first mouse, and I was exceedingly proud of having got out the first crystal structures. But inevitably the results with the spectrometer, especially the solution of the diamond structure,[67] were far more striking and far easier to follow than the elaborate analysis of Laue photographs, and it was my father who announced the new results at the British Association, the Solvay Conference and at lectures up and down the country and in America, while I remained at home. My father more than gave me full credit for my part, but I had some heartaches. I knew, however, that, if it had not been for our collaboration, I could have made little progress.

[65] Edward Victor Appleton (1892–1965); famous for his work on radio waves in the atmosphere. Awarded the Nobel Prize in Physics in 1947.

[66] DISR, Department of Scientific and Industrial Research; at the time, the main science funding body.

[67] Published by WHB and WLB in *Proc. R. Soc. Lond. A* 1913 **89**, 277–291.

Poor Common Salt!

'Some books are lies frae end to end' says Burns. Scientific (save the mark) speculation would seem to be on the way to this state![. . .] Professor W.L. Bragg asserts that 'In Sodium Chloride there appear to be no molecules represented by NaCl. The equality in number of sodium and chlorine atoms is arrived at by a chess-board pattern of these atoms; it is a result of geometry and not of a pairing of the atoms'.

This statement is more than 'repugnant to common sense'. It is absurd to the n. . . th degree, not chemical cricket. Chemistry is neither chess nor geometry, whatever X-ray physics might be. Such unjustified aspersion of the molecular character of our most necessary condiment must not be allowed any longer to pass unchallenged. A little study of the Apostle Paul may be recommended to Prof. Bragg, as a necessary preliminary even to X-ray work, especially as the doctrine has been insistently advocated at the recent Flat Races at Leeds, that science is the pursuit of truth. It were time that chemists took charge of chemistry once more and protected neophytes against the worship of false gods; at least taught them to ask for something more than chess-board evidence.

Figure 18 Extract from a letter to *Nature* in 1927 by Professor H.E. Armstrong.[68]

Cecil Hopkinson's reaction to my new ideas about X-ray diffraction was typical. He was tremendously excited, and I can hear him now exclaiming and muttering as he tried to thrash out the method of analysis for himself. He was the warmest-hearted and most loyal friend it was possible to imagine. We were at that time occupying the same set of rooms in Trinity. This went right against college regulations, but as a scholar I had a large room with two bedrooms and somehow Aunt Evelyn managed to wangle permission for Cecil to occupy the second bedroom, which was really little more than a box room. Cecil at that time was carrying out research into the shock waves when a projectile hits a plate, under the direction of his brother Bertie Hopkinson, the Professor of Engineering.

While at Cambridge I demonstrated in Searle's famous Part I practical laboratory. Searle was a really extraordinary character. He thought

[68] Henry Edward Armstrong (1848–1937) was an English chemist. He worked on the chemistry of naphthalene derivatives. He is remembered today largely for his ideas and work on the teaching of science.

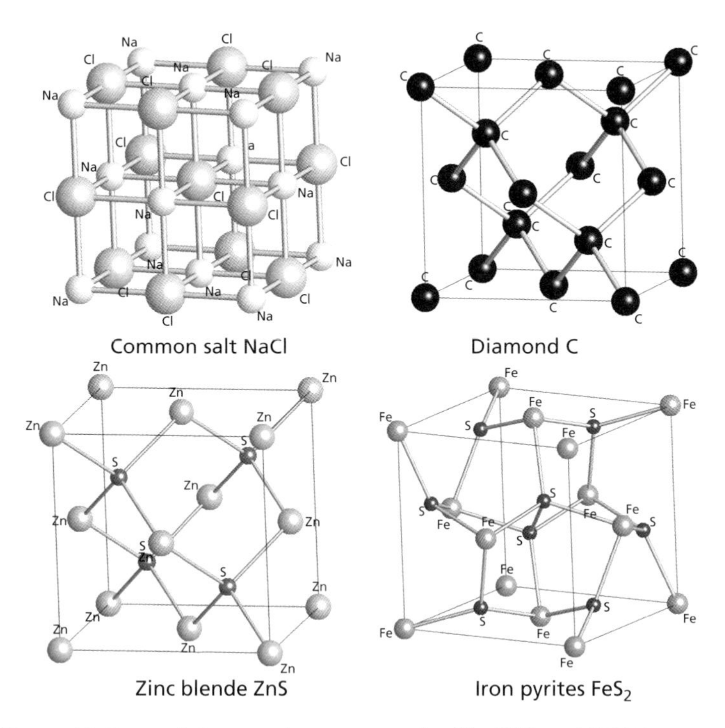

Figure 19 Some of the crystal structures solved by WLB and WHB between 1913 and 1914. (Courtesy: the Royal Institution London)

less than nothing of the research side of the laboratory; to him teaching was its sole important function. He had come to the Cavendish in 1888 about the time my father left to go to Australia. As I have said earlier, his lectures were very dull and I think they were so because he despised research and so all the new advances physics was making. Time had stood still since 1880 as far as he was concerned. He took infinite pains over the practical class, writing a booklet for each experiment and supervising the work of each student personally. On the whole I think it was a good discipline and wish I had taken Part I Physics at Cambridge, though his experiments were perhaps too regimented. He was a terrific tyrant, though really very kind-hearted. I remember his shouting to me down the length of the classroom, 'Bragg, come here and see what this fool has done. This is the kind of thinking you have got to look out for.' And another time, holding up the large sheet on which the students' experiments were registered, 'Bragg, what do you think these marks are? Human tears,' he said, pointing scornfully to a wretched girl sobbing her heart out over his desk. Another famous remark was overheard by R. W. James. A man continued to get inexplicable results with his galvanometer which puzzled even Searle for a bit, but he guessed the trouble rightly and said to the man, 'Do you wear a truss?' He had health crazes. At one time he would go round the class asking each student, 'How're your bowels?' At another, he unexpectedly clapped glasses from an optician's box in front of their eyes and asked if it improved their sight. He had started early in J. J. Thomson's time. In the Second World War he was brought in to demonstrate again, and it was extremely hard to persuade him to give up when the war ended. He must have been about eighty, when I, seeing his bicycle propped against the lab wall, admired its streamlined mudguards and he said, 'I reckon I get three or four extra miles an hour from those.' He had been a great racer in his young days. When he taught me as a student he was one of the quite senior members of the staff and we were all in great awe of him. There are endless Searle stories.

2.4 War!

Then, out of the blue, came the war. At one moment we were in the midst of the usual carefree easy-going Long Vacation, and then we were plunged into a strange new world. At that time my brother Bob and I were in a kind of Territorial unit called King Edward's Horse. It was composed of men who had come originally from the Dominions and was first called 'The King's Colonials'. My cousin, Steenie Squires, had got into the unit on the grounds that his mother was Australian. We were mounted infantry, trained in the tactics found useful in the Boer War. The emphasis was on marksmanship, riding and care of our horses. We were in sections of four, and on exercises three of us dismounted and delivered a withering fire on the enemy while the fourth held the horses. We then mounted again and proceeded to the next strategic point. It must have been an expensive unit to run, since we had much practice in the riding school, including jumping and vaulting, and took our horses to camp each summer; the expense was borne by the Dominions. Bob went into camp with this unit at the outbreak of war, but I applied for a commission and trained in Cambridge while waiting to be posted. A small group of dons sorted us out and, as I had experience both of riding and mathematics, it was decided that the right billet for this combination was the horse artillery, and I was sent to a Territorial battery, the Leicestershire RHA.[69] Bob also eventually applied for a commission and joined a field battery which was quartered in Leeds.

I was very much a fish out of water in the battery. The officers were hunting men, who talked and thought horse to the exclusion of most other interests. It was strange to see how many horses were needed for a four-gun battery. Six horses per gun, six per ammunition wagon, all the ranks mounted, the officers, doctor, vet, bugler, servants, grooms, farriers and so on, together with the horses for the ammunition train, came if I remember rightly to about 240 in all. We were quartered for the first year of the war at Diss in Norfolk. This is an attractive town, centred round a small pretty lake. The horse lines were in the northern outskirts of the town, and in winter they were distributed among barns in the neighbourhood. At first I shared a wonderful billet with Reg Whitehead. I cannot remember the name of our host, but I recall that he made Gossling's Pig Powders. Our hostess was reputed to be the finest cook in Norfolk and I can believe that this was so. One builds up a fine appetite when out all day on horseback, and we did justice to her dinners. I particularly remember a partridge each on toast. They were such terrific banquets that we found it difficult to rise in order to ring the bell for the next course, so we tied a piece of string to it and rang it from the table. Alas, the good lady found it too much effort to feed us, and we had to find another billet.

I hold that the finest food is to be found in such country households or farms where the lady of the house cooks herself. It is far better than the productions of many famous restaurants or hotels to a gourmet whose palate is unspoilt. A fowl from the estate, peas picked that morning in the garden, new potatoes, home-cured bacon and ham and home-baked bread are far better than the grandiose food of the restaurant, where the staleness of the raw materials is disguised with cream, butter, cheese and those wretched ersatz cultivated mushrooms which are not the same species as the true field mushrooms and are quite tasteless. One can tell the difference by the contrast in the ravishing smell of cooking in such a household, and the repulsive odour of garbage which rises from the street grille ventilating the kitchen of many a hotel or restaurant.

[69] Royal Horse Artillery.

I was fortunate in having another magnificent billet for part of the time I was there. A grand old lady (called Mrs Taylor?) occupied a Georgian house with chains on white posts in front.[70] I kept my two horses, looked after by my groom Staniforth, in the stables. I breakfasted alone, real porridge with cream and brown sugar, and generally a fish dish plus an egg or a meat dish, all off priceless Nanking china. In the evenings I dined with her and her husband, who was very old and frail. They had a very ancient parrot, which had eaten all his feathers except those on his head which he could not reach. The dear old man suffered attacks of wind after dinner at rather frequent intervals, which would have been easier to ignore had not the parrot repeated each eructation with the greatest fidelity.

Although I only served with the battery for a year, I have regularly been invited to the annual reunions and recently attended one of these occasions. It is wonderful how time softens all things. Although I know I had many shortcomings as an officer, I found that after nearly fifty years I had become a legendary figure! I was also delighted to find how many names I could remember, including those of my groom Staniforth and my servant Cobley.

Then one day in August 1915 I got a letter from the War Office instructing me to report myself to Colonel Hedley.[71] The proposition he put before me was this. The French had started experiments with a method of getting the positions of enemy guns by measurements made on sound waves. Major Winterbotham, RE, who directed survey and similar activities at the front, was convinced that this method might be useful and had persuaded the army authorities to set up an experimental section under the direction of an officer who had scientific training.[72] Colonel Hedley had got my name, I think from Cambridge, and asked me whether I would take on the assignment. I was thrilled. Coming back along Whitehall I realised what was meant by 'walking on air'. To have a job where my science was of use, after feeling so inefficient in the battery, seemed too good to be true. I think the battery was equally thrilled to part with me, from the warmth with which I was urged to take on the

[70] This was probably the Manor House, Mount Street, in Diss, owned by the Taylor family.

[71] Colonel Walter Coote Hedley (1865–1937), of the Royal Engineers.

[72] Major Winterbotham is Harold St John Lloyd Winterbotham (1879–1946).

assignment. Incidentally, when writing up an account of sound ranging at the end of the war, I saw the early correspondence. The RA had clearly been very doubtful that the method would be of any use but finally agreed (in so many words) that, if an officer were sent out who was not needed for anything else, they would raise no objection. Another young officer, Harold Robinson, was detailed to do the experiments with me.[73] Robinson was on Rutherford's staff at Manchester in peace time and was serving in a heavy coastal defence battery. We were sent out to France together, to report to Colonel Jack; at GHQ, Jack was in charge of the field survey sections, one of which was attached to each army.[74] They did surveying, maps and printing. Later, the sections included sound ranging and flash spotting (locating enemy gun positions by sighting on their flashes). Our section was attached to Jack's office at St Omer while still in the experimental stage.

We were first sent to a French section in the Vosges. The French were trying out several methods. This section used a recording apparatus designed by Lucien Bull and made in the workshops of his laboratory at the Institut Marey in Paris.[75] It was a beautifully designed and accurate recorder, employing a six-string Einthoven galvanometer.[76] The six microphones which registered the report of the gun were connected, one to each string, and the time differences read on a running cine film which recorded displacements of the strings. The French section was in a quiet part of the front, so quiet indeed that I cannot remember a gun firing during the two or three weeks we were there, so our instruction was rather theoretical. We lived in a skiing club's chalet, and near us was a French battery, with the officers of which we messed. I remember how the washing used to be spread out in the

[73] Harold Roper Robinson (1889–1955) was a British physicist who collaborated with Lord Rutherford. He was Head of the Physics Department at Queen Mary College, London, and then Vice-Chancellor of the University of London from 1954 to 1955.

[74] This is Lieutenant Colonel Ewan Maclean Jack, Royal Engineers.

[75] Lucien Bull (1876–1972) was a French physicist of Irish extraction who worked on electrocardiography and recording heartbeats. He remained a close friend of WLB for many years after.

[76] Willem Einthoven (1860–1927) invented a string galvanometer and the first useful electrocardiogram; he was awarded the Nobel Prize in Physiology or Medicine in 1924.

gun emplacements to dry, and once or twice when a plane came over there was a general rush to get it in.

When we came back to St Omer to report to Colonel Jack, I heard from my parents that Bob had died of wounds at Gallipoli, very soon after the landing.

I was then sent to Paris to take over a recording set, mounted in a lorry, and Robinson and I set up the first sound ranging station near Kemmel Hill, just south of Ypres. We had six microphone stations in a row about a mile behind the front, connected by lines on poles to the recorder which was stationed in a village called La Clytte. It was there that I heard of the award of the Nobel Prize to my father and myself.[77] I remember that the friendly priest, in whose house we were quartered, brought up a bottle of Lachryma Christi from his 'cave' to celebrate the occasion.

At first we had many disappointments, and our results were of little value, though we did not admit this. Our microphones were of the wrong kind for sound ranging on guns. When a gun fires a shell

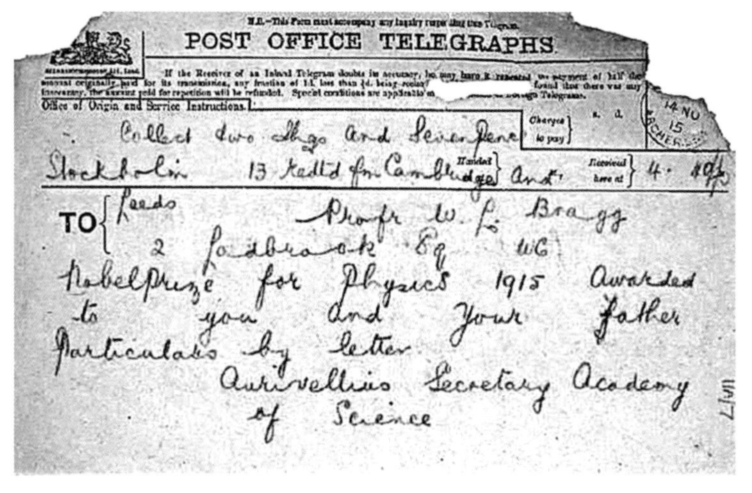

POST OFFICE TELEGRAPHS

Office of Origin and Service Instructions.

Collect two shillings and seven pence

Stockholm 13 Handed in for Cambridge

TO Leeds Profr W. L. Bragg

Nobelprize for Physics 1915 Awarded to you and your father

Particulars by letter

Aurivellius Secretary Academy of Science

Figure 20 An important telegram arrives. (Courtesy: the Royal Institution London)

[77] This was in 1915. They are the only father and son team to be jointly awarded a Nobel Prize, and WLB at the age of 25 remains the youngest recipient ever of a science Nobel Prize.

with an initial velocity greater than that of sound, the shell creates a wake or 'shell wave' which is heard as a loud crack. The phenomenon has become familiar now that aeroplanes pass the 'sound barrier'. This crack is far more intense than the low-toned 'wuff' which is the true gun report and the one on which the sound ranging measurements must be made. Our microphones, of a type used for long-distance telephony, were far more sensitive to the shell wave than to the gun report, which tended to be completely obscured by the loud crack. They were also sensitive to rifle fire, traffic noises and dogs barking. Howitzers make no shell wave because of the low initial velocity, but most of the activity was that of gunfire, which defeated us.

It was a year before we solved this problem. In our quarters at La Clytte, the privy was reached through the kitchen, and as the house in winter was hermetically sealed, when one sat down one closed the only aperture between the house and the outer world. There was a six-inch gun battery behind the house, and the crack of the shell wave was deafening. What we noticed, however, was that the shell wave only affected our ears, whereas the succeeding low gun report lifted us somewhat and let us drop again. Clearly the gun report created a large pressure difference. Then, a sapper joined my section, named Corporal Tucker, who had been a lecturer at Imperial College and while there had done research on the cooling of fine platinum wire by air currents.[78] His results hinted at a solution of our problem. We sent to England for some Wollaston platinum wire and when it arrived we bored a hole in the side of an ammunition box, stretched the fine wire across it and arranged that it should be in one arm of a balanced Wheatstone bridge, with our galvanometer string recording any change of resistance. The gun report created a draught through the hole, and one could see the red-hot platinum wire blinking. I shall never forget the occasion when we first tried out the new device in our dugout. A German field battery started up. The shell waves produced hardly any effect because their vibration was so rapid, but the gun reports made large clean 'breaks' on the cine film due to the deflection of the wire.

[78] William Sansome Tucker (1877–1955) was the inventor of the 'Tucker' microphone.

From this time on, sound ranging produced exceedingly valuable results. It recorded not only the position of the gun, but also the corresponding shell burst, so that one could tell on what target the hostile battery was ranging, and from the time of flight one could tell the type of battery. The great disadvantage was that it could only work when the wind was light or was blowing from the enemy towards us (an easterly wind). It was as hopeless to try to work when the wind was westerly as it would be to see through a thick fog. When the sound is travelling against the wind, it is thrown upwards. However, conditions were ideal in foggy weather, when other means of location failed. Before some of the main attacks, the sound rangers had mapped out all the opposite battery positions with an accuracy which was verified when we advanced. Towards the end of the war there were about forty sections covering the whole front. When we captured a German order of the day, it showed how much the Germans respected our sound ranging:

> In consequence of the excellent sound ranging of the English, I forbid any battery to fire alone when the whole sector is quiet, especially in east wind. Should there be any occasion to fire, the adjoining battery must always be called upon, either directly or through the Group, to fire a few rounds.
>
> June 23rd, 1917

It might be added that by this time we had no difficulty at all in sorting out the records when two batteries were firing at once. Our microphones were evenly spaced along the arc of a circle, and it was easy to tell by eye which 'breaks' belonged together because they formed a regular curve on the film. One's eye soon became very expert in seeing these curves and spotting which were due to enemy guns.

I think that a main reason for the rapid development in an efficient method of sound ranging was that it all took place in the front line. We had at first no experimental station in England. We gave each section a mechanic with tools and a small lathe, so that it could try out gadgets on the spot. We arranged a meeting of heads of sections about every two months, when experiences were exchanged and new devices described. Thus, experiments were carried out by men who knew the front-line conditions and could properly assess the advantage of any new device. We generally had our meetings at

Doullens, and they were followed by a dinner which was rather an orgy. There were hardly any scientific schemes in the First World War, in complete contrast to the many developments in the Second World War.[79]

The amplifying valve was just coming into its own and was used for radio communication and for overhearing enemy telephone conversations in the front line. There were geological experts who advised on wells and water supply. There was little else, and almost no demand for scientists for special duties. When we were increasing the number of sections, I went to the base camp at Rouen to get recruits. I had only to say, 'B.Sc.'s,[80] one pace forward' to get dozens of men stepping out of the ranks. These men were put through training tests and we selected the best. I remember after one such selection the sergeant coming to me and saying, 'Could we not perhaps pass Jackson, sir, he says he comes from the Royal Observatory at Greenwich and is the man who found another satellite of Jupiter.' I think he had been failed for faulty arithmetic, but on the basis of this claim we let him through. He was John Jackson, who later became the Astronomer Royal at the Cape![81]

Shortly after America came into the war in 1917, Professor Trowbridge of Princeton University was posted to us to study the British method of sound ranging.[82] The American army had decided to adopt our method, after comparing it with others. We trained some American officers in our school. Trowbridge later became a great family friend. My wife and I stayed with the Trowbridges in Princeton in 1924, and he was godfather to our daughter, Margaret. He was a modest, wise and most likeable man with a wonderful sense of humour. One

[79] WLB and his sound ranging team can be credited for playing a major part in shortening the Great War as their microphones pinpointed the German artillery so closely that they were effectively disabled. An excellent and thorough account of his contribution can be found at <http://rsnr.royalsocietypublishing.org/content/59/3/273.full>.

[80] That is, those with Bachelor of Science degrees.

[81] John Jackson (1887–1958) worked on stellar parallax.

[82] Augustus Trowbridge (1870–1934) carried out research on infra-red radiation and built a machine to make optical gratings, the 'echellete'. Curiously, his biographical memoir for the National Academy of Sciences of the USA in 1937 makes no mention of WLB's role; see <http://www.nasonline.org/publications/biographical-memoirs/memoir-pdfs/trowbridge-augustus.pdf>.

rather awful thing happened. Soon after we had shown our secret methods to the Americans, a complete description of our sound ranging method was published in an American journal. We never found out how the leak occurred; I am sure it was not Trowbridge, who was discretion itself. There was also some 'trouble' after the war, as is shown by the following extract from a letter written to me in March 1920 by Cooke.[83] Cooke was also a professor in the physics laboratory in Princeton; he had enlisted in the British Army early in the war and later became a sound ranger:

> Well, he gave an illustrated lecture on S.R. here in which he got off the usual dope—French showed it was feasible, British brought method and organization to perfection, Americans copied and added nothing. Next morning the New York 'Tribune' had a long account of the lecture in the best U.S. journalistic style—British attempted crude but unsuccessful experiments, but it remained for the Americans, particularly Princeton men, to bring method to technical perfection. War won by this means. Trowbridge was madder than he had ever been in his life. They refused to publish a denial. He ran down the reporter and found he hadn't attended the lecture! He got his denial published on a threat of publishing an account of the whole business in the New York 'Sun' through his friend Munsey, the editor.

The opposite numbers of the sound rangers were the 'flash spotters'. A section had three or more observation posts, furnished with theodolites. When an observer saw a gun flash, he pressed a button which gave a signal in the other posts; this helped the posts to concentrate on the same gun. The intersection of their bearings gave the gun position. Flash spotting conditions in general were favourable in clear westerly weather, when sound ranging would not work, so the two branches fitted in well together. Harold Hemming stood in the same relation to the flash spotters as I did to the sound rangers.[84] We both had a period at the front and then were pulled back to GHQ to supervise the scientific side of the work generally.

[83] Hereward Lester Cooke; expert on acoustics. He supervised Arthur Holly Compton who gained the 1927 Nobel Prize in Physics.

[84] Henry Harold Hemming (1893–1976) was a Canadian artillery officer serving with the British Third Army.

Here is a *Times* report of an article in the *Geographical Journal* in January 1919 by A. R. Hinks, then secretary of the Royal Geographical Society:[85]

> Sound Ranging and Flash-Spotting
>
> Mr. Hinks remarks that the most elegant development of all scientific warfare is in sound-ranging. It will never been known how many people thought of its elementary principles, and we do not at present know whether the enemy thought of the idea independently, or whether he heard somehow that sound-ranging was practised on the other side. So far as we know, his method of putting the idea into practice with stop-watches was but a poor thing, and was little credit to him as a scientific soldier. In a captured order, signed by Ludendorff, reference is made to the fact that the English have a well-developed system of sound-ranging, and insisting upon the importance of capturing the self-recording apparatus used by us. Ludendorff added that precautions should be taken to camouflage sound, but as to these precautions, Mr. Hinks remarks that the enemy will be disgusted to learn that if you fire six guns at once it may be that the positions of the six guns are determined at once when you have really good and skilled operators. The enemy got his sound-ranging going about two years too late, having until near the end only the elementary method of stop-watches. At flash-spotting—another recent development—he was better. Both sound-ranging and flash-spotting, however, require the maintenance of a complex telephone system, which is far from easy, especially when Tanks are about. According to Mr. Hinks, the German, in his official reports, was eloquent on the delay due to the destruction of his connexions by enemy fire.

I had a school under my charge at GHQ in which we trained the recruits, both officers and men. My ally in setting up the school was R. W. James. James and I were in the same Part II class at Cambridge. Just before the war he volunteered to go as a physicist with Shackleton's expedition to the Antarctic.[86] War broke out while the expedition was in the South Seas, and the party was of course ignorant of the war's existence since they had no radio communication. James played a

[85] Arthur Robert Hinks (1873–1945) was a British astronomer and geographer, credited for measuring the solar parallax.

[86] Ernest Henry Shackleton (1874–1922) was an Anglo-Irish explorer who participated in four British expeditions to the Antarctic, of which he led three.

notable part in the adventures of the party. The ship was trapped in an ice sheet and sank, leaving the party stranded on an ice floe, which slowly drifted north. Their chronometers had run for too long without a check and were unreliable, so they had no idea of their longitude. A copy of the Nautical Almanac had been saved from the wreck, and with its help James mastered the method of using the occultation of stars by the dark limb of the Moon in order to determine the true time and hence the latitude. This enabled the party to know when they were nearest to Elephant Island. The boats were launched and a successful landing made on a minute strip of beach. Shackleton made his epic boat voyage to South Georgia to get help, leaving the main party on Elephant Island, where they had to subsist for some months on penguins and seals until a steamboat could rescue them. They lived under an upturned boat, and James relates how in the middle of the night he heard a comrade mutter in his sleep, 'Turn it over on the other side and it will be done nicely,' so obsessed were they with dreams of food.

Very soon after his return to England, James volunteered and, hearing about our sound ranging, managed to get attached to my unit at Kemmel Hill. We worked together a great deal—he was a tower of strength. When I went to GHQ, he joined me there. We had got so used to working together that, when I was appointed to the Chair at Manchester after the war, I asked whether James could join me as a lecturer. James had profound influence in building up the 'Manchester School' of X-ray analysis after the war.

One of my great pleasures when I came home on leave was taking out my sister Gwendy. She was seven years old when the war started, seventeen years younger than me. It was the greatest of fun to give her fun. I remember taking her to a revue where the inimitable George Robey appeared, and Gwendy was weak with laughing and giggling. There was an Australian soldier sitting behind us, and he said to me, 'I came here to see this revue, but I have spent most of my time watching your little girl!' It is strange how much anyone involved in a war turns to children. When units of the BEF were quartered in Cambridge at the onset of the war, due to go out to France shortly, the men borrowed prams complete with babies and took them out for walks.[87] It was the same in France and Belgium where any children were available.

[87] BEF, British Expeditionary Force.

Figure 21 The Sound Ranging Team, Section France. Left to right: J. P. Gott, William Tucker, Lucien Bull, A. H. Atkins, WLB, Captain T. E. Paris, unknown. (Courtesy: the Royal Institution London)

Anyhow, Gwendy and I would go on expeditions together or go out rowing. She had rather a serious time with no other young family, and it was so very pleasant to make her enjoy herself.

After the armistice in 1918 I stayed for a while in Colonel Jack's headquarters at Campigneulles, near Montreuil sur Mer, in order to collect material for a handbook on sound ranging. Then, early in 1919, I returned to Cambridge. Just before the war I had been elected to a Lectureship and Fellowship of Trinity College, and I started my new duties of supervising students and giving lectures at the Cavendish Laboratory. Soon after my return, however, Rutherford was incited to take charge of the Cavendish Laboratory, which J. J. Thomson had resigned. I was invited to go to Manchester in Rutherford's place and accepted, so I had only a brief time at Cambridge. The college was very generous to me. In the earlier part of the war it had made up my lieutenant's pay to the stipend which I should have received as a lecturer at Trinity, and this went on till I became a major and my army pay became the greater. I wish I had stayed longer at Trinity to repay this generosity, but a chance like Rutherford's chair at Manchester had to be seized. In this brief interlude at Cambridge, however, the most important event of my life occurred, because I met for the first time my wife-to-be, Alice Hopkinson. I had heard a good deal about her. She was first cousin of my great friend, Cecil Hopkinson. Her father, Albert Hopkinson, had been a doctor in Manchester. Alice was always referred to by the relations as an extremely pretty girl, a statement generally accompanied by a slight shaking of the head. She was then up at Newnham College, reading history and having a whale of a time. A large number of naval officers had been sent to Cambridge after the war, most attractive young men who gave endless parties with their war gratuities, and the undergraduates were many of them demobilised officers, so that the girls at Newham and Girton were in great demand. We met first at a *thé dansant*, a form of entertainment then popular but now I believe obsolete. I think it was on the fourth occasion we met that I proposed. But Alice was having much too good a time, she was only nineteen, and she had an ambition to complete her university studies.

2.5 Manchester Days

Alice's family moved from Manchester to Adams Road, Cambridge, just about the time I joined the staff at Manchester University, in October 1919. Her father had a general practice in Withington and was dearly beloved by his patients. When later he came to stay with us in Didsbury he seemed to be stopped by every third person in the street to talk about old times. He had a serious illness at the end of the war, largely due to overwork, and decided to give up his practice. In Cambridge, he was appointed demonstrator at the Anatomy Laboratory. All doctors who trained at Cambridge for a whole generation knew and loved 'Hoppy'. I cannot help thinking that he stood out in contrast to other demonstrators because he had had a long practical experience.

I was only twenty-nine when I took up my duties as head of the physics laboratory in Manchester and was handicapped by not having first served my apprenticeship in a junior position in a department. Further, we had forgotten most of our physics during the war; I think everyone had the same experience. Not only was it hard to remember physics but also the textbooks bored us profoundly after the excitement of the war and sent us to sleep. Finally, the students were largely older men who had been demobilised and were a tough crowd. The staff consisted of Evans,[88] Florance,[89] Robinson, James, Tunstall,[90] and me. We had a strenuous and difficult time getting the laboratory going again and training ourselves to teach. It was not made easier by a vile series of anonymous letters which started soon after I came to Manchester. These letters and postcards were directed mainly against myself and James and abused us bitterly as incompetent and useless. They were the worse because it was clear that the writer had access to laboratory gossip and knew of every mistake we made: there was just a small element of truth in every criticism. This experience made me understand the deadly nature of anonymous letters. At first one ignores them, but the effect is cumulative and in the end it drove me into what was really a nervous breakdown. Curiously enough, I recovered when the letters began to attack my father and Rutherford as well.[91] Even in my state, I realised the absurdity of such vile travesties. I shall never cease to be grateful to colleagues in Manchester who understood the trouble and helped me through the worst of it. My recovery was hastened also by James and myself getting some first-rate results in our research.

As a bachelor in Manchester, I kept house with Douglas Drew,[92] who had been a sound ranger and who was a lecturer under Conway in the Classics department.[93] We were looked after by old Charlotte, an arrangement made by my mother. It was interesting to live in such close contact with an Oxford classicist; it was almost as if we talked different languages and was very stimulating. I was much attached to Drew.

I was elected a Fellow of the Royal Society in 1921. Among the letters I received was one in a handwriting which made my heart

[88] Evan Jenkin Evans (1882–1944) was Assistant Director of the Manchester Laboratories from 1919, moving to Chair of Physics, University College of Swansea, in 1920.

[89] David C. H. Florance worked as a student under Rutherford at Manchester and later became Professor of Physics at Victoria University College, New Zealand.

[90] Norman Tunstall; lecturer in Physics at Manchester.

[91] It seems that these letters stopped when a certain member of staff left to go elsewhere.

[92] Probably D. L. Drew; Professor of Greek in Swarthmore College. Formerly Lecturer in Classics in the Victoria University, Manchester.

[93] Robert Seymour Conway (1864–1933) was a British classical scholar and comparative philologist.

turn over. It was Alice writing to congratulate me. We had never met or written to each other since I left Cambridge. My troubles in Manchester had been such that I think I felt my chances in that direction were nil; it seemed all wrong to drag such a gay and popular girl into a world of stress. But—I replied, asking when I could come to see her, and arranged a tea date in Newnham. The train from Manchester on that day only got in a quarter of an hour before the tea appointment, and I considered this was cutting things too fine, so I remember I stayed in London the night before in a hotel so as to run no risk of being late. We became engaged the next day.

Alice's parents were always dears to me. Her mother was utterly outspoken with no inhibitions whatever and endless moral courage. She had perhaps inherited a good deal of the aristocratic instincts of her German forbears, who were reputed to ride their horses through the windows of tradesmen who annoyed them by sending in bills; but in her case the same self-confidence and courage ran entirely in good channels. She was an ideal mother-in-law, always taking my side, though it was not often necessary to take sides. My father-in-law was a really good man. It is remarkable what an effect such a man has on all who come into contact with him. I would talk with the shop people in Cambridge, sometimes pretty worldly people with an eye for the main chance, and quite a different tone would come into their voices when they mentioned 'Dr Hopkinson'. I think sinners are the best judges of saints. They both had the saintly quality of thinking little of themselves and thinking other people much better than they really were, so that they had to try to live up to what was thought of them.

I had some grand holidays after the war with George Thomson on his yacht *Fortuna*. We sailed her by degrees from Harwich round to Dartmouth. She had no auxiliary engine, so it was quite tricky sometimes getting into or out of harbour. We had, of course, endless adventures, as one always does when sailing. We stayed for a few nights with the Goodharts at Frinton, where they had taken a house for the summer belonging to a Mrs M—. It was a very grand house, rather like an American one in a film, complete with swimming pool in the garden. I was put in a very small bedroom in the back region with a label on the door, 'Mr. M—'s bedroom'. I longed to follow up this interesting sidelight on the M— family.

Figure 22 J. J. Thomson and family. The young G. P. Thomson at his father's shoulder. (Courtesy: Patience Thomson)

When we were on the south coast, Arthur Goodhart joined us for a time.[94] He and George were then Fellows at Corpus, and Arthur Goodhart is now Master of University College in Oxford. We were sailing in the Solent, I being the helmsman, on a day of very light airs. I kept a good lookout ahead but was surprised by a loud rushing sound which bore down upon us from astern. It was a channel buoy; we were actually moving through the water more slowly than the counter current of the tide and were drifting backwards on to it. Goodhart, with perhaps memories of the tactics in the Backs, put out a leg to fend the buoy off. Our boat was four and a half tons, and the buoy perhaps twenty; it was very lucky that his leg was not trapped when we struck the buoy a glancing blow.

When we were in Poole Harbour, Goodhart was very keen to see an 'English watering place', so we all trekked along to Bournemouth for a really grand dinner. Alas, we looked so disreputable that none of the grand hotels would admit us, and we had to eat in a small café.

[94] Arthur Lehman Goodhart (1891-1978) was an American-born British jurist and lawyer, and a Fellow of Corpus Christi College, Cambridge. He was the first American to be a Master of an Oxford College (University College). He was married to Cecily Carter.

I also had two climbing holidays with Cambridge friends. The first of these was in the Lake District, where we stayed at the Wastwater Hotel and did a number of climbs on Great Gable and elsewhere. I never became good enough to lead, but I liked the thrill of the climbing. In another summer we went to Skye, dividing our holiday between Sligachan and Glenbrittle, and climbed in the Coolins. Sligachan had a civilised hotel with quite a number of guests. Glenbrittle on the other hand was a tiny place reached by a rough track over the moors, just one farm and the postman's house where we stayed—Aston,[95] Milne,[96] Darwin and I. At Sligachan there was a rather unattached, no-longer-very-young lady from New Zealand and, to our horror, when we went to the tiny four-roomed house at Glenbrittle, she came too. We had hoped to be alone, and we tried various devices to get away. It was as usual very wet, there being about one fine day in three and, as we had to tote everything in a rucksack, we only had two suits of clothes. We put on the wet one in the morning and changed to the dry one in the evening. The lady used to follow us when we set out in the morning, but we could usually shake her off by walking through a deep river in the glen. When we came first to breakfast Charles Darwin used to try discouragement by blowing tobacco smoke over her bacon. The kind-hearted Aston was the only one who behaved like a gentleman to her. The climbing in Skye was very interesting, the rock being so crystalline and rough that one can stick to quite steep slopes. The scenery was wonderful too, with the contrast of the jagged, black main Coolin range with the smoke-red Blaven.

In the summer of 1921, when Alice and I were engaged, we had a holiday with my parents in Brittany, where they had taken a house near St Briac. The Queen of Romania also had a villa near us, and the royal party used to bathe on the same beach. My mother happened to have a white anchor embroidered on her bathing dress which looked not unlike a crown at a distance, and the photographers used sometimes to sneak up and take us under the impression that they were photographing royalty. I remember once that one of the ladies in waiting fainted on the beach, and one of the royal scions poked her

[95] Francis William Aston (1877–1945); awarded the 1922 Nobel Prize in Chemistry for the invention of the mass spectrograph at the Cavendish Laboratory, Cambridge.

[96] Edward Arthur Milne (1896–1950) was a British astrophysicist and mathematician. He was an early researcher on the expanding universe concept.

Figure 23 Alice's favourite umbrella pines at Valescure. (Sketch by WLB; Courtesy: Patience Thomson)

with his toe as she lay and said something which we interpreted as 'she will come round soon', which I thought was a superb example of princely reactions. Then I went to stay with Alice's people, who were having a holiday at Ullswater in the Milcrest Hotel.[97]

We were married on 20 December 1921, at Great St Mary's in Cambridge. Alice's bridesmaids were Enid, Gwendy, her cousin Alice Hopkinson, Felicity Hurst and Molly Thynne, and Vaughan Squires was my best man. We went for our honeymoon first to Rockdunder near Wrington in Somerset, lent to us by Alice's Aunt Monica Wills, and then to Valescure, near Fréjus on the Riviera. We had a delightful room in the hotel, and the weather was so balmy that we were generally able to breakfast on its balcony. It was Alice's first Christmas away from her parents and, to make the break seem less, I dug up a juniper tree in the hills, we bought decorations for it in Nice, and so we had a real Christmas tree in our room. We explored the hills behind the hotel for ever greater distances till we began to be familiar with the paths through the woods and finally made a long excursion to the highest ridge of the Esterel range and saw snow mountains, the first Alice had ever seen. Alice had only once been abroad before, when she went with her family to Brittany, and everything was new to her. We had lunch with the de Broglies on our way through Paris, and the Duchess had promised Alice that when she woke up after her night in the train she would see

[97] Originally a temperance hotel; now the Best Western Glenridding Hotel. In 1955 the hotel was the base for Donald Campbell's attempt at Ullswater at the world water speed record, at which he succeeded, thus making him the first man through the 200 mph water speed barrier.

'arbres exotiques'.[98] There they were, the palms and all, when I pulled up the blind of the sleeper: a great moment. We stayed at Avignon on the way back, and there I had evidence of Alice's decided character. Our hotel bustled us away in the morning, saying that there was barely time to catch the Paris express. We did the last packing in a great hurry, leapt into a carriage, the horse was lashed, and away we sped to the station, which was at a very short distance away. Of course we found when we got there that the train was not due for another half an hour, and Alice began to remember details she had not packed. So we marched back to the hotel.

I was very anxious about the house I had bought for us in Didsbury, as I had to choose it alone. Alice was at that time absorbed in the final examination for her Tripos and, as houses were hard to come by, I had to snatch at the chance of one before Alice could come to see it. Charlotte was in charge while we were away on honeymoon, but I foresaw

Figure 24 'The hotel at Valescure, where we spent our honeymoon', 1921. (Sketch by WLB; Courtesy: Patience Thomson)

[98] Louis-César-Victor-Maurice, Sixth Duc de Broglie (1875–1960), was a French physicist. who made important contributions to X-ray diffraction and spectroscopy. He was married to Camille Bernou de Rochetaillée (1888–1966) in Paris.

trouble if she were there when we returned and begged my mother to find someone else to help Alice in the house. I do not think Mother quite appreciated the psychological importance of doing this—anyhow, Charlotte was still there when we returned from our honeymoon and it was very uncomfortable for Alice. Charlotte used to go about the house, muttering, 'Poor Mr. *Villy*', profoundly distrustful of leaving me to Alice, who had absolutely no experience of housekeeping and cooking, though she soon learnt it with characteristic thoroughness. Fortunately, Charlotte departed after a few days, and we had for a time a rather unsatisfactory mother whose daughter also lived in the house. Then we got Sarah as a cook. Sarah was minute and came from a Cumberland mining family. She was engaged without an interview, and Alice was staggered when this tiny creature with an almost incomprehensible dialect appeared on our doorstep. She was blunt and 'jannock', characteristically North Country and a 'treasure'.[99]

It was not an easy time at first. Manchester was a dreadfully dirty and ugly place with a vile climate, foggy and drizzly. It was worse then than it is now, since it has gone in for smoke prevention. Through almost all the winter the pavements were wet, there being never enough sun to dry them, and so many of the days were dark. When her family had migrated from Manchester to Cambridge, Alice had thankfully supposed she had left such a clime for good—and then I bore her back there! Then again, she had such a riotous time at Cambridge with all the young people, and here she was married to a professor. We went to formal parties of the middle-aged, where Alice as the new bride wore her wedding dress and was taken in by the host. At one of these Sir Edward Donner, a leading Manchester figure but then in advanced years and whose sight was failing, led Alice into the china cupboard instead of the dining room, followed by some of the guests.[100] One hostess held regular tennis parties, and I remember that when Alice and I turned up at one of these she said, 'Of course, we held our young party yesterday.' Oh dear!

Popular girls have a gay life before they are married, and young husbands feel acutely how hard it is to give them anything which can

[99] 'Jannock' is Northern dialect, according to the dictionary: pleasant, outspoken, genuine, straightforward, generous.

[100] Edward Donner (1840–1934), First Baronet, was a banker and a philanthropist.

correspond. The husband comes home tired after a day's work and can at first do so little to amuse a wife who has spent rather an empty day looking forwards to fun when he gets back. It is perhaps different now, when there are no servants, and the young wife looks after the house and often has a job too. In our early days maids were regarded as part of the normal scheme, and indeed my mother was shocked that we only had one.

Then again there is a difficulty peculiar to scientists. When such a one is absorbed by some problem of research, he gets sunk into a deep well of concentration and outside events produce no impression—he really is 'lost to the world'. He is then a dull companion. This is hard on his nearest and dearest, who cannot resist pulling him out of his well, from time to time, to see if he is still there, and it takes a long time for him to get back to his problem again. I know I was often torn between the research and Alice, feeling that I was failing her wretchedly as a companion when I was hot on the chase.

In our first married year we had an Easter holiday at the British Camp hotel near Malvern.[101] It was a lovely spring holiday, with the woods and fields around full of wild daffodils. Alice encouraged me to take up watercolour sketching again, of which I had done a little owing to Mother's inspiration when I was a boy. Again Alice showed her spirit in not putting up with things, where I would have meekly not made a fuss. There was something wrong with our bed, as I think the springs had come through the mattress, and our first night was very uncomfortable; so at two in the morning we arose and raided the sitting room downstairs. We took away every cushion we could find, some thirty or more, and built a wonderful mound of them on our bed on which we reposed peacefully. If I remember rightly, we said nothing but let the pile in the morning tell its own story. It really was completely effective; the management was thoroughly upset and apologetic. In the summer Alice and I had a holiday on the Broads, which she really enjoyed. I was somewhat anxious about the camping in the boat, but any primitiveness was compensated for by Alice's intense love of country and of water.

In the autumn we paid a visit to Sweden. As my father and I had been awarded the Nobel Prize in 1915, in the middle of the war, it had not been possible for us to go to Sweden for all the ceremonies and

[101] Now the Malvern Hills Hotel.

festivities which take place when the King distributes the prizes. So when I was asked to give a lecture in Norway, I asked if I might come to Sweden as well, and this visit was made the occasion of a series of receptions for us by the Swedish scientists, though the programme did not include the usual ceremonies in the palace. We went first to stay with the Bohrs in Copenhagen. At a large dinner party they gave, it suddenly became clear that I was expected to make a speech, and Alice was on tenterhooks till the ordeal was over because she had never heard me make one.

We then went on to Stockholm and were met at the station by the great Arrhenius himself, the famous Swedish physical chemist, who had a dinner party for us and took in Alice to sit with him.[102] He started the conversational ball rolling by saying, 'I am—years old, I have married two wives, I had—children by the first and—by the second. How about you?'

Alice replied, 'I am twenty-two years old, I have had only one husband, I have been married for seven months and as yet I have no children.'

We also saw much of Benedicks, the metallurgist, whose young assistant, Westgren, was our guide around Stockholm.[103, 104] Benedicks' room in his laboratory was extraordinary. The walls were covered with small drawers, and in them was a supply of every item one could imagine; there were dozens of drawers with different kinds of screws, others with nuts and bolts, foil, wire of all gauges and so forth. Westgren said of him, 'His soul, it is full of little *veels*.' When Benedicks gave a dinner party for us, he was very anxious that all should go well. He warned me that I should be expected to make a speech and that I should start with 'the position I occupy on the left of my hostess gives me the right to reply for the guests'. 'They will all wonder', he said, 'how you came to think of such a clever remark.' He went on, 'I do not know the arrangements for dinner—that is the province of my wife—but I think there comes at one time a little fruit, and the little fruit will to you a signal be.' Sure enough, about the entrée stage

[102] Svante Arrhenius (1859–1927) was awarded the 1903 Nobel Prize in Chemistry and became Director of the Nobel Institute in 1905.

[103] Carl Benedicks (1876–1958) was a Swedish professor and metallurgist.

[104] Arne Westgren (1889–1975) was a Swedish pioneer in the use of X-rays in the study of metals.

Benedicks, perspiring profusely, made his speech, including a somewhat dubious reference to my joint work with my father as an effort of 'The Father, The Son and The Holy Ghost'. After another course or two I got restive, as it seemed so rude to leave his kind words unanswered, but each time I tried to rise to my feet he waved me down in great agitation. However, at last the little fruit appeared and I rose with thankfulness, but could not help maliciously delaying the 'right and left' bit, despite Benedicks' frantic signals. Of course I brought it off in the end and all was well. We then went on to Oslo, where I was to give a lecture to the Anglo-Norwegian Society. I realise now it was a bad popular lecture; I had not learnt the right level at which to aim. We had a holiday in the Norwegian countryside, above Lake Myosen, in lovely sunny still weather with the grass brilliant green and the trees yellow, red and purple. We elected to come back over the mountain railway to Bergen, but this was a great mistake. Alice felt too miserable with all the twists and turns to be able to enjoy the extraordinary way in which the line goes along almost perpendicular precipices. It was very rough too coming back from Bergen to Newcastle in the *Jupiter*.[105]

This Scandinavian visit sent up the value of my shares as a husband because I was able to show Alice that being married to a scientist offered certain advantages in the way of welcome in foreign countries. It was sad that our visit did not include the reception at the palace, but on the other hand, if I had gone to this directly after the war, I should not have had Alice to come with me. My father never went to Stockholm to give his Nobel lecture as I did. Why this was so I have never been able to understand.[106] The Swedes were very disappointed according to Barkla, who also got his prize in the war and afterwards went to be honoured.[107] The Swedish papers remarked on our youth, especially on my being accompanied by such a very young wife, and they were also astonished that I gave my Nobel lecture in a lounge suit. It had never occurred to me that I ought to appear in a morning coat.

[105] Operated by the Bergen Line and Nordenfjeldske.

[106] Apparently WHB refused to go, saying 'Germans will be there!', thus reflecting his bitterness at losing his son Bob during the war.

[107] Charles Glover Barkla (1877–1944) was awarded the 1917 Nobel Prize in Physics for his work on X-ray spectroscopy.

It was in this year that I started the research work which led to the series of 'B. J. B.' papers. James, Bosanquet and I made a series of experimental investigations into the optics of X-ray diffraction on a sounder quantitative basis.[108] This in its turn made it possible to attack much more complicated crystal structures, because quantitative measurements gave far more information. So we started the 'Manchester School' of X-ray analysis.

In 1923 Alice and I had our Easter holiday at St Ives in Cornwall. Alice had chosen St Ives for a holiday because she had been there before and was very keen that I should like it. In the summer we went to another place known to the Hopkinson family, Morwenstowe, on the north coast of Devon. We had a gentle holiday because Alice was expecting her first baby, Stephen, who was born that November. We had by that time acquired a car, a two-cylinder Wolseley. It was a very small two-seater car (it was rather risky to cram in three people because on occasions this broke a spring) which did over fifty miles to the gallon. This first car was a great joy. We blithely set off for Devon, not realising how long and steep some of the hills were, and it was rather a marvel that we had no mishaps. The back-wheel brakes, which were the only ones, were worked by cables, and when the car went over a bump these cables were apt to vibrate, so that the brakes were intermittently rammed on causing terrific jars and jolts. On the way back to Manchester, suspecting something was amiss, we took the car to the Wolseley works, and it was discovered that three of the five bolts holding in one of the back wheels had sheared through. Even the works experts were somewhat shattered, especially when they heard we had negotiated Porlock Hill.

Alice had a very bad time when Stephen was born. Our doctor was her father's former partner, Dr Stocks: I believe he was brilliant, but I think he lacked common sense. I realise now that I should have insisted on his calling in a specialist when it was clear that the birth was going to be a difficult one, but of course I had no experience. The doctor and nurse were both completely worn out when the baby finally appeared, and Alice was left alone to pass the night in a sad state of exhaustion when a minute angel in the form of Sarah appeared in the small hours to comfort her and give her tea.

[108] Charles H. Bosanquet had been a sound ranger in the First World War.

In 1924 we had great adventures. I had been invited to give a series of lectures in the summer vacation at Michigan University, Ann Arbor. The British Association was holding its summer meeting at Toronto in Canada, so we combined the two events. Stephen, then eight months old, was parked on the kind grandparents in Cambridge with his nurse, Miss Graesser. Ann Arbor is a small town, centring round the large university. We were given rooms by one of the chemistry professors, Smeaton.[109] I was billed to give a course of, I think, four lectures a week on crystal analysis, and four on X-ray spectra. The former I could do without too much strain, but the latter were difficult, as spectra were not really in my line. I was too inexperienced to realise that eight lectures a week, of a new kind, were too many; I ought to have held out for the one set only. However, I struggled through.

We loved Ann Arbor. It was very hot, but there was a river for bathing and canoeing, there was an interesting country round, and we played golf and tennis. The faculty members were very kind to us, but on one occasion Alice got into trouble. One of our great friends was Dean Kraus of the graduate school, and he took us to a 'ball game' in Detroit.[110] Alice was invited by a reporter of the students' paper to give a Britisher's view of such a game and did so in very racy terms. She was subsequently called on by a very formidable lady, who said she was speaking for the faculty wives generally and informed Alice that her story was much too undignified and deplored by them all.

After some six weeks at Ann Arbor, we next were off to Toronto, and a friend had offered to drive us to Detroit, where we could board the train. We travelled in those days with enormous cabin trunks, golf bags, tennis rackets, hatboxes, etc. and, as all these could not get into the friend's car, we arranged for an agency to take them to Detroit and produce them at our train. After a day seeing Detroit, we went to the station—no trunks! As we had booked our sleepers and had to produce all our belongings for customs at the frontier, I was in despair. Our kind friend took us to a number of 'depots' in Detroit and to our joy we finally found our belongings in one that was miles from the station. As the sympathetic railway official said, 'We sure had a

[109] William Gabb Smeaton (1874–1959) worked at the University of Michigan from 1902 to 1942.

[110] Edward Henry Kraus (1875–1973) was a mineralogist and the recipient of the Roebling medal from the American Mineralogical Society in 1945.

howl coming'. When I have a nightmare, it is very often about travel plans going hopelessly wrong and I think this started with the Detroit incident.

After the British Association meeting in Toronto, we had a choice between a tour across the continent to Vancouver or a holiday in the Canadian woods. We chose the latter. I think we were almost the only people who did so. We went first to Bigwin Island, in the Lake of Bays. The hotel owned and used the whole island.[111] There was a large central building with the usual recreation rooms, candy stores, tobacconists, news kiosk and so on. The bedrooms were in separate one-storey buildings, and the restaurant was built on piers over the lake. University students served in the restaurant, and we would meet the young lady who brought our breakfast at the dances in the evening. Half an hour after each meal started, further entrance to the restaurant was barred and anyone who came late had to eat cold food in a side room. A golf course had been carved out of the woods; there were tennis courts, motor boats, sailing boats and canoes. Most of the guests were Americans, and two things about us puzzled them. They could not understand our pleasure in going out in a small sailing boat, when one could roar along far faster in a speed boat. They also could not understand our habit of going for long walks around the island alone; they concluded we must have just married because we had this strange taste for being alone together. One day when I was sketching I heard a loud shriek from Alice nearby and running up I found Alice and a mink eying each other. I said, 'I am sure it's not dangerous—don't be frightened.' Alice said, 'You don't understand, it will just go round my neck!'

Then we went on to a very exciting trip. We went to the Algonquin Hotel in Algonquin Park, and after staying a night or so we got a guide—Jack—two canoes and enough food for a ten-day trip and made a long round in the lakes and woods. When we had to make a land passage from one lake to the next, the procedure was as follows. First, the guide and I took a canoe each on our heads and walked along the track through the woods, with Alice carrying the eggs and

[111] The hotel is probably the Bigwin Inn, opened in 1920 but now closed since 1960. From 1941 to 1945 the Dutch Royal Family summered on the island when they were exiled in Ottawa. Beatrix of the Netherlands and her family lived in private cottages. The Rotunda was used to store the constitution of the Netherlands.

the fishing rods. We then came back and took on our heads the tents and blankets and then came back a third time for the food. A three-mile portage thus involved a fifteen-mile walk, carrying some seventy pounds for nine of the miles. It was surprising how rapidly one became fit. On the last day, when most of the food had been eaten, I was able to carry a canoe and a pack, so that we only needed one journey. Jack paddled one canoe with most of the gear. Alice and I were only just able to keep up with him paddling our almost empty canoe together. It was an unforgettable experience. The lovely lakes, the quiet, the solitude (we only saw two other parties in the ten days) and the bird and animal life were fascinating. Practically every night we had a wolf chorus. A scout wolf would see our camp fire and howl, replied to by the rest of the pack. Alice had qualms about wolves and made the guide promise that we should camp on an island each night; as Alice's understanding of geography is not very strong, Jack had little difficulty in persuading her that he was carrying out her instructions. He also assured her that wolves did not eat humans unless the month had an 'r' in it, it then being the last half of August. Alice found it hard to believe that the wolf calendar was so accurate. Actually, I believe no case has ever been known of a wolf attacking a human being in Canada. A paper once offered a large reward for an authentic story, but it was never claimed. I thought that the loon cries were even more eerie than the howls of the wolves. The beaver colonies with their dams were very interesting. Jack 'guided' in the summer and worked as a telegraphist on the railway in winter. He had never been away from the northern woods. He marvelled at our ignorance of the country. On the other hand, he just could not bring himself to believe facts which were familiar to us. For instance, I remember telling him that, in mid-ocean, wave crests were often a quarter of a mile or so apart. This he found quite incredible.

It was in this year that I analysed the structure of aragonite, a considerably more complex crystal than any hitherto attempted. Alice thought the aragonite pattern was so pretty that she embroidered it on the front of a nightdress. It and calcite are different crystalline forms of calcium carbonate ($CaCO_3$), and the relation between the two forms proved to be very interesting. We were becoming more confident in tackling complex crystals, partly because of our quantitative techniques and partly because we were getting experience of the way in which the atoms packed together. This work on aragonite led later to my calculating the double refraction of calcite and aragonite, with a quite fair agreement between theory and experiment, although the assumptions on which the theory was based were imperfectly true.[112] Hartree had by this time come to Manchester and helped the X-ray work by calculating the scattering factors for the atoms.[113]

I remember the years best by the holidays and the visits abroad. In 1925 Alice and I attended the Solvay Conference in Brussels, the first conference to which Alice had been.[114] Old Madame Solvay was very kind to the ladies, although they were not officially there. They were provided with transport for visits to places of interest in Belgium, they had a box at the opera at their disposal, and we found at the end of the stay that all their hotel bills had been paid. 'Even our baths have been paid for!' exclaimed Peggy Rideal.[115] Flowers were sent to Madame Solvay as each new evidence of her kindness was disclosed, but the last discovery of complete hospitality demanded, the ladies felt, a large plant in a pot. I was proud to be able to take Alice to this conference where we were given such a royal time.

In the summer we stayed with Alice's Aunt Monica Wills. When Uncle Harry Wills was alive they always took a grouse moor for the autumn, and Aunt Monica was keeping up the family tradition in renting the laird's house with shooting attached at Aviemore in the Spey valley. The shooting party consisted of Aunt Monica, a lady missionary from India, a headmistress recovering from sleeping sickness, the mother superior of a convent, Alice—who was then expecting her next baby (David)—together with her father and mother, Stephen (then two) and myself. The poor head gamekeeper felt sorely tried to justify his existence. I was the only member of the party whom he could look after, and I had done very little shooting. We went on

[112] Double refraction refers to the splitting of a light ray into two perpendicularly polarised components on passing through certain crystals. This property is known as birefringence.

[113] Douglas Rayner Hartree (1897–1958) was a mathematician and physicist who carried out numerical calculations in the field of atomic physics.

[114] The Solvay Conference held in Brussels was founded by the Belgian industrialist Ernest Gaston Joseph Solvay (1838–1922) in 1912; the most famous physicists of their time were invited to it. Solvay was known for developing the ammonia–soda process for making sodium carbonate.

[115] Peggy Rideal (Margaret Atlee, née Jackson) married Eric Keightley Rideal (1890–1974), a celebrated physical chemist.

hunting roe deer with a rifle and a sinister knife, which was attached to the keeper's belt, but thank God we never had a shot. The deer were, quite unnecessarily, wary. The head gamekeeper was terribly pleased when I shot a woodcock, and there were a few other victims to my gun. He also instructed Alice and me in fly fishing. It was lovely country, and the 'Doune' was a fascinating old house.[116]

Then we went from there to a holiday specially arranged for us by Alice's Aunt Mabel Hopkinson, at Portsalon on Lough Swilly in Donegal. Poor Aunt Mabel was very disappointed that David was on the way, so that Alice could not be very active. It was fun to see Aunt Mabel, from Belfast, up against the Irish. We went by train to Londonderry and were to be met by a car sent by the hotel. Sure enough, a smart large Wolseley was waiting outside the station, and Aunt Mabel said, 'They have sent us a nice car, dears. I hardly expected anything quite so large, but then it is a very pleasant hotel to which I am taking you.' So we packed our trunks, when a dear old gentleman came up and said he thought there must be some mistake as it was his car. We then noticed behind it a dreadful old Model T Ford, with mudguards like banana leaves. The driver had pieces of string in his pocket and tried to lash our trunks on the mudguards. Luckily it turned out that the old gentleman was going to the same hotel, and he offered to take all our baggage. Then again, if we arranged for a car to take us anywhere, we had to allow for its arriving two hours late: one hour because the countryside took no notice of summer time, and the other the obvious hour of normal lateness. Breakfast was billed for nine o'clock. Aunt Mabel used actually to enter the dining room at nine o'clock and sit there in stony silence. At about a quarter past anxious faces peered round the door, and at half-past perhaps food began to arrive. There were two hills not far away, Muckish and Errigal, and we planned an excursion up Muckish. The driver of our car said he was indeed glad we were not going up Errigal, which was the second highest mountain in the world (the highest of course being Everest). Aunt Mabel was an ardent golfer, with a very characteristic game. She could not drive more than about one hundred yards, but all her shots were dead straight and she never foozled. It was rather terrifying to play her as partner, as one's mistakes showed up so glaringly. At one time my game was improving quite fast, but the paralysis of my left hand

[116] The Doune of the Rothiemurchus is the laird's family home.

was even then overtaking me and was disastrous for golf, where the left arm is so important in the swing. There was a pleasant rough golf course at Portsalon, and the local youth caddied for us. I remember asking one of them what he did and he said, 'In the summer I caddy'.

'And what do you do in the winter?'

'There's no work in the winter.'

David was born on 1 March 1926. We had a silly Welsh monthly nurse who overfed him with rich food, and he became very ill. The poor baby was only skin and bone when we had our summer holiday at Abersoch in Caernarvonshire. Both sets of parents in turn came to stay with us there. Alice lost her gold wristwatch on the beach; her father characteristically borrowed a rake from the farm and painstakingly searched for hours till he finally raked it up from the sand.

The next year we had the first of a series of family holidays in Pensarn, near Abergele in North Wales. The boys loved this place. It consisted of two rows of gaunt houses, practically all of them boarding houses, and a railway station. But there was a good beach for them, and the railway line was a very busy one, so that Stephen could indulge his growing interest in engines. I remember later that he went with us to a preparatory school where we had some thought of entering him. The headmaster took us round, and on each dormitory door was the name of a famous man. 'Nelson, do you know of him, boy?'

'No, Sir.'

'Shakespeare, have you heard of him?'

'No, Sir.'

'Ruskin?'

Here Stephen's face lit up and Alice and I hoped our progeny was at last going to distinguish himself. 'Yes, Sir, the name of an engine which runs through Pensarn.'

The river Elwy, inland, had pools and pebble beaches where the boys played. By this time we had a grand car. Alice never knew what present to expect from Aunt Monica—it might be a very big one or just a tract. On this occasion she opened an envelope and out came a cheque for £350 with which we were to buy a car. In those days, a car for £350 was a very grand car indeed and we bought a magnificent Humber, which was our pride and joy and envied by our university friends.

In the autumn we went to Italy and had a most interesting time. It was the occasion of the centenary of Volta's death, and the Italians had arranged a gathering in Como and a subsequent visit to Rome.[117] Mussolini was in his heyday and we were treated royally.[118] We were given our fares to Italy, railway passes anywhere in Italy, and all our hospitality for both Alice and me. We went first to Lenno on Lake Como for a fortnight or so and explored the towns around the lake and the villages in the hills. Then came the conference in Como, attended by scientists from all over the world. The Italians politely asked their foreign guests to give their papers first, and I remember Aston saying on the third or fourth day, 'Let's take this day off, it is only the Italians giving papers now.' He was a single-minded man. Anyhow, Aston, Eddington, I think Fowler, Alice and I went off bathing in the lake.[119] The bathing raft had a slide, and Aston thought this looked rather splintery and dry, so he made Alice go down first. Sure enough there was a tearing noise, and Alice declared that she just could not come out of the water. Eddington most gallantly removed his tiepin and closed the rent under water, so that Alice was able to emerge.

Alice, like many other members of the gathering, was taken ill at Como, we thought because of some lemonade which had been served at one of the parties. The stouter spirits who had restricted themselves to alcohol were not affected. She was so bad that we feared she would not be able to go on to Rome but she just managed it. We stayed there in the Hotel des Ambassadeurs. It was a very grand hotel and we tried to preserve the labels on our bags for as many years afterwards as possible. Meetings and festivities were continued. One of the secretaries, De Bosis, gave us a romantic party in quarters which he had inside the fortifications built by the invaders of Rome, I think in the fifth century.[120] De Bosis lost his life later, being shot down in a plane from which he was distributing anti-Mussolini pamphlets. Mussolini gave us a party. The ladies were asked whether they would rather have Mussolini or the Pope and plumped for Mussolini (the Pope meant special dresses). Mussolini received us and had something to say to

[117] Alessandro Giuseppe Antonio Anastasio Volta (1745–1827) was an Italian physicist known for the invention of the battery in the 1800s.

[118] Benito Amilcare Andrea Mussolini (1883–1945), known as 'Il Duce', was the leader of the Italian National Fascist Party, ruling the country as Prime Minister from 1922 until his ousting in 1943.

[119] Arthur Stanley Eddington (1882–1924), renowned British astrophysicist, wrote on the theory of relativity.

[120] Lauro de Bosis (1901–1931) was an Italian poet and actor.

all the guests in their own language. I was very impressed. We had a tour to Ostia, the old port of Rome. Against our friends' advice (it was a very hot day), we bathed after lunch. Whether it was this or something we had eaten, I do not know, but that night Alice and I were bowled over by the most terrible internal trouble. I have never experienced anything like it. I recovered, but Alice remained very under the weather and we were unable to use the passes all over Italy which were burning in our pockets. We had meant to see Naples, Florence and Venice, but we had to return tamely to England. Poor Alice continued to be unwell; she had paratyphoid and a long spell in bed. Nowadays we should have been inoculated before the visit and taken drugs with us, but such precautions were then not available.

The research in the physics laboratory was all this time acquiring momentum. We had gone on with the investigations on strength of diffraction, and these were paying big dividends. Darwin had developed the theory of X-ray diffraction in two brilliant papers before the war, which had appeared so long before anyone was able to test his equations that they had almost been forgotten. We based all our work on these two papers. On the one hand, the quantitative work led to estimates of the scale of atomic structure, to measurements of thermal vibration by James and his co-workers, and finally to a brilliant direct measurement of zero-point energy by James, Waller and Hartree.[121, 122] On the other hand, they enabled us to be far bolder in tackling difficult structures. We published a paper by 'Bragg, Darwin and James' which summarised our knowledge. Looking for suitable material on which to try out our new methods, we selected the silicates. They are available as well-crystallised material, and they have quite complex structures. I did much of the early work with West.[123] The final result was a general survey of the mineral kingdom and brought order into a very complex story. Silicates in particular had been very hard to classify on the basis of their chemical composition, and mineralogists had been

[121] The residual energy when the temperature is at absolute zero. In quantum mechanics it is the energy of the ground state.

[122] Ivar Waller (1898–1991), Professor of Physics at Uppsala, developed a theory of the thermal vibrations of atoms with Peter Joseph William Debye (1884–1966), a Dutch physicist and physical chemist who was awarded the 1936 Nobel Prize in Chemistry.

[123] Joseph West; student of WLB, and a sound ranger.

guided by the external form of their crystals. The mineralogists as it turned out were completely justified. Our work led to its being possible to classify the silicates largely by the geometry of their structure. One could draw the main silicate types on the back of an envelope; they corresponded to the various ways in which one could join tetrahedra made of oxygen atoms by their corners. I have always thought of this work on the silicates as one of the most satisfactory pieces of research I have helped to do—it reduced such a welter of information to such a very simple scheme.

I was invited to be Visiting Professor at the Massachusetts Institute of Technology in 1928. I went there early in the New Year and at first stayed with Duane in an apartment he rented in the Faculty Club.[124] I shall never forget my arrival at 'MIT'. The president was then Dr Stratton,[125] and the Professor of Physics was Norris.[126] Stratton was worried about the institute because the professors were so heavily engaged in industrial problems and indeed earned far more than their salaries by acting as consultants to firms. The situation had got out of hand and the interests of MIT were sacrificed. Norris spent most of his time investigating bricks; his office was full of them. When I presented myself to him he took me along the corridor to a room on whose door my name had been freshly painted. He showed me the efficient arrangements for hanging up my coat. We opened drawers and found stationary of every kind, pencils of all colours, stamps, etc. We pressed a bell, a beautiful blonde secretary appeared, and Norris waving his arm towards her said 'She's yours'.

Then he said, 'I have forty young assistants and I would be grateful if you would introduce them to original research.' Having been provided with the pencils and the secretary, it was up to me to do the rest. Actually I think I did quite well. I gave lectures on crystallography as part of the curriculum and set an examination paper at the end in approved style. The lectures were well attended and were, I believe, popular. I remember one young lady coming to me after class and saying, 'I have made the cutest little model of that last structure

[124] William Duane (1872–1935) was a Harvard physicist who had worked with Marie Curie.

[125] Samuel Wesley Stratton (1861–1931); administrator in the American government; physicist and educator. He was President of MIT from 1921 to 1930.

[126] This may have been Charles Head Norris of MIT.

you showed us out of toothpicks and chewing gum.' I talked about research problems with some of the young men, the professor's son, Buerger,[127] and above all Bert Warren.[128] He and I worked together on some measurements on diopside made by James and which I had brought with me from Manchester.[129] Warren had the brainwave which provided the solution and it turned out to be a key silicate structure because it showed the silicon–oxygen tetrahedra joined by corners to form long strings. This was the break for which we were looking; it provided the clue to an understanding of silicate formulae in general. I think I really had a part in starting Warren on the fine research work in the X-ray field which he has followed up ever since; so I did at any rate 'introduce one student', as Norris had asked.

Alice came to join me, I think, about the beginning of April. She had a dreadful voyage out. It was extremely rough and the boat was coated with ice when it arrived in Boston Harbour. This was quite an occasion for me. Kind friends in Boston whom I had met first at a tea party offered me the use of one of their four cars, a large Buick. Duane coached me in driving and I passed my test. I had kept this a secret from Alice. I planned to meet her at the docks and, as the Boston streets are narrow, complex and jammed with traffic, I walked over the course the day beforehand to make sure I knew all the one-way directions. Well, the boat came in hung with icicles, I hung over the barrier, and eventually Alice came down the gangway bearing a bag and a ukulele. I dashed forwards to greet her, but a typical American official of the be-capped, square-featured, chewing variety held up a large hand and said, 'Stop! She may have the plague.' Alice had to go through quarantine first. Then I led her to the car and we packed her bags aboard. Alice said, 'Where is the driver?' and this was my great moment—'My dear, I am the driver!' To drive a car abroad seemed more of an adventure then than it perhaps seems now. Poor Alice was still rather wan and thin after her bout of paratyphoid.

[127] Martin Julian Buerger (1903–1986), Professor of Mineralogy at MIT. Invented a type of X-ray camera called the precession camera, which gave an undistorted picture of the X-ray diffraction pattern from a crystal.

[128] Bertram Eugene Warren (1902–1991) was a crystallographer who became a specialist in diffraction from liquids and non-crystalline materials.

[129] Diopside is a pyroxene mineral with composition $MgCaSi_2O_6$. The solution of its structure was a particular triumph in its time.

On our first night we had dinner with Stratton and his social secretary Paris in the lovely President's House at MIT. We were to meet Governor Fuller, whose name was so familiar because of the Sacco–Vanzetti case.[130, 131] America was then in the Prohibition era, and we were astonished that Stratton, entertaining the governor of the state, should provide wine and other drinks at the dinner. I think there was a polite convention that such drinks were remnants left over from the pre-Prohibition days, but of course this was impossible after so long a lapse of time. Alice was very upset because, at such short notice, she had a desperate time finding someone to do her hair, which was rather wild after the voyage, and she was very disappointed with the result.

We greatly enjoyed Boston. I had acquired a small apartment in the Hotel Lincolnshire just off the Common. We had a sitting room, bedroom and minute bathroom which was so small that the bath bore the same relation to an ordinary bath that an upright piano does to a grand. It was a kind of vase, and by judiciously filling it half way up the water rose to one's chin when one sat down. Alice was at first very worried at the temperatures in all the rooms, with windows closed. I had got used to it, and in a week or two Alice became quite accustomed also. Duane had made us members of a club on the Common and which was noted for its good food and it was fun to dine there. I would set off in the morning for my tasks at MIT and Alice generally joined me there for lunch in the refectory. We had a great time with the car, exploring the neighbouring country. At first it was very cold and snowy and we had to drive with chains on the wheels, but after two or three weeks came the dramatic change into spring, with the temperature soaring up into the seventies. We visited the picturesque New England villages and picnicked in the woods. Most of the 'colonial' villages were now occupied by Poles. Warren was our great friend and we went out with him at weekends. Poor Warren—on one occasion the highlight was to be cooking hot dogs in the snow. We unpacked rolls, butter, mustard and then—no hot dogs; he had left them in the icebox at home. He took a long time to recover from this incident. We paid a visit to the Eves in Montreal, and

[130] Alvan Tufts Fuller (1878–1958) was one of the wealthiest men in the United States at the time and was Governor of Massachusetts from 1925 to 1929.

[131] Nicola Sacco and Bartolomeo Vanzetti were Italian-born anarchists who were convicted of murdering two men during the armed robbery of a shoe factory in South Braintree, Massachusetts, United States in 1920. They were executed in 1927.

then had a skiing holiday with their relations the Pilchards in the Laurentian Mountains; this was Alice's first experience of skiing.

We returned to England in midsummer, and I believe it was later that year that I had a sailing trip with George Thomson and Blackett on the west coast of Scotland. We started from the Clyde, went through the Crinan Canal and then on to Mull. The tide races were terrific; it was fortunate that we had a powerful motor in our yacht. We found some very remote and romantic anchorages by the shores of the many small uninhabited islands in those parts.

It was about this time that crystal analysis began to take on a new complexion, and the advance stemmed from this crystal of diopside, $CaMg(SiO_3)_2$, which I have already mentioned. In the earlier days we used to determine the structure by measuring crucial reflections which were sensitively dependent on the positions of some atom or another and so fixed its coordinates. In other words, we made a selection of reflections and measured those carefully. West and I had gone about diopside in another way. The kind Professor Hutchinson, in the Mineralogical Laboratory at Cambridge and who always helped me so much, had provided sections of diopside perpendicular to its main axes.[132] West and I mounted these sections on the X-ray spectrometer and, by turning them round in their own plane, we obtained a complete series of measurements from all the crystal planes parallel to each axis or around each 'zone', as it is termed. The new feature was the completeness of the survey. This was the material on which Warren and I worked, and on which the solution of the diopside structure was based. When we came to write up our paper, I was able to show that this survey of reflections could be used, for instance, to pin down the positions of the silicon atoms in a direct way. Now we had already used the orders of reflections from a given plane to form a one-dimensional Fourier series which represented the density parallel to that plane.[133] This procedure was first proposed by my father in his Bakerian lecture in 1915. I tried hard to see how all the reflections around a zone could

[132] Arthur Hutchinson (1866–1937) was Professor of Mineralogy at the University of Cambridge.

[133] Fourier series, named after the French mathematician Jean Baptiste Joseph Fourier (1768–1830), are used in crystallography to add together the contributions from diffracted X-ray waves in order to generate the positions of atoms in a crystal. This was a most important contribution made originally by WHB to crystal structure solution.

be used to build a two-dimensional Fourier projection on a plane, but I missed the answer. It was provided by my father. He had been thinking about the same problem and wrote to me on my return, showing how the Fourier series should be formed but asking my advice as to whether the coefficients should be intensities or amplitudes. I replied that they should be amplitudes and that I had sets of them obtained from diopside ready for trial. I put these coefficients in the Fourier series, summed it up and, behold, there were the calcium, magnesium, silicon and oxygen atoms clearly shown by peaks in the Fourier summation. I wrote to my father with much excitement and proposed that we should write a joint paper about it. He insisted that I should write it up alone, but I have always regretted that I agreed to do so. The idea as to how the Fourier series should be summed was entirely his, and I think he felt sad that the first paper describing its use should appear over my name. He should have published the idea, and I should have published its application to diopside. It was in this year that West and I published a paper on the analysis of structures with many parameters in which we summarised all we had learnt about techniques.

Warren came to work in Manchester in 1929, and I believe it was then that Waller from Uppsala also came. We had a galaxy of talent in Manchester during these years. Waller was a fine theoretical physicist, and he collaborated with James, who did the experimental measurements, and Hartree, who calculated the structures of atoms, to make the famous direct measurement of zero-point energy in rock salt. Even at absolute zero, the atoms contain a certain amount of vibration. Zachariasen came to work then.[134] He was an incredibly rapid worker; we expected a new structure from him once a fortnight, and he did much to build up the survey of silicate structures. Warren had pushed on from the study of diopside to other pyroxenes and amphiboles. Later two very famous theoretical physicists were at the same time in the Manchester Laboratory: Peierls and Bethe.[135, 136] Bradley

[134] Fredrik William Houlder (Zak) Zachariasen (1906–1979) was a crystallographer famous for his work on the structure of glass.

[135] Rudolf Ernst Peierls (1907–1995) was a German-born British physicist who worked in many areas of theoretical physics, including semiconductors. He was active in the British nuclear programme.

[136] Hans Bethe (1906–2005) was a German-American nuclear physicist who worked on the Manhattan Project. He made many contributions to theoretical physics and was awarded the 1967 Nobel Prize in Physics for stellar nucleosynthesis, the process by which the natural abundances of the chemical elements assemble in the cores of stars.

had started his work in the alloy field; he was the leading man in the application of X-ray methods to alloy structure, having served his apprenticeship with Westgren in Stockholm.[137] Practically all the X-ray crystallographers in this country either received their training at Manchester or in my father's laboratory at the Royal Institution, or they are students of this first generation.

We were invited to go to Holland in the spring of 1929. The students of the Dutch universities and of the technical college of Delft asked me to give lectures at each centre, and this was combined with a visit to the Philips Company in Eindhoven. We stayed in Utrecht with Professor Cohen who was later killed by the Germans in the war,[138] with the Fokkers near Delft,[139] and the Zeemanns in Amsterdam.[140] We should have stayed with Coster in Groningen, but his children were having the mumps, so we were put up in a hotel.[141] The hotel was incredibly clean; it looked as if every scrap of floor, wall, ceiling and furniture were polished every day. We enjoyed the strange Dutch breakfasts in which cheese played a large part, and the excellent coffee. While we were there the students somehow discovered that it was my birthday, and a deputation presented me with a large 'Groningen cake'. The Zernickers had a charming little daughter called Wout and we went on a walk with her to study newts in a pond.[142] The students

[137] Albert James Bradley (1899–1972) was WLB's first research student and worked with him in Manchester and later in the NPL and Cambridge on the structures of metals and alloys using diffraction from powders, as opposed to single crystals. He was the first to solve the structure of γ-brass. Sadly, in later life he suffered from a form of paranoia, which meant that he could no longer continue in research. A good account of his life by Henry Lipson can be found in *Biographical Memoirs of Fellows of the Royal Society*, **19**, 116 (1973).

[138] Ernst Julius Cohen (1869–1944) was a Dutch-Jewish chemist working on the allotropy of metals in the University of Utrecht. He was elected a Foreign Member of the Royal Society in 1926. He died on 5 March 1944 in a gas chamber at Auschwitz concentration camp.

[139] Adriaan Daniël Fokker (1887–1972) was a Dutch physicist and musician. He studied with Albert Einstein, Ernest Rutherford and with WHB.

[140] Pieter Zeemann (1865–1943) was a Dutch physicist who shared the 1902 Nobel Prize in Physics with Hendrik Lorentz (1853–1928) for the discovery of the Zeemann effect, where spectral lines are split under a magnetic field.

[141] Dirk Coster (1889–1950) was Professor of Physics and Meteorology at the University of Groningen. He was one of the co-discoverers of the element hafnium.

[142] WLB is probably referring to the family of Frits Zernike (1888–1966), a Dutch physicist who was awarded the Nobel Prize in Physics in 1953 for the invention of the phase contrast microscope.

gave us a very happy and interesting time. The summer holiday was spent at the boys' favourite, Pensarn.

Mother loved life so much and it was tragic that she died at a comparatively early age. I remember her saying once, 'I have just heard you all laughing, so I cannot really be so very ill.' She loved arranging the lives of others, a very Todd trait, it being clearly understood that no responsibility was incurred. I remember other quaint traits. She had a great liking for beer with her lunch, but we were stopped with loud exclamations if we tried to fill her glass to the top. Two three-quarter glasses were her limit. She always knew exactly which bit of a joint she wanted, and when Dad was carving she used to jump up and point with her finger to the right part. I used to get impatient with her assumption that everything was perfect, and to explode into frankness. Bob managed her much better than I did by playing her innocent games of make-believe. Mother loved the grandchildren, Stephen and David, and one could always make her forget her pain and become quite animated by telling her of their doings and sayings. I went with her to the centre where she had deep X-ray treatment, and I remember how gallantly she waved to me through the window of the cubicle where I was watching, although the treatment made her very ill indeed. Poor Dad could not bear to talk about Mother's illness—his face became blank when it was mentioned. I remember the surgeon taking me aside once because he thought that he had quite failed to convey to my father the hopelessness of the case, and he asked me to help. Of course this was not so; it was just that my father so completely concealed his feelings.

My dear mother had some very unusual traits of character. She had grown up in the rather haphazard way peculiar to the large Todd family, and as she told us herself, she had resisted with complete success any attempt to educate her. In consequence she had no exact knowledge of any subject, and in fact had ideas which sometimes astonished one by their strangeness. For instance, she was convinced that chops came from a sheep's cheek, because as she said, animals licked their chops. Nothing could persuade her that Cambridge was not south of London. I remember once, at the British Association meeting in Glasgow when my father was president, that Mother told us that while we were in a meeting she had found out exactly how to go to a highly recommended restaurant. My father in a fit of naughtiness said, 'Well, Gwen, you lead the way and we'll follow.' Of course Mother had not really taken in where it was; it was a painful episode. She was always quite vague as to what anyone's post was, as to what was going on in the world, as to rules and regulations of any kind. Yet she was convinced that she had a vast fund of knowledge on all these subjects and carried off any situation with a superb air of mastery. She could sense halfway through a sentence from the listeners reaction when she was on the wrong tack and change what she was going to say in a flash. Then again, it always had to be quite clear that any family decision was not hers, but that she was giving way to the wishes of others and doing what she thought they wanted, although actually Mother very effectively guided family decisions in the direction she thought best. She never liked to buy anything for the house, or any clothes, unless some member of the family was with her who could afterwards be said to have made the choice. Yet her taste was very good indeed, and she was always perfectly turned out.

Mother could never bear to be alone. If my father was out in the evenings, some faithful retainer such as an aunt was brought in to keep her company. She could not bear an admitted failure; we always had to agree that everything had been done in just the very best way. Her talents, which were great, were natural and untaught. She was a brilliant watercolour painter and could draw with her brush in an assured way which was astonishing. The painting of a rough Australian field which I have is a good example, and there is also the family group at Port Elliott. She could sew beautifully and design clothes. In Adelaide, she was the leading spirit in planning occasions of a kind we do not seem to have nowadays, and which I am at a loss to describe by name. There were parades, masques, formal dances, allegories and pageants, I suppose in aid of good causes. Mother designed all the dresses for the girls representing different flowers, for instance, or the boys' herald costumes, as well as the decor of the hall. Her fertile imagination seemed to have no end, and it was all done by the light of nature. She loved parties and occasions of all kinds, and her great endeavour was to 'make them go', which she certainly did.

In 1930 we had a spring holiday at Manesty Farm on Derwent Water. This was the first time that the boys had seen hills. Stephen, then six and a half years old, furnished himself with a stout stick—and almost as soon as we arrived he started out uphill like the hero in 'Excelsior'. It was a pretty place, with a birch wood running down to the lake and in which the boys could play, and a boat in which we all went fishing.

This was a sad year because I had some kind of nervous breakdown when we returned to Manchester. It was precipitated by worry about an invitation I had received to go to Imperial College, though I expect the root causes were much deeper. When I knew the invitation was on the way, I went to stay with Dad and Gwendy at Watlands, so that I could discuss my decision with Dad, but it was one of those times when it was quite impossible to get him to talk. I decided not to go and then found that Alice, though not thinking my decision wrong, had thought of the offer as a possible way of escaping from Manchester. I am very bad at thinking out all angles of a problem, and it was borne in on me (or I thought it was) that possibly this was the last chance to take a post elsewhere and that I had committed both of us to a life in Manchester. I had been swayed by my deep gratitude to Manchester for the way I had been supported and tided over difficult times, but I thought I had made a decision which was very unfair to Alice. A victim of a nervous breakdown feels so very guilty. I was in a bad way and caused Alice much distress. She tackled my father when we were staying at Watlands for the summer holiday, and between them they made a plan for my taking a term off and spending it at Munich in Sommerfeld's department.[143]

We went in January 1931 and had three months there, staying in a boarding house in Ohmstrasse run by two dear old ladies, the Frauleins Lammers and Junker. Nearly every weekend Sommerfeld used to take a party of students to ski in the mountains, but alas, Alice could not join these parties because she was then expecting Margaret. Sommerfeld's assistant, Sellmeyer, had boarded up a section in a byre in the mountains, occupied by the cows in the summer. Mrs Sellmayer went with us to cook, and we set out first on the mountain railway and then in Indian file up the mountainside with Sellmeyer holding aloft a blazing torch to guide us. In the byre, Sommerfeld slept in the crib and the rest of us in an enormous bed which held eight like the great bed at Ware. Preparing Sommerfeld to go skiing was a great sight. He was a short man and fat and, by the

[143] Arnold Johannes Wilhelm Sommerfeld (1868–1951), Head of the Theoretical Physics Institute in Munich, made important contributions to the quantum theory of atoms. He was approached by von Laue with a request to try to diffract X-rays by crystals but originally refused. Sommerfeld supervised more doctoral students who went on to gain Nobel Prizes than anyone else since.

time all the leggings, sweaters, coats and scarves had been donned, he was practically spherical in shape. It was the duty of Sellmeyer to ski slightly behind the Herr Professor and to stand him upright again when he fell over; it was merely a matter of rolling him into the correct orientation. In the evenings the air in the small room was thick with smoke and the odours of cooking, and we were dropping to sleep after a day in the open air. But this did not deter Sommerfeld at all; when he had got his cigar going well he would say, 'Now we will discuss the Ramsauer effect', or some equally stiff problem of theoretical physics.

Alice and I much enjoyed Munich and it did a great deal to put me right again. We were there during 'Fasching', the carnival time, which was still kept with traditional festivities. The shadow of Hitler had not yet fallen on the land. A wonderful dance was held in one of the Munich halls, with the room decorated by myriads of coloured balloons and everyone in a fancy dress. The opera, too, was delightful in the perfect little Mozart theatre. I am not musical, but even I could appreciate these occasions. Ohmstrasse leads on to the Englischer Garten, which had been planned by Count Rumford and was a very picturesque large park in which we walked.[144, 145] We also went on expeditions such as to the Zugspitze and to Garmisch-Partenkirchen. It was only sad that Alice could not ski.

After Margaret's birth on June 23rd of that year we went to Pensarn for our summer holiday. Alice could not leave Margaret, so I had to set out in the Humber car with Warren, Waller and I think an Austrian student called Struntz to find accommodation in a boarding house. I am not the best interviewer of landladies and rather let the family down, and also Alice had a painful time after Margaret's birth, but still we had a good holiday; Pensarn was always such a favourite with the boys and, like all children, they enjoyed going to a place they knew well.

The year 1932 was noteworthy for a visit to Russia. The plan had been mooted at an evening party in Cambridge at which Kapitza was

[144] It was in this park that von Laue said he first had the idea of diffraction of X-rays by crystals.

[145] Count Rumford, Benjamin Thompson (1753–1814), American born, made important contributions to the notion of heat. He was active in the American War of Independence.

present.[146] Though he and his wife Anna had their home in Cambridge, and children at a Cambridge school, Kapitza tempted providence by going regularly to Russia every summer. On this occasion there was to be a conference in Ioffe's laboratory in Leningrad, and Kapitza had promised to make arrangements for our attending it.[147] The party included R. H. Fowler, Dirac and Gurney, together with the Kapitzas and us.[148, 149] Our visas were obtained, and accommodation booked on a motor vessel going to Leningrad though the Kiel Canal. Alice was very alarmed when she heard we were going in a motorboat, picturing it as one of the seaside variety, but actually when we boarded it at Hays Wharf it was a reasonably sized ship, though it rolled abominably in the North Sea. In our party, Anna Kapitza and I were the only ones who were not sick; I had been pretty well hardened by being cook on our sailing trips.

Life on the boat was not quite orthodox. Fowler organised deck-tennis, and the only space where there was room for a court was on the bridge. The officers most complaisantly let us play there, only asking us to pause when they wished to take a sight. If the quoit was thrown somewhat high, it struck the wire running to the funnel and turned on the blast of the whistle. Halfway across the Baltic, a target was erected on deck and the stewardesses had rifle practice. There were not many stewardesses allotted to the cabins and we had pretty well to fend for ourselves. I used to fill Alice's hot-water bottle from a large samovar near the dining room. There was only one plug to several bathrooms and after my bath I pocketed it to hand over to Alice. I sometimes forgot I had it till the others tracked it down.

We arrived in Leningrad to be met by the Kapitza parents on the quay. We had no Russian money and I ill-advisedly traded some

[146] Piotr Leonidovich Kapitsa (1894–1984) was a Soviet physicist who worked on liquid helium. He spent several years at Cambridge with Rutherford (who called him 'Crocodile'). He was awarded the 1978 Nobel Prize in Physics for his work in low-temperature physics.

[147] Abram Fedorovich Ioffe (1880–1960) was a prominent Russian/Soviet physicist. He received the Stalin Prize in 1942, the Lenin Prize in 1960 (posthumously) and the Hero of Socialist Labour in 1955.

[148] Paul Adrien Maurice Dirac (1902–1984) was a brilliant theoretical physicist famous for his work on quantum mechanics. He shared the 1933 Nobel Prize in Physics with Erwin Rudolf Josef Alexander Schrödinger (1887–1961).

[149] Ronald Wilfred Gurney (1898–1953) was a research student of WLB. He discovered how α-particles decay.

pounds for roubles at the exchange bureau, only getting a few roubles to the pound. When we got to the hotel, our Russian friends pressed hundreds of roubles into our hands, saying, 'You will want something for the trams.' We stayed in a large but derelict hotel reserved for foreigners. At that time everything was very shabby. The people looked ill and tired and the children, owing we were told to there being no soap, suffered much from impetigo and had their faces bound up. It was extraordinary how quickly one's scale of values altered. It seemed amazing to see flowers in some of the squares, and when we saw a racing four on the river it seemed a marvel. The trams were very crowded, and we had such a time with fleas that we looked like bramble golf balls. The food was also very queer. The soup had a piece of meat in it which we swore appeared night after night and which we attributed to a mammoth. But it was good in the hotel as compared with what was obtainable for private use. We had breakfast with a young English couple doing research with Ioffe—it was porridge with the husks of the grain and eggs that were very far from fresh. They had given us of their best, but it was all we could do to eat it politely. Our scientific friends were very kind. One trip which they took us on was to Tsarskoye Selo, the former residence of the Tsars and about twenty miles from Leningrad.[150] Things seemed to happen so queerly in Russia. Gurney was always getting lost and on this occasion he had gone out to buy stamps and not returned when we were due to go to the station. Cries of 'my god, where is Gurney?' were heard on all sides from our Russian friends. Finally we tumbled into a car and were driven at a great pace to the station, only to meet Gurney on the platform. He said he had arrived back after we had left the hotel and had taken a tram.

The old palace at Tsarskoye Selo was fantastic in its sumptuousness and barbaric taste. In one room the walls were covered with amber, in another, with Old Masters cut to fit like bricks. In Queen Catherine's boudoir we were shown a chandelier and asked to admire it. We said it was elegant but small and were told it was made entirely of diamonds. The propaganda was very clever. In the entrance hall was what appeared at first sight to be a large dais, with Queen Catherine on it accompanied by Potemkin and surrounded by courtiers, all in

[150] Tsarskoye Selo means literally 'Tsar's village'.

their original dresses.[151] Looking closer, the dais was seen to be composed of serfs, with bowed backs on which the court stood. In another of the most sumptuous rooms was a small table with a knout on it and a photograph of a man being knouted—no words. Afterwards we went to the pathetic modern palace, all ugly Victorian, just as the royal family had left it with even the children's toys lying about. Alice has written a diary of this time in Russia; she saw much more than I did because I spent so much time in meetings. I was asked to give a talk to the Russian Academy and afterwards taken home with Alice in the car. We asked if we could stop on the way to our hotel at a flat where Lady Muriel Paget was quartered—she was doing relief work for distressed governesses and jockeys stranded in Russia after the revolution and had asked us to pay her a visit.[152] When we got there, with the two scientific friends who were kindly taking us home, they went as white as sheets. It transpired that Lady Muriel had as guests that evening two members of the secret police, and I suppose our friends thought they would be marked down as fraternisers with foreigners. I hope no harm came to them.

Alice and I had an invitation from Monkhouse of Metropolitan Vickers to stay with him at his dacha near Moscow on our way home, and Kapitza had managed to get tickets for the whole party on the Moscow sleeper by some wonderful staff-work.[153] The rest of the party stayed in a Moscow hotel. The dacha seemed very civilised after our experiences in Leningrad. The main pipe supplying Moscow with water ran not far away, and Monkhouse had drilled into it and joined it up to his house, where we had a plentiful supply. We were very impressed because he always slipped a pistol into his pocket when we went on any expedition in the car. Moscow was also very shabby, and the street cars had people hanging on by every projection. We were fortunate to link up with Monkhouse, because he helped us to get back to England. The State had ordered that no foreigners were to be allowed to attend the conference in Ioffe's laboratory, but by the

[151] Prince Grigory Aleksandrovich Potemkin-Tavricheski (1739–1191) was a Russian military leader and a favourite of Catherine the Great.

[152] Muriel Evelyn Vernon Paget (1876–1938) was a British philanthropist and humanitarian relief worker.

[153] Allan Monkhouse worked many years in the Soviet Union but was accused of spying for British intelligence in 1933 and put on trial.

time this order was issued we were already on the high seas heading for Leningrad—so there we were. This had the advantage that, not being there officially, we were allowed to roam around where we liked without a guide. But we had no tickets back home. I arranged to collect roubles by a promise to write an article for a Russian scientific periodical. When I got home I handed the money over to Monkhouse who produced the tickets—but I have often wondered how far my roubles went; I expect he paid out of the kindness of his heart. So we made the long journey back, passed at the frontier from the dirty untidy Russian side into the spick and span Polish station, so through Warsaw and Berlin to home.

Monkhouse told us very interesting stories about the end of the war. He had lived for a long time in Russia and played a part in the development of its electrical industry. When the revolution came, he organised the escape from Moscow of the English colony stranded there. He hired two goods trucks in which the party lived during the long journey to Vladivostok, foraging for food and persuading the railway authorities at each centre to attach the trucks to a train going the next stage of the journey. He told us that by good fortune a party of Cossacks travelled for a good part of the way with them, and the Cossacks had been so rowdy that the transport managers were only too anxious to pass the party on. The train was once raided by bandits, the English had very nearly fallen foul of the Cossacks and they had in fact endless adventures. His story of their finally rolling into Vladivostok station, to find a party of sailors from one of our warships to receive them, was very moving.

The famous trial of Monkhouse and his colleagues for espionage and sabotage took place shortly after our stay in the dacha. We met Monkhouse again on his return to England and heard his account of the trial. He was convinced that the Russians never believed in his guilt. When arrested, he badly wanted time to think, so he demanded that he should have a hot bath before being taken away. They were supposed to be on trial for their lives, but while the proceedings were on he went to his offices and to our embassy for part of each day. When we were staying with him we were looked after by his very efficient Russian secretary, a former princess, who seemed devoted to him. We were very surprised that she was one of the principal witnesses for the prosecution, describing a number of occasions when she had overheard Monkhouse and Thornton plotting against the

Soviet State.[154] We asked Monkhouse about this, and he said it was one of the brightest features of the trial. She knew their engagements, of course, and was careful to give as the dates of these occasions those times when it could be easily proved that Monkhouse and Thornton were a thousand or more miles apart. Monkhouse said, 'Once when giving her evidence she gave me a wink—a great moment.' I asked whether her ruse had not been discovered, and Monkhouse said the Russians had tumbled to it. She was punished by being forced to type for the secret police and having her theatre permits cut down to one a month.

Before going to Russia, I had a cruise with Blackett, Eric Holmes and Garnett.[155] Blackett had hired an ex-pilot ship at Penzance, and we took with us a former coastguard as crew. She was an old-fashioned boat of twenty-seven tons, with a deep straight stern and wonderful at sailing on a course when the helm was lashed. We set out for Cork and were three days at sea, and I had evidence of Blackett's powers as navigator. He had a sextant and a watch, the latter of which he checked by the time signal every evening. It was hazy weather with the sun just visible. Yet when we first saw land through the mist, the Cork Harbour entrance was over our bowsprit. We had rather a fright halfway across because the boat began to take in water faster than we could pump it out. Finally we discovered that Holmes, who was suffering from seasickness, had opened the stopcock to flush the lavatory and forgotten to close it again, so that water was pouring in at a great rate. The old coastguard lived in the forepeak and had brought a sack of provisions with him so as to be independent. He never went on shore and only emerged from his retreat when on duty. Once I persuaded him to share in our evening meal and, as he handed his plate back he said, 'A pretty drop of stew, sir.' A homing pigeon landed on our boat when we were well out at sea and seemed to adopt us. It flew off each morning in Cork Harbour and returned at night. We only lost it when we moved to Kinsale Harbour.

Early in 1933 we moved from our house in Pine Road, Didsbury, to Windy Howe at Alderley Edge, which we were able to afford because

[154] Leslie C. Thornton was Monkhouse's chief engineer and made a signed confession to working for British intelligence.

[155] William Garnett (1850–1932); professor and educational adviser, specialising in physics and mechanics and taking a special interest in electric street lighting.

Alice had received a legacy when Aunt Monica died. We rented Windy Howe from the Kings. Alice and I loved the new home. We had a very interesting garden, looked after by a fine old fellow, John Massey, who was a great expert and a secretary of the annual flower show. He had started work as a boy in the copper mines on Alderley Edge, which was at the top of the hill up our lane. The copper was extracted from the mined sandstone with sulphuric acid, and the hillside was covered with the golden heaps of the treated sand. The mine was a very old one, dating we were told back to Roman times, but it became uneconomic towards the end of the nineteenth century and was finally abandoned when a main boiler burst. John was deaf owing to a box on the ears he had received from one of the miners as a boy. The garden was really lovely, all the more so because we had come from the sad dirty one in Didsbury where few things would grow. Soon after we settled in, we made a visit to Spain, and I remember well our arrival home, seeing as we came up the drive snowdrifts of arabis and all the blossoms, a kind of fairyland. John had a large tin tray he put in the drawing room in late summer and filled it with chrysanthemums with gigantic heads. I remember the greenhouse was filled with schizanthus the first year we were there. There were not just a few struggling violas around the rose beds, but hundreds of them—and the roses! Then again, our house, which was on a rise, looked out over the Cheshire plain, with the mountains of Wales in view on a clear day, and to pull up the blind in the morning and see the plain lit by the early sun was something I have never forgotten. As we were high up, we were above the fogs which settled on the low-lying land around Manchester, on just those clear sparkling days of blue sky in winter and which made all around us so beautiful. Altogether it was as if we had been transplanted to a new world.

We went to Spain at Easter time. I had been invited by Professor Cabrera in Madrid to spend three months there as a Visiting Professor, but I could not take so long a leave from Manchester University, so it was arranged that West should go for three months to represent our school of X-ray analysis and that I should come for a final three weeks to top up the course.[156] West had great trouble in starting his lectures. Whenever he asked Cabrera about arrangements, Cabrera

[156] Blas Cabrera y Felipe (1878–1945) was a Spanish solid-state physicist who introduced crystallography to the Instituto Nacional de Física y Química.

would say, 'My dear fellow, it is early days as yet, we will think about it soon.' West, a conscientious fellow, finally came to the end of his tether and said at one of the laboratory gatherings, 'Tomorrow at eleven o'clock I shall go into the main lecture hall and commence a course on X-ray analysis.' He carried out his promise and started lecturing in a loud voice to a completely empty theatre. After a while startled heads appeared around the doors, an audience gathered and all went swimmingly. But things did happen there in an approximate way. I was billed, towards the end of our stay, to give a lecture to the Spanish Academy of Science, and Alice and I duly turned up at the appointed hour which was nine o'clock in the evening. We were met by Cabrera who said, 'The French company is appearing at the theatre, and tonight they are acting *The Rape of Lucrece*; would it not be a fine idea if we went to see them; they are said to be very good?' Alice said that she appreciated his plan, but that she thought she ought to stay for my reception by the academy and my lecture. Cabrera said, 'Oh, that's all right, your husband can come too, we shall be back in plenty of time—they will all wait for us.' So off to the theatre we went. As time passed I got more and more uneasy and kept suggesting to Cabrera that we really ought to return to the academy, but he said, 'No, no, we must wait, we are just coming to the crucial moment of the play.' Finally we came back, and the only gesture of mild reproach I saw was made by an old gentleman who pulled his watch out of his pocket and rotated it by its chain in front of Cabrera. In Madrid we had a friend called Arris, whom we had known in Manchester—he was a representative of an English company making, I believe, adding machines, and he and his family were very kind in taking us about. Arris and I played golf at the country club near Madrid, and I shall always remember it because there was a notice at the entrance to the swimming pool, 'Spitting and Making Love Prohibited.' Arris was a terrific smiter and caused cries of appreciative amazement from the spectators as we drove off. Dolly Arris and Alice used to go shopping together and Alice once asked her what it was the men on the benches were saying as they went along the boulevards. Dolly said, 'Oh, they are saying, "I wish I could win the lottery",' the idea being that if they were in such a fortunate position financially they might be able to make an offer to the ladies which would prove acceptable.

Poor West, the most modest of men, once had an uncomfortable time with Madame Cabrera, who asked him about his wife and how many children he had. West replied that he was not married, and Madame Cabrera said, 'Ah, yes, I understand, you just have a little lady friend.' Dinner was so late in our hotel that we never once achieved it. We had a kind of glorified high tea between seven and eight o'clock and then went to bed about ten, long before the dinner hour. Cabrera and his son took us on excursions to Asturias, Toledo, and up the Guadarrama hills where people were still skiing. The new university town was just being built; all the trams bore notices of the lottery which was financing it. The university was later the site of bitter fighting in the civil war. A slight embarrassment towards the end of our stay was that I could not pin down getting my money, which I was counting on for the hotel bill; I had to be somewhat firm.

I had the operation on my arm in 1933; luckily it was my left arm, so I was soon able to play tennis again. In the early summer the Goodharts invited Alice and me, with David, to stay in a house they had rented at Folkestone for Easter. Their eldest boy, Philip, was about David's age. David was thrilled because an extra nurse was hired to look after him and dry him after bathing. Later in the summer we had a holiday with Alice's parents at Woolacombe in North Devon. Margaret was then just over two. She was a most determined little girl and very keen on scrambling on the rocks. We had by this time acquired a cine camera and some of my first pictures are of Margaret making sandcastles and doing rock climbing, the boys looking after Margaret in the sea, and the boys surf-bathing and riding, as well as of the dear grandparents, who loved this holiday.

I was invited to go to Cornell University at the beginning of 1934 as Visiting Professor in the Baker Laboratory (Chemistry). My duties were to give a series of lectures and to write a book. I went out alone in January, leaving Alice with the children. I was offered hospitality in a students' 'fraternity' called Telluride, where I had a room and all my meals with the young men. Fraternities differ in their character: this one was of a distinctly highbrow type. It was said that a girl once exclaimed to a friend, 'I've just remembered that I have been asked to a dance at Telluride in a month's time and I have not started doing my reading for it yet.' Most of the boys had been to a boarding school in the western desert, founded by a man who had made his money in gold at a place called Telluride, and they had gone on with scholarships to this fraternity in Cornell University. They were extremely kind hosts. I wondered how it happened that whenever I was at a loss, to get a sleeper on a train, see someone in Cornell, buy something in Ithaca, etc., there always seemed to be a student at my elbow to help me, and I only later realised that it was a kind of fag service; they took it in turns to be there to look after me. I was allowed by the students to attend their weekly meetings; in alternate weeks we had an evening when the house went into committee to consider domestic problems, and a meeting at which speeches were made. About three of the fraternity each made one of these speeches for ten minutes without notes. He could speak on anything he liked and I remember once 'The Nature of God' was followed by 'The Price of Electrical Power in Ithaca'. It was an excellent training. Two things about their regulation of their affairs interested me. The number of freshmen taken on in their first year was twice as great as the number of second and third year men, so there was a weeding out of half the young aspirants at the end of their first year. They were sized up during the year by their elders and voted on at the end of the year. I felt it was rather an ordeal for the freshmen. Another cast-iron rule was that if any two members quarrelled both had to leave the fraternity, without any questions as to who was the guilty party. In consequence there was very little open friction. They had two other guests beside myself: a dear old professor who was a leading expert on witchcraft and who was so anxious not to abuse the hospitality of the young men that he insisted on sleeping on a sofa, and a Professor of 'Pomology'. I asked what pomology was and he said, 'Everything except bananas', and added that there were seven professors of pomology at Cornell. It ought to be said that there was a large agricultural college attached to the university. It was extremely cold when I arrived at Cornell, about 40°F below zero, and one had to be careful. It was quite an ordeal walking from the fraternity to the laboratory, and many of the students were going about with bandages on their heads because they had incautiously allowed their ears to get frozen.

My lecturing duties were very pleasant. I had a convenient number in my class and they were all keen. I gave a series of lectures on crystal structure. I was also offered laboratory facilities in case I wished to do research, and had an excellent assistant in Hughes.[157] But I decided that six months was not long enough to get research going, so I

[157] Edward Wesley Hughes (1904–1987) was an American chemist and crystallographer who introduced into crystallography the method of least squares to handle the large amount of data involved in the refinement of crystal structures.

devoted all my spare energies to writing my book, and Hughes helped me with my lecture demonstrations and later in getting the book ready for the press. The book was *The Atomic Structure of Minerals*, published by McGraw Hill in New York. It was all written in four months, because I had every afternoon free and could get down to it without interruption in my bedroom at Telluride. If only one can manage it, the ideal way to write a book is to have a sabbatical six months or a year. The total time spent if one tries to write it in one's 'spare time' is many times greater. The material for a book can be collected in spare time, but the writing ought to be done in one spurt, so that one need not think of anything else and can remember what one has written. Otherwise, so much has to be rewritten. The figures took over a year to prepare; I left the rough drafts with Hughes and he faithfully attended to their being drawn properly. The book remained quite a popular textbook for many years.

An interesting evening during my time there was the pouring of the glass for the 200-inch reflector for Mt Palomar, at the Corning glass works. Special arrangements were made for visitors to pass through the factory and witness the operation. It was interesting that the suitability of the molten glass in each ladle was judged not by pyrometers or other gadgets but by an 'old man', an experienced hand who judged its fluidity and temperature and allowed it to be poured or sent it back to the furnace. Actually this pouring was a failure. In order to give the mirror strength and lightness, it was shaped like a waffle, with the cavities in the back being formed by brick pillars on the floor of the basin into which the glass was poured. One of these pillars came adrift, the bricks floated up to the surface and the operation had to be repeated.

Alice came to join me just as the first signs of spring were appearing. I met her at Boston and we went first to Charlottesville in Virginia to stay with one of the university professors. Alice, I knew, would need a rest after the Atlantic crossing, and so I begged our host to arrange a quiet time for us both. When we arrived we found that every moment of every day had been planned, speeches here and there, visits round the laboratories and to other colleges in the neighbourhood, parties, etc. I rebelled and said we must have one half-day with nothing whatever to do, and I do not know to this day whether I did right. Frantic telephonings, we could not but be aware, went on in our host's house, and at some of the places we visited people were clearly angry at having their plans changed.

We then went to Cornell, where for the first three days we stayed in Telluride. It was a house rule that no woman should stay there for more than three days, a rule made, we were told, because once the wife of a professor had attempted to mother the boys. So I had boldly rented an 'apartment' in a big block on the edge of the campus, consisting of sitting room, dining room, kitchen, bathroom and bedroom. It had tables, chairs and beds but nothing else, and Alice was rather appalled when she saw its bareness. The faculty, however, helped with sheets and blankets, and we bought all our cooking stuff and crockery from Woolworths. Most of it just lasted our three months' stay. We had tremendous fun looking after ourselves and shopping. We hired one of the students to come in from time to time to do the heavy cleaning. I remember he asked whether we were paying him by the time or the job, and when we said 'by the job' he dashed around at enormous speed. He was a track-runner. He paid his way through college by acting as butler in the home of a rich Ithacan citizen. Towards the end of our stay he pressed me to talk to a student society of which he was a member. I tried to get out of it, saying that our last days were full of engagements, but he said, 'I found your engagement book in your bureau, and you are free on [here he cited the evenings]', so there was no way of escape.

We had a Ford V8 car and made many excursions. The country of the Finger Lakes around Cornell is very beautiful. The ice dammed back the lake at one time so that it was nearly at the level of the flat hilltops which are now high above it, and as the ice melted successive beaches were formed which are very noticeable as a series of benches. At the same time the small streams meandering gently to the original lake started emptying themselves down the steep slopes of its shores as the lake subsided and cut back a series of narrow canyons, with a waterfall at the head of each. It was, in fact, the same history as that of Niagara Falls, on a much smaller scale. We used to take our supper out and explore these canyons, lovely in the evening quiet, finding the flowers of the brief American spring: trilliums, Dutchman's breeches, violas, columbines and a pretty trailing arbutus with pink flowers. The hillsides were ablaze with dogwood, and there were squirrels, chipmunks and woodchucks to watch, as well as many birds in which I was very interested. We liked Cornell the best of all the university centres where we made prolonged stays.

The young men in the Telluride fraternity paid Alice a great compliment towards the end of our time. Every two years they had a kind of 'May Week', a long weekend when they invited their girls to stay in the fraternity and had a series of dances and other festive occasions. The girls occupied the bedrooms, while the young men managed somehow in the attics. I think they took to Alice when we had our three days stay, and they invited her to stay and to chaperone; I joined in the festivities but went home to our apartment each evening. I had heard a great deal about this weekend in the fraternity committee meetings, which I was allowed to attend. If I remember rightly, a member who contributed fifty dollars was allowed to invite his girl. Some of the men could not afford this; they paid ten dollars and had certain limited rights and duties. They could come to all the parties but a certain secret sign was arranged so that if the entertainer of a girl felt he just must be free for a while, he gave the signal and it was the duty of the ten-dollar subscriber to cut in and take the girl off his hands. Then we voted on how much 'dog' we could afford—for instance could we run to 'corsages'. I strongly suspect that, whatever was decided in committee, the fraternity steward, who was old in its service and had experience of many such occasions, did just what he considered appropriate. The young men were the slaves of the girls, waiting outside their bedroom doors to take them to breakfast in the morning and faithfully attending them all day. The place was lit in the evening with pink-shaded lights of low luminosity, and the girls looked like houris. It was quite a shock to see them as 'co-eds' on the campus afterwards, spectacled perhaps and with their hair screwed back, not made up and in sensible clothes. I remember a remark by a co-ed with whom I was dancing one evening; I had politely asked her what course she was pursuing and she said she was 'majoring in philosophy and sanitary engineering', which had always seemed to me the ideal preparation for married life. Alice attended the law courses while in Cornell. I do not think I have mentioned before that Alice was a student at Lincoln's Inn and had successfully passed all the bar examinations except her finals. She did one examination, roughly speaking, between each baby and was due to take the finals after Patience until the war made it impossible.

The British Association met that summer in Aberdeen, and we had a holiday first at Ballater on the Dee. The boys, now eleven and eight, were excellent walkers over the hills and we climbed most of the heights in the neighbourhood. We had got to know the Lord Provost of Aberdeen through making arrangements for the BA meeting, and his son Hugh, who was about Stephen's age, stayed with us for a while. One day we climbed Mount Keen with the Slaters (he was Professor of Physics at the Massachusetts Institute of Technology) and I remember vividly their astonishment at hearing Stephen and Hugh conversing together in Latin on the way down.[158]

I gave the public evening lecture at that BA meeting on silicates, and I called it 'The Ethics of the Dust' after Ruskin's book on minerals. I experienced a curious coincidence. Bunn was then in charge of crystallography in the ICI works at Northwich in Cheshire, the place where Brunner and Mond had made their first soda when they acquired the rights to develop the Solvay process.[159, 160] They had bought a school at Winnington, and the soda was made in the former stables. I asked Bunn to help me with photographs of crystals for the Aberdeen lecture, and one day, when I was having tea with him in the lovely old house used as a clubhouse by the staff, I asked him about the origin of a curious semicircular depression in the lawn. It was, he told me, a kind of open air theatre when Winnington was a girls' school, and it was there that Ruskin, engaged as a part-time master, gave talks to young ladies about minerals; afterwards, he published these talks as *The Ethics of the Dust*.

Another incident I remember at Winnington was being shown by Freeth a specimen of a new type of substance which had just been made in one of their high-temperature, high-pressure vessels.[161] Freeth, who was the head of the research department, was always interested in high-pressure chemistry. He said, 'I have no idea if it will ever be of any use commercially, but it seems to have some interesting physical properties.' This was polythene!

[158] John Clarke Slater (1900–1976) worked on the theory of the electronic structure of atoms, molecules and solids, and on microwave electronics.

[159] Charles William Bunn (1905–1990) worked as a crystallographer for many years at Imperial Chemical Industries (ICI). His obituary can be read at <http://scripts.iucr.org/cgi-bin/paper?S0021889890013048>.

[160] The Solvay process, or ammonia-soda process, is the major industrial process for the production of soda ash (valued primarily for its content of sodium carbonate). The ammonia-soda process was developed into its modern form by Ernest Solvay during the 1860s.

[161] Francis Arthur Freeth (1884–1970) was a British industrial chemist. He was manager at ICI Winnington.

It was around this time that the work on alloys in the laboratory advanced very rapidly. It was Bradley's genius in the main which inspired it. It is really extraordinary to look back and see how much was accomplished by Bradley in a decade, and it was tragic that his illness cut short such a brilliant output. Goodeve once said to me that, when he took up his post as director of research of the Iron and Steel Institute and set himself to read up the recent fundamental work on the physics of alloy structures, it was 'all Bradley'.[162] Bradley was a wizard at sorting out complex ternary equilibrium diagrams and the structures of complex alloy phases. Bradley's work dovetailed in with Hume-Rothery's theories of alloy structure.[163] Again, Sykes was then in the research department of Metropolitan Vickers and investigating the possibility of using iron–aluminium alloys for the wires in electric fires.[164] He found to his surprise that the resistance varied by a factor of two depending on the heat treatment the wired received. Bradley investigated the phenomenon and found it to be due to the difference between an ordered structure of the alloy when cooled slowly and a disordered structure when it cooled rapidly. The former has much lower electrical resistance. Bradley described his results at a colloquium in the physics department, and I put forwards some general qualitative ideas about the nature of the order–disorder change. E. J. Williams was present at my talk and subsequently came to my room next morning with the precise mathematical theory which he had worked out that night.[165] Williams and I published a series of papers on the 'order–disorder' phenomenon, and these attracted a good deal of attention. We thought at the time that our ideas were original, because I had consulted metallurgist friends as to whether any previous

[162] Charles Frederick Goodeve (1904–1980) was a Canadian chemist and a pioneer in operations research for the British. During the Second World War, he was instrumental in developing the 'hedgehog' antisubmarine warfare weapon and the degaussing method for protecting ships from naval mines. He was Director of the Iron and Steel Research Establishment.

[163] William Hume-Rothery (1899–1968) was a distinguished British metallurgist, known in particular for his set of basic rules describing the conditions under which an element could dissolve in a metal, forming a solid solution.

[164] Charles Sykes (1905–1982), a metallurgist, worked at Metropolitan Vickers before going to the NPL.

[165] Evan James Williams (1903–1945) is primarily known as the discoverer of pi–meson decay; he published over thirty academic works, bringing a semiclassical approach to modern physics.

work had been done in this field which was new to us, and they all replied in the negative. I found afterwards that Johannson and Linde in Stockholm had arrived a year or two before at the same main equation.[166, 167] We, however, pushed the analysis much further and I think it is fair to claim that the Bragg–Williams papers inspired the application of the order–disorder principles to a number of other types of phenomenon. Sykes obtained leave from Metropolitan Vickers to devote all his time to a study of the changes in the alloys and built a most effective apparatus for plotting the variation of specific heat with temperature and which tied up with the theoretical treatment. Both Bethe and Peierls, who achieved worldwide fame as theoretical physicists, were then in my department and they worked on the order–disorder phenomenon. Bethe in particular gave a most elegant statistical treatment and developed the idea of 'local order'. Incidentally, Sykes' contributions were later recognised via his election to a Fellowship of the Royal Society. Altogether, it was a fine corpus of work.

I gave the 1934–1935 Christmas Lectures at the Royal Institution. Alice suggested the subject, electricity, and the experiments were planned by Kay and me. William Kay was the wonderful laboratory steward at the Manchester physics laboratory, known all over the country for his cleverness, keenness and likeable personality.[168] He had served under Rutherford, who had tried hard to persuade Kay to go to Cambridge with him, but fortunately for me Kay would not desert his native heath. Kay was up to all the students' tricks. For instance, if a lighted candle figured for some reason in a lecture, Kay always cut off the top half of it, just as the lecture started, because the students were sure to have spat on it to prevent us lighting it. I always remember too how I would say in the lecture on magnetism, 'Suppose we take a bar of iron which is red-hot', and on the dot Kay would issue from the door to the preparation room with a red-hot poker and respectfully present the cool end to me—I did not even

[166] Carl Hugo Johansson (1898–1982) was a Swedish physicist. From 1946 to 1959 he was head of the laboratory Nitroglycerin AB, researching into explosives.

[167] Otto Jonas Linde (1898–1986) was a Swedish physicist who became Professor of Solid State Physics at the Royal Institute of Technology in Stockholm.

[168] William Alexander Kay (1879–1961) was Chief Steward of the Manchester Physics Laboratory from 1908 until his retirement in 1946, having served under Rutherford and WLB.

have to look round. He threw himself into everything with the utmost keenness, and he took immense pride in the preparations for the Christmas Lectures. We had great fun making a moving cartoon of the lines of force as two charged spheres approached and receded from each other, for like and unlike charges. Instead of drawing a series of pictures of the lines, we made one picture for like charges and one for unlike charges, and blackened in alternate strips between the lines. One had only then to slide this diagram towards and away from the cine camera and arrange that simultaneously the charged spheres should move so as to cover the points from which the lines appeared to diverge. The result was most realistic. It is surprising how a stock of experiments, made up for a special occasion like this, can be used again and again for other talks. Thirty years later many of them are still proving to be useful.

We had one experiment in which we placed Stephen on a platform supported by a wire suspended from the roof of the theatre and showed that he could be pulled round by an electrified ebonite rod. In order to get the twist out of the wire we weighed Stephen at school in Manchester and left a corresponding weight on the platform overnight. Just as a check, I weighed Stephen again just prior to the lecture, and to our astonishment he was nearly a stone heavier after Christmas. Both sets of scales were vouched for as accurate. I afterwards wrote up these lectures as *Electricity*, published by Bell, and the book had a good vogue.

Windy Howe had a billiard room, and I used this for my literary labours. The chapters were arranged around the table with the appropriate diagrams, photographs and references in the middle, and as I worked through the chapters I moved my stool on from one chapter to the next under the bright overhead lights. It was an excellent scheme because nothing had to be put away, one just put the cover over the papers at the end of the day, and everything was handy. David had to write an essay (both boys were at a local prep school, Harden House) on 'Our Home' and it started, 'Our Home has a billiard table where Daddy works and a cellar in which he keeps his beer.'

We sent the boys to Harden House after much cogitation about preparatory schools. Harden House was not ideal, but the mathematics was well taught by a former Manchester student of mine, and Stephen duly got a scholarship at Rugby, where we had enrolled him with

Vaughan when he was a few months old.[169] David stayed on there till we left Manchester and then went to 'The Downs' near Malvern, of which Geoffrey Hoyland was headmaster, and after it to Rugby too.[170]

We had our 1935 Easter holiday at Bull Bay in Anglesey. Margaret was four then and very loyal to and proud of 'my boys' as she called her brothers, whom she tried to copy in everything. Patience was on the way that summer and Alice, not able to do anything strenuous, packed me off for a walking holiday with Ritchie.[171] He was an old friend of my student days in Cambridge and was then Professor of Philosophy in Manchester after having been lecturer in anatomy, a remarkable transition. We went via Cork to Killarney and set out on our tramp from there. We were fortunate to have fine weather. Getting accommodation for the night was a chancy business because possible places were so few and far between, and when one is walking with a pack it is not much help to be told at the end of a long day that there might be a place ten miles further on. We were nearly stuck at McGillicuddy's Reeks. The hotel refused accommodation, I fear because we looked so disreputable, and we went off to the village shop a mile away where there was an extremely fat but very kindly lady behind the counter. She said she thought the postmistress could take us in but 'if she can't, come back here. My husband's away for the night and it's a big bed, and you but two slips of lads.' I have never been sure whether I am glad or sorry that we did eventually find quarters with the postmistress; it would have been an interesting and I am sure quite innocuous experience. We stayed there some days to walk on the Reeks, and then went on via Waterville, Sneem and Kenmare back to Killarney. A memorable holiday because the country was so beautiful and fresh, although it was August, and the sea coast was magnificent.

[169] William Wyamar Vaughan (1865–1938) was the headmaster of Rugby School from 1921 to 1931. He died after an accident at the Taj Mahal in 1937 where he broke his leg, which was subsequently amputated.

[170] Geoffrey Hoyland (1889–1965?), married Elsie Dorothea Cadbury (1892–1971) in 1919, and became headmaster of the Downs School in 1920. He used the family wealth to improve the school. He built new buildings and introduced an innovative curriculum emphasising science and arts. He supervised the construction by pupils of a miniature railway which exists to the present day (see http://audensociety.org/13newsletter.html for comments by Stephen Sender on Hoyland's private life).

[171] Arthur David Ritchie (1891–1967) was a British philosopher.

Patience was born on 11 September. She arrived as a very perfect and mature baby and McKassack, our doctor, was much impressed. We had the dear Nurse McRoberts as midwife. She was very old and her arms were like sticks but she could do anything and never seemed to need sleep. She had an ideal way with distraught husbands and used to set me to boil large amounts of water in the dead of night, I think for no purpose but to ease my mind.

There was considerable excitement in the local garage at the time of Patience's birth. We still had the grand Humber car which Aunt Monica had given us, now about nine years old, but it was an open car and although the hood and window fittings were good it was a draughty car when Alice and I had to go into Manchester for dinner parties in winter. We decided on a second saloon car for Alice, and I promised her one on the taxation basis of one horsepower for each pound of baby plus an extra horsepower. The garage produced likely cars and had bets on the result. In the end Patience tipped the scales at nine pounds and it should have been a 10 h.p. car, but so attractive a specimen at 12 h.p. was pressed on us that I fell for it. In those balmy days before the war we had a large house and garden, two cars, four maids and a whole-time gardener; everything was so much cheaper.

So far as I remember, 1936 was an uneventful year. I was busy writing *Electricity*, and the research was going well. The most exciting aspect was perhaps the work on 'order–disorder' with Sykes, Bradley, Williams, Bethe and Peierls. We much enjoyed the country walks around Alderley Edge. Though it was so near Manchester the intrinsic beauty of the country could not be spoilt. The fields were full of Star of Bethlehem in spring and the rare water-violet, a kind of primula, grew in the ponds at the end of our lane. Wood warblers and garden warblers were common in the trees on the Edge. Our garden was pretty and varied, and we had many parties on our tennis lawn at weekends. I bought John Massey a petrol mower because many of the lawns were so sloping and he was no longer young, but he never got quite used to controlling the pace with the throttle and tried to hold the mower back or push it forwards, with the result that it often ran away with him under apple trees and the like. In the summer we took the boys on the Broads in a sailing yacht, starting at Oulton, and we had a grand time. It was a moot point whether to make them wash up; we felt it was good for morale if they did so, but often as the bowl of dirty water was chucked over the side there was a glint of metal and another knife or fork went to the bottom of the river. We had a small sailing dinghy as tender, and the boys had much fun with it.

2.6 A Brief Interlude

In the summer of 1937 I was invited to succeed Petavel as director of the NPL at Teddington.[172] I remember my father telephoning me to say that a letter making the offer was on the way, and my going out into the garden to tell Alice. I was overjoyed to have the offer, because it meant that Alice could go to the south of England; it was a tremendous relief to me to have the chance to make a move. I was only fearful that she would not like our official residence, Bushy House. She first saw it with Sir Frank Smith, the Secretary of the Department of Scientific and Industrial Research, under which the NPL came, and when I met her in London after this visit and found that she had really fallen in love with the old house, I was quite overcome![173] I had felt so very remorseful about keeping her so long in Manchester.

Bushy House is very large. The director has the top two stories for his residence, and the vast ground floor and basement are used by the laboratory. When the laboratory was founded, everything was accommodated in the old house. It was built in the early eighteenth century, and at one time belonged to Lord North, of America fame. Later it was used by William IV to house Mrs Jordan and her numerous progeny, and when he died Queen Adelaide adopted it as her residence. It was, I believe, derelict when Queen Victoria assigned it to the newly founded laboratory in 1900. Alice and I got the help of the Worthingtons in choosing the decorations, carpets and curtains, and in the general planning of the flat. The kitchen was moved from a part of the house forty yards from the dining room to an adjacent position. The rooms really looked very lovely indeed when they were finished, as the ceilings, doors and windows were so elegant to begin with. And then, almost before we had settled in, I was invited to succeed Rutherford at the Cavendish.

[172] Joseph Ernest Petavel (1873–1936) worked on the combustion of gases, designing the Petavel pressure gauge. He made important contributions to aeronautics.

[173] Frank Edward Smith (1876–1970) was a physicist who wrote papers on electrical units.

Frank Smith was very generous in his attitude to the situation. He had been one of the electors to the Cambridge chair, and he urged me to accept it although it would mean leaving the NPL a year after our going there. My father was against the move. I think at bottom he never felt quite happy in the Cambridge atmosphere. The High Table at Trinity with the witty asperity of the Fellows' conversation made him feel uncomfortable. After much deliberation we decided to go. Quite apart from the fact that the Cavendish chair was the greatest honour a physicist could be offered, I believe I felt that the academic atmosphere was one in which I had more chance of doing something constructive. Of course I never had a chance to do anything at the NPL because the move to Cambridge was decided upon so soon after our arrival there, and I did not want to start any plans which I could not see through to completion. But whether, if I had stayed, I could have been effective, I have often doubted. The NPL is a vast organisation, requiring a very clear head and gifts of statesmanship if all its ramifications are to be run properly, and this is a field in which I have never shone. At the same time the director must have scientific imagination, because it is only too easy for such a place to lose contact with reality and pursue lines of work which do not lead to useful results. I might perhaps have made my contribution in this last way, when experience emboldened me to take a firm line. Undoubtedly, many things were still being done which had long ceased to be useful.

Not only did our impending departure make our time there an unreal one, but also war was threatened. It was the time of Munich. We all thought war inevitable, and we all breathed sighs of intense but uneasy relief when Chamberlain averted the crisis. I have a vivid memory of the awful feeling at that time that we were caught quite unprepared—that we had been hiding our heads in the sand—that at last England had got herself into a position from which no escape was possible. One feels ashamed now to have doubted the tremendous power of recovery which the country possessed. I have never felt proud of that year at the NPL. I feel that my judgement went to pieces with the newness of the job and the threat of war and, though I think I was not alone in this, I lost a sense of proportion. I felt that the NPL was a very obvious target and arranged for a series of shelters to be made, *not* the thing to bother about at that crucial time.

We had a brief time of what might be called normal life at Teddington. It is a strange place. The director's quarters are very lovely indeed, and he has about seven acres of garden and a tennis court. Petavel had planted many thousands of daffodils in the grounds which made a wonderful sight in spring—in fact one got rather satiated with daffodils. Beyond the ha-ha at the bottom of the garden stretched Bushy Park with its fine trees and open stretches. Margaret, then seven, went riding in the park, and she and Patience had a huge garden in which to play. The children, however, never liked life there. The ground floor was used by the laboratory, and they felt they had no privacy when playing in the garden. The feeling of having a place to oneself, however small, is immensely important to a child. It is reported that when Bullard was the director, and his children were being very noisy in the garden, he leant out of the window and shouted, 'Do not make such an infernal din, you might wake someone up.'[174] Alice loved the graciousness of the rooms. In particular, she had a small room of her own, off the drawing room, which had a view over the park and garden. We put an immense amount of work into decorations and furnishing of the house, and it was very sad that Alice was borne away so soon.

In the winter of that year we had a skiing holiday at Zell am See in Austria. It was an attractive little town, and there was a funicular to the heights above. Alice spent part of the time in a ski school, and we had expeditions on the hills. Mark joined us there, and I had an expedition with him up the glacier and to the summit of the —, about 12,000 feet high.[175, 176] We spent the first day climbing up to a hut at the foot of the glacier; we stopped for the night at that hut and then did the ascent the next day. The — is a peak with a flat space on top some ten yards or so in each direction, and then a very steep fall on every side. It ought to be straightforward to stand upright on a space ten yards across, yet to the inexperienced it is extraordinarily difficult. We had a marvellous view into surrounding countries. On the way down Mark spread a rug on the glacier, in a place where we could see a mile or two in every direction, and while we had our lunch he asked me what possibility there was of his getting a post in England. I was too dumb to

[174] Edward Crisp Bullard (1907–1980) was a distinguished British geophysicist; he was appointed Director of the NPL in 1949.

[175] Herman Francis Mark (1895–1992) was an Austrian-born chemist who worked on the structure of polymer fibres. He made important contributions to the understanding of the molecular structure of cellulose. His father was originally Jewish but converted to Lutheranism upon marriage.

[176] The names of the mountain or the glacier were not given in WLB's manuscript.

realise the urgent significance of his enquiry. There was a sense in Austria of impending trouble, but I did not guess how near the Anschluss might well be. Mark, owing to his Jewish connections, was in danger and was taking all the precautions he could for a safe getaway. He told me that his rucksack always contained enough to live on for a week. When the Anschluss came he made a good escape. He was very popular among the students as he was a brilliant footballer, quite apart from being such a likeable person, and I expect he had many friends. He was allowed to leave the country in his car, and he had arranged, he told me later, that the tools in his toolkit should be made of platinum. They came to see us at Bushy House when they reached England. Mark first found a post in Canada, I think helped by my father, and later went to the Brooklyn Technical Institute in the United States.

In the spring we had a New Forest holiday near Beaulieu. The Malcolm farm had a very large number of cats, which had a passion for David; an odd half-dozen or so used to visit his bedroom in the morning. The stay was also notable for a remarkable creature which Patience reported she had seen. It flew, it ran and it hopped. It ran down a hole in the ground and then flew up a tree. It made an extraordinary selection of noises which Patience repeated with verve—in fact, Patience had us all guessing. She was then three. In the summer we had a holiday at that delightful place, Studland in Dorset, where we rented a small house.

During that summer of 1938, Alice became heavily involved with the Women's Voluntary Service (WVS) association. Lady Reading engaged her help in the head office in Tothill Street, from which she was organising the service all over Great Britain.[177] Alice was given the task of starting the organisation in north-west England and went to Manchester and other places to find women who would take responsibility for their areas. She went every day to Tothill Street from Bushy House. I believe it was through Sylvia Fletcher-Moulton that Lady Reading and Alice were brought together.[178] When we went to

[177] Stella Isaacs, Marchioness of Reading, Baroness Swanborough (1894–1971), née Stella Charnaud, was a British aristocrat and philanthropist. She was the founder and chairman of the WVS.

[178] Sylvia Fletcher-Moulton (1902–1989) was a magistrate and a member of the Royal Commission on Penal Reform, of the Department of Inquiry into the Sunday Observance Acts, of the Council of Tribunals and of the Governing Body of Girton College, Cambridge. She was awarded a CBE in 1961.

Cambridge Alice had perforce to give up her work in the central office, but after an interval she became head of the WVS in Cambridge and had a very large organisation for canteens, clubs, salvage and many other duties which devolved on the WVS. This was the real start in any big way of the public work of which Alice has done so much ever since. She had an organiser in every street of any importance in Cambridge, and she got to know the town, and the townspeople got to know her in a way which was of the greatest importance in her later position on the town council, as Mayor, as Justice of the Peace, and on hospital and school boards.

2.7 Return to Cambridge

We moved to Cambridge in October, Alice's father having found a house for us in West Road, one of the few of its kind in Cambridge. It was early nineteenth century—cream brick, with a slate and lead roof, high ceilings in the main rooms, and extensive outbuildings. We had one and three-quarters of an acre of garden. The house belonged to Caius College and had been rented to Mrs. Wollaston, whose husband was shot so tragically by a demented King's undergraduate.[179] The Wollaston children were very sad that the house was being taken over by newcomers. They made the young Singers, neighbours in the house called *Fenella*, which abutted on our garden, swear eternal enmity and wrote 'Death to the Braggs' in candle smoke on the whitewashed ceiling of the cellar. This house was the family home for fifteen years. Margaret and Patience grew up in it and the boys spent much of their holidays there. They brought their friends to West Road, which became quite the centre for young people in Cambridge; it is surprising how many we have met since who remember the place with affection. It was a place of which we became very fond, and to the children it has always figured as the family home. It was so large and rambling that it was impossible to keep it warm in winter. It was a difficult place for Alice to run because it had so many rooms, and it was expensive to keep up, but we loved it.

[179] Alexander Frederick Richmond 'Sandy' Wollaston (1875–1930) was a British medical doctor, ornithologist, botanist climber and explorer. He was murdered by a student, Douglas Potts.

Figure 25 A drawing of 3 West Road, Cambridge, where we lived from 1938 to 1952. (Sketch by WLB; Courtesy: the Royal Institution London)

Figure 26 Quy Pond, near Cambridge—an old coprolite digging and a favourite bathing pool, July 1948. (Sketch by WLB; Courtesy: Patience Thomson)

War again loomed over us in the summer of 1939. Cockcroft, who was then head of the Mond Low-Temperature Laboratory and had been appointed to the Jacksonian Chair at about the time we came to Cambridge, was the originator of a scheme for making physicists available should war break out.[180] It was rather an uphill task to get the services to accept them. The Royal Society had made itself responsible in 1938 for compiling a list of scientists, with notes about their lines of work and experience: the 'Register of Scientists'. I remember a meeting presided over by my father at which the register was described to the heads of the army, navy and air force research departments, and how these heads in turn said effectively, 'You do not seem to realise that, once the war starts, research stops. We may need perhaps a dozen scientists but not the hundreds on your "Register".' (Six months after the war started, these same scientists were being clamoured for.) Not deterred by their lack of interest, Cockcroft proposed that he and I should make a tour of the service laboratories, armed with a list of what we considered were the hundred or so brightest young people in the universities' physics departments. When we made this tour, the only body which 'played' was the radar research establishment which we had visited at Bardsey. As a consequence, during the summer of 1938 over a hundred clever young physicists helped as a vacation task to man the radar stations along the coast. Although when war broke out all these men did not go immediately into radar research, most of them found their way eventually to Christchurch, Malvern and other radar establishments or worked on it at universities. I am sure that this early contact with radar was a major factor in the unique development of radar in this country. Practically all the fundamental ideas which made our radar so powerful came from university research students, the outstanding example being the magnetron developed at Birmingham by Randall working under Oliphant.[181, 182] Not only the practical experience but also the personal contacts made in 1938 gave us at least a year's start.

[180] John Douglas Cockcroft (1897–1967) was a physicist who shared the 1951 Nobel Prize in Physics with Ernest Walton (1903–1995) for splitting the atom.

[181] John Turnton Randall (1905–1984) was a physicist and biophysicist who after the war led the team working on DNA at King's College London.

[182] Marcus 'Mark' Lawrence Elwin Oliphant (1901–2000) was an Australian-born physicist and a student of Rutherford; he worked on nuclear fission and the development of the nuclear bomb.

Even in 1937, while at the NPL, I had been invited to report on the equipment and tactics of the sound ranging sections in the army, because of my experience in the First World War. I have always felt remorseful for my line of action at that time. I was appalled by the 'whiskers' which had grown on the sound ranging gear. We had left it at the end of the First World War stripped of everything except essentials and made it as simple a procedure as we possibly could. I imagine that in time of peace those engaged in research and development for a war of a kind difficult to foresee and which might or might not come have a desperately difficult task. They have to think up gadgets to show they are being active, but the result in the case of my beloved sound ranging was that it seemed to be hung around with as many gadgets as the White Knight. I felt, when later I had experience as a consultant scientist during the war, that there had often been a wrong placing of emphasis in planning research in peacetime. Thinking up gadgets is of restricted value between wars, because it is so very hard to foresee quite what will be needed. On the other hand, one found in war that knowledge was lacking about quite fundamental subjects which could well have been investigated thoroughly and in an unhurried way in peacetime. For instance, for ASDIC (echo-sounding for submarines), information was needed about the transparency of water for different frequencies, temperature gradients in the sea, reflections from the bottom and from water disturbances, etc., all bits of fundamental knowledge which had to be got in a hurry in difficult wartime conditions. The same was true for the physical effects in the neighbourhood of a ship, pressure disturbances, noise, magnetic field, etc., which imagination might have foreseen as possible effects to be used for mines. To return to sound ranging, I made the mistake, which I have often repented, of throwing my weight about far too much. A little gentle pressure would have been much more effective. My violent attacks only put up backs generally and made the powers-that-be concerned to defend their subordinates. Once the apparatus got to the front and was used in the field, it reverted very much to the simple affair we had used in the First World War.

It was in 1938 that my association started with the protein research which was to lead to such spectacular results twenty-five years later. Perutz was then working in the Cavendish: he had come there from

Vienna, where conditions were becoming very difficult.[183] He showed me some very fine X-ray diffraction pictures which he had obtained with haemoglobin crystals. Some fortunate intuition made me feel that this line of research must be pursued, although it seemed absolutely hopeless to think of getting out the structure of so vast a molecule. It was at that time difficult to get a grant for a refugee student. I think there was some regulation by which we had to prove that the work for which we wanted the money could not possibly be done by a British subject before we could get it for an alien. I wrote to the Rockefeller Foundation in New York asking for a grant of £375 a year: £275 to pay to Perutz (I think he was then about 24) as my assistant and £100 for an X-ray tube. This was granted and it is interesting to compare the £375 which then supported the research with the annual budget of the MRC Unit for Molecular Biology at Cambridge under Perutz's direction, which has developed from this modest start. In 1962 it was £375,000 per annum. Perutz could do little during the war, which started so soon afterwards. For part of it he was interned, and when free again he worked on an amazing scheme for making a large floating ice platform in the mid-Atlantic, to be used as an air base.[184] His particular contribution was to make stronger ice by reinforcing it with sawdust. Canada was to provide both the freezing of the ice and the sawdust from its lumber mills using a battery of refrigerators! The project was not absolutely mad—it might have come off; Louis Mountbatten was very interested in it.[185] The protein research of which I must give a connected account later in this history was often interrupted and often almost given up as impossible; but yet it went on in a sporadic way, with just enough success to keep a feeble flicker of interest going, till finally it flared up into a blazing success in the 1950s.

Before Rutherford's death, a sum of money had been collected to build a considerable addition to the Cavendish Laboratory. Baldwin

[183] Max Ferdinand Perutz (1914–2002) was an Austrian-born molecular biologist who shared the 1962 Nobel Prize in Chemistry with John Cowdery Kendrew (1917–1997) for solving the structures of myoglobin and haemoglobin.

[184] This was Project Habakkuk.

[185] Lord Louis Mountbatten (1900–1979), First Earl Mountbatten of Burma, was a British statesman and naval officer, an uncle of Prince Philip, Duke of Edinburgh, and second cousin once removed to Elizabeth II. He was assassinated by the Provisional IRA while in County Sligo, Ireland.

was then Chancellor of the Exchequer, and the story is that he was given a list of eight powerful industrialists in alphabetical order, to whom he was to appeal for the £90,000 which was estimated the new building would cost.[186] Being a lazy man, he only wrote to the first on the list, as that person ended up giving the whole amount; hence, the building was called the Austin Wing. Holden was the architect, and plans for the building were complete when I took up my post in 1938.[187] Cockcroft was the moving spirit in all the designs, but no building had started. War was again threatening and we only got permission to carry on with the building by promising that it could be used by the services if hostilities broke out. It was occupied during the whole war and indeed for some time afterwards by an army ballistics unit and a navy signals unit. Fortunately, I bought all the furnishings for the meeting rooms, offices, museum and library before the outbreak of war, helped by Joan Worthington. They were magnificent, such as we could never have got after the war, and we were allowed to retain the library as a place in which we deposited the curtains, carpets, tables, chairs, etc. Holden designed the building in a very practical way. The strength was in the outer walls and corridors and in the reinforced concrete floors. The interior walls separating the rooms were very light and could easily be pulled down and erected in a different place with breeze blocks. The window plan was quite regular; hence, the whole scheme was very elastic, and frequently during my time we enlarged rooms, subdivided them or altered the layout completely as new needs arose. One large vertical shaft conveyed the services from floor to floor, and the piping and cables were in the open along the corridors, their colour scheme creating quite a pleasant aesthetic effect. Again and again, we had reason to be grateful to Holden for such a sensible building.

It had risen some fifteen or twenty feet from the ground when I had a letter from Lord Austin to ask when he was to lay the foundation stone![188] We were very ashamed, but Holden proposed a solution. The

[186] Stanley Baldwin (1867–1947) was First Earl Baldwin of Bewdley. As a member of the Conservative Party, he served as Prime Minister on three occasions: 1923, 1924 and 1935.

[187] Charles Henry Holden (1875–1960) was an award-winning architect known for his designs for several London underground stations.

[188] Herbert Austin (1866–1941), First Baron Austin, was a British designer of automobiles and the founder of the Austin Motor Company.

steps up to the entrance had a low flanking wall, and Holden suggested that we should make the foundation stone part of this wall. Austin was delighted with the idea, and after the ceremony he said to Holden, 'I really am pleased with this very original departure from the usual way of putting a foundation stone in the wall.' Holden let us down completely by saying, 'Oh, you heard about that, did you? We forgot all about the bally thing.'

When I came to the Cavendish, Feather,[189] Dee and Bretscher were the leaders in the nuclear laboratory,[190, 191] which revolved around the high-tension apparatus purchased from Phillips in Holland, and the cyclotron.[192] Cockcroft was in charge of the Mond Laboratory,[193] built originally for Kapitza, and had with him Pippard,[194] Shire and Schoenberg.[195, 196] Ratcliffe ran the ionospheric laboratory on the rifle range which he had inherited from Appleton.[197] They had a hut in the middle of the range, and the electrical supply came on cables strung along the hedges. Once, one of the horses which grazed on the range, annoyed with the obstruction, seized a cable in its teeth and fell down dead—according to the owner, it had been an extremely valuable horse. The whole laboratory had of course centred round Rutherford

[189] Norman Feather (1904–1978); British nuclear physicist who worked on the secret Tube Alloys project to develop nuclear weapons. Together with Egon Bretscher, he suggested that the 239 isotope of plutonium would be a suitable choice for sustaining a nuclear chain reaction.

[190] Philip Ivor Dee (1904–1983) was a British physicist who worked on airborne radar during the Second World War.

[191] Egon Bretscher (1901–1973) was a Swiss physicist who worked with Rutherford at the Cavendish Laboratory from 1936.

[192] A nuclear particle accelerator in which charged particles are accelerated in a spiral arrangement from the centre outwards, making it a compact machine compared with a linear accelerator.

[193] Named after Ludwig Mond (1839–1909), a German-born chemist and industrialist who became a British citizen.

[194] Alfred Brian Pippard (1920–2008); British physicist who worked on electronic properties of metals and superconductors. He was Cavendish Professor from 1971 to 1984.

[195] Edward Samuel Shire (1908–1978) was a British nuclear physicist. During the Second World War, he worked on proximity fuses for bombs.

[196] David Schoenberg (1911–2004) was a British physicist, born in St Petersburg, Russia. He worked on low-temperature physics and magnetism at the Mond Laboratory.

[197] John Ashworth Ratcliffe (1902–1987) was a British physicist who worked on radio waves in the ionosphere.

and had been devoted to nuclear research, with the exception of the Mond Low-Temperature Laboratory and the ionospheric work under Appleton. He left shortly after I came, to succeed Sir Frank Smith as Secretary of the DSIR. The Jacksonian Chair thus became vacant and, at one of my first committee meetings in Cambridge, Cockcroft was elected to succeed Appleton. R. H. Fowler had the Plummer Chair of Theoretical Physics. Bradley, who had been with me at the NPL, came as Assistant Director of Research and at some point Lipson joined the team.[198] Orowan, who had come to England as a refugee, found a home in the Cavendish.[199] So with Perutz, Bradley, Lipson and Orowan I had a group working on X-ray crystallography and metal physics in which I could take an especial interest.

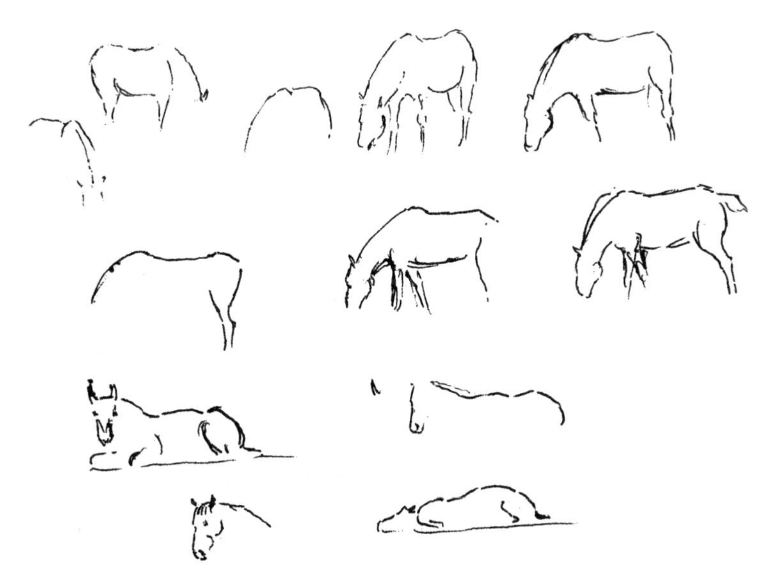

Figure 27 Rough study of horses. (Sketch by WLB; Courtesy: Patience Thomson)

[198] Henry (Solomon) Lipson (1910–1991) was a British physicist who worked as a crystallographer with WLB in Manchester and followed him to the NPL and Cambridge.

[199] Egon Orowan (1902–1989) was a Hungarian-born physicist and metallurgist.

At the Cavendish dinner in December 1938, Lord Austin was our chief guest. He stayed with us, and I remember that the boys were very worried because our car was a Morris.[200] He was the kind of great industrialist sometimes pictured in fiction, apparently with no interest outside his work. He told us he had not had a holiday, even on Sunday, for I think he said twenty years, and his health was quite broken down. At the dinner, Bhabha unveiled an oil painting which he had done (without a model) of me, in the grand eighteenth-century style, with velvet curtains partly concealing a vista.[201, 202]

It seemed only a very brief interlude before the war started. It was declared while we were having a holiday in a house belonging to Hugh Lyon, in the Barmouth estuary.[203] We had promised to vacate the house if war came, because Lyon wished to evacuate his family there. For some curious psychological reason, everyone felt that their families were safer in any place other than at home. Sharing this illusion, we parked the children first with Enid and Malcolm at Newport in Wales,[204] and then with the Cawleys in Herefordshire,[205] while Alice and I returned to Cambridge. Some evacuations were of course sensible. I remember seeing all the trains roll in full of children from London. But the conception of what bombing would mean was fantastically inaccurate quantitatively. There were stories going round that there was not enough wood on London to make all the coffins which would be necessary.[206] The fact that there was almost no way

[200] Morris Motors Ltd began in 1912 and manufactured automobiles under the Morris name until 1984, when the Austin brand took over.

[201] Homi Jehangir Bhabha (1909–1966) was an Indian nuclear physicist and founder of the Tata Institute of Fundamental Research and the Bhabha Atomic Research Centre. He participated in the British Tube Alloys project and in India called for the development of nuclear weapons.

[202] This painting (slightly damaged) is currently located in the museum storeroom at the Cavendish Laboratory.

[203] Percy Hugh Beverley Lyon (1893–1986) was a British poet and was Headmaster of Rugby School from 1931 to 1948.

[204] Enid and Malcolm McGougan. Enid was WLB's sister-in-law.

[205] Robert Hugh Cawley (1915–1954), Second Baron Cawley, and Baroness Vivienne Cawley, née Lee (1879–1978), lived at Berrington Hall, Herefordshire; the hall now belongs to the National Trust.

[206] During the Second World War, John Desmond Bernal carried out research on the effects of enemy bombing on animals and people. Analysis of bombing on Hull and Birmingham showed that aerial bombing caused little actual disruption.

of stopping a bomber from attacking a particular target led people to think that every target would be attacked, quite forgetting that the bombers available could only deal with a minute percentage of the possible targets. It really was fantastic, the contrast between the belief that practically all London was doomed, with the reality that the number of deaths from bombs throughout the war was less than the number of road accidents. On the other hand, the discomfort and dislocation due to bombing were very real. Strangely, Cambridge was the first place in Great Britain to have a serious bomb attack.

Most of the researchers in the Cavendish at once left for some form of war work and those who remained were involved in schemes of some kind for the services, so normal research stopped entirely. Teaching went on with increased activity. Queen Mary College and Bedford College were evacuated to Cambridge, and we solved the problem of the shortage of teaching staff by pooling our classes and resources. Robinson, Head of Physics at Queen Mary and my old colleague of sound ranging days in the First World War, billeted with us.

Alec Wood was a great stand-by;[207] he excelled at lectures for Part I Physics, and dear old Searle who really should have retired long ago, was in his element; he came back to take charge of the practical classes for Part I. He celebrated his eightieth birthday, I believe, during the war, and it was extremely difficult to get him to relinquish his classes when the staff returned at the end of hostilities. When Ratcliffe was due to return and take over he wrote to me to say how essential it was that Searle should have disappeared from the scene before he organised the classes again. Shortly after Ratcliffe's return I went over to see how the Part I practical class was getting on. All the students were clustered at one end of the big room, listening to an exposition—by Searle. Ratcliffe stood in solitary glory at the other end.

The new Austin Wing was occupied by service research departments, and we had to cram our students, augmented by our guests from Queen Mary and Bedford, into any accommodation we could find. Radar was becoming so very important that there was a great demand for trained men, and we had a special radar laboratory. The Mond was devoted to very interesting research for the Ministry of Supply. It started with the development of a fuse which could be set

[207] Alec Wood is probably Alexander Wood (1879–1950), who lectured on sound and optics.

electrically as the shell left the muzzle of the gun: this never became an operational success, but in the course of the investigation Pippard and Schoenberg developed a most ingenious and simple device for measuring the accelerations to which a fuse was subjected during the passage of the shell up the bore of the gun. They were then able to match these accelerations quantitatively by attaching the shell to a block of steel and firing a rifle bullet at the block, thus obviating the slow and expensive necessity of tests by firing and recovering shells on a range. This method was taken over in a big way by the services; it was a real winner. Vick came regularly to the Mond to get a report on their progress.[208] It was the first time I had met him and I came to realise what a gift he had for organisation of research.

I remained based in Cambridge during the war, except for the eight months I spent in Canada as Scientific Liaison Officer in 1941. My main concerns, apart from running the Cavendish, were sound ranging and the Admiralty ASDIC. The headquarters for sound ranging research and teaching were on Salisbury Plain, and my great friend from the First World War, Atkins, was in charge of the sound ranging section, and I stayed with him frequently.[209] Harold Hemming, who had been my opposite number in the First World War, concerned with flash spotting, was also quartered there. Sound ranging had really hardly altered at all since the First World War. The one big step forward which we envisaged even in 1918 was a 'radio link' from the microphones to the recording apparatus, but this had never been developed into a practical form in the intervening years. I still think it might have been done, if sufficient intelligence had been applied to the task; a radio link would have enabled sections to work without the miles of connecting wires,

[208] Francis Arthur Vick (1887–1998) was a British solid-state physicist. During the Second World War, he worked at the Ministry of Supply. In 1966 he was appointed Vice-Chancellor of Queen's University, Belfast. His letters and papers are kept at the University of Warwick.

[209] Arthur Henry ('Tommy)'Atkins (1887–1946) was heavily involved in artillery activities during the First World War. He headed sound ranging sections, particularly the 'S' section for some time, and he was prominent at WLB's sound ranging conferences. From Nov. 1917 he had an important training role at the Overseas Artillery School on Salisbury Plain. He returned to the Western Front in March 1918 to undertake a compilation-of-intelligence role during the German offensive and was also involved in the allied technique of flash spotting. He trained sound rangers in the Second World War (private communication from John Jenkin). He remained a lifelong friend of WLB.

which they had to lay across country to their microphones, and which were at the mercy of transport, tanks and scroungers who were not averse to picking up a length of such good cable for their own purposes.

It was not anticipated before the war that sound ranging would be of much use, as it was doubted whether there would be time to install sections in a 'war of movement'. The station has to survey six microphone positions over a base of some 9,000 yards, to establish a headquarters for the recording gear, and run separate double cables to the six microphones and to the two observation posts in front of the base, which set the recorder going when a gun fired. This would be at best a matter of one or two days. Actually, sound ranging proved to be a winner. If there were a hold-up in an advance, it would take two or three days for our guns to establish themselves and fix their positions, and during this time sound ranging could establish its survey and base. Since technical advances had almost eliminated the flash from guns and, as in poorly surveyed and featureless country like North Africa, air observations were almost meaningless, sound ranging was often the only means of finding enemy battery positions. So sound ranging, linked to the same survey as the gun positions and thus independent of maps, was the main basis for counter-battery tactics. The accuracy was probably less than in the First World War because the ranges were greater. When trying to silence an enemy battery, however, shells anywhere near the position have a deterrent effect. If a shell bursts within a quarter of a mile of a gun position, gunners are apt to say, 'My God, they have discovered our position.' At any rate, sound ranging must have been appreciated because, in both the Sicilian and the D-Day landings, the units were sent in, I was told on D-Day plus three.

An interesting development of sound ranging was the location of the origin of the V2 rockets towards the end of the war. They travelled at supersonic speeds and rose a hundred miles in the air. The measurements were made on the 'shell wave' produced by the rocket going faster than sound, and this noise appeared to come from a point in the trajectory. By spacing microphones at the corner of a square, it was possible to get a line showing the direction from which the shell wave came, and so intersecting the trajectory. A grid of such squares gave a series of lines from which the trajectory and hence the firing point could be deduced. It was a strange form of sound ranging. There was no need of a forwards observer, because the rocket exploded a minute or so before the sound of the shell wave reached the microphones; there was time for an observer in London to telephone headquarters and set the recording apparatus in motion. The 'base' in England stretched from the south coast to the Wash, and the links were made through our telephone system. Headquarters were established in a school under the shadow of Canterbury Cathedral. Taylor and Clews were the moving spirits in directing the investigation.[210, 211] The method gave a position accurate to a mile or more in breadth and perhaps ten or twenty miles in range; it was, I suppose, regarded as of tactical importance to know that the rockets were coming, for instance, from the neighbourhood of The Hague. Later, when the enemy were driven out of Belgium, Taylor and Clews established their recording station in Malines and used the Belgian telephone network. It was an uphill task because the Belgians had so successfully sabotaged their own lines when the Germans were using them. The V2s were at that time falling on Antwerp.

It was very interesting to act as adviser to the navy. In the ASDIC system, an emitter is housed in a water-filled 'dome' beneath the bottom of the ship. The piezoelectric effect is used to send a short burst of water-borne sound waves, the wavelength being so short that the beam is only a few degrees in angular width.[212] After the emitter has sent out its 'ping', it is switched over so as to be used as a receiver to pick up an echo from a submarine in the neighbourhood. The direction in which the submarine lies is deduced by swinging the beam round and noting the direction from which an echo is received, and the range to the submarine is deduced from the time between 'ping' and the receipt of the echo. A hunting drill helps the destroyer to close in on the submarine and drop its depth charges. The method was initiated during the First World War, and my father's group at Parkeston Quay developed it, but it never reached the stage of actual use in that war. It reached a practical stage in the interwar years. There was an unhappy

[210] Taylor is Major P. W. E. (Bill) Taylor.

[211] Charles John Birkett Clews (d. 2010), a major during the war, was associated with identifying the location of the V2 launching sites. In 1952 he was appointed Professor of Physics at the University of Western Australia. During his appointment, he became Deputy Vice-Chancellor and had a road named after him. He died in Oxford in 2002.

[212] The piezoelectric effect occurs when the application of an electric field on a certain material causes it to become strained, thus changing its shape. If a high-frequency oscillating field is used, then the material responds accordingly, creating ultrasonic waves. The effect also works in reverse, converting the reflected sound waves effectively into electrical signals.

situation at the beginning of the war, because the chief scientist who had been responsible for the developments in peacetime was thought, perhaps with some justice, to have become so set in his ideas that he would not willingly try out new schemes or accept suggestions from the university recruits to the station. A. V. Hill in particular was very worried about the situation, and so the chief scientist left his post.[213] I was asked to take the place of the chief scientist but felt uncertain of my powers as an administrator of so large a centre.[214] Roberts was then asked and, after consulting me, he accepted.[215] He became a very good head, and he also had the help of Anderson, who was his second in command and succeeded him.[216] The station was at Fairlie on the Clyde during the war, and I went there regularly.

I liked working with the section. Naval enterprise seems to me to have a really practical bent even in peacetime, I suppose, because the investigators are up against the sea and all the difficult problems it imposes. The naval investigators were very pleasant people to work with. Besides Roberts and his second in command, Anderson, there was Vigoreaux,[217] who was adept in electrical circuiting, Norman Astbury (brother of the Astbury at Leeds [Bill Astbury][218]) and Fisher, who were first-rate applied mathematicians,[219, 220] Alexander,[221]

[213] Archibald Vivian Hill (1886–1977) was a British physiologist known for his work on muscles. He shared the 1922 Nobel Prize for Physiology or Medicine for his study of the production of mechanical work by muscles. During the Second World War, he was a member of a committee overseeing the development of radar.

[214] This would have been HM Anti-Submarine Experimental Establishment, Fairlie.

[215] John Keith Roberts (1897–1944) was an Australian-born physicist who worked on adsorption of gases by metal surfaces.

[216] J. Anderson became Superintendent Scientist at Fairlie.

[217] P. Vigoreaux worked at the NPL.

[218] William (Bill) Thomas Astbury (1898–1961), originally a student of WHB, was a British physicist who initiated the use of X-ray diffraction on biological molecules and took the first X-ray photographs of DNA. In 1938, he was the first to introduce the term 'molecular biology'. WHB proposed him for Fellowship of the Royal Society: he was elected in 1940.

[219] Norman F. Astbury, was the younger brother of Bill Astbury, with whom he shared a love of music. He was elected Director of the British Ceramics Research Association in 1969.

[220] J. W. Fisher (d. 1954) was a mathematical physicist and a medical doctor.

[221] Possibly E. A. Alexander, who later worked on jet engines in the Combustion Division at Rolls-Royce.

who was an expert on supersonic waves, and my friend Ben Browne from the geophysics laboratory in Cambridge.[222] I would go at regular intervals, spend one or two nights at the hotel near Fairlie and talk with the researchers about their problems. I find it hard to estimate how much I helped. Only the people on the spot could appreciate the practical difficulties, and such help as an outsider could give came from a knowledge of the man to consult about this or that special point. I sometimes found that years later some suggestion I had made had apparently borne fruit. But quite apart from direct help, I think the researchers liked talking about their problems to someone who understood and could appreciate their work. I continued this link after the war, when the station returned to Portland Head, near Weymouth, and I served as an adviser for over fifteen years.

Roberts died while at Fairlie, a very sad loss. He had tuberculosis as a younger man and had recovered, but it attacked him again. He had a hard and sometimes distressing time at Fairlie, the latter because some members of the staff were very loyal to the head who had left and they resented Roberts's coming. I do not think, however, that his strenuous time was the cause of the recurrence of the tuberculosis, and even if it lowered his resistance, he would, I am sure, have wished to use his last years for such an important and stimulating task.

I had gone over to France just before the end of the 'phoney war', when the advance of the German armies began. Hodgkinson and I flew over with a gunner, Colonel Paterson, to meet colleagues in Paris and inspect a sound ranging section at the front.[223] We were much feted, taken to a cabaret in Paris and given such a tremendous lunch in a small inn near the sound ranging section that my memory of the occasion is somewhat dim. I remember the beams of the inn were festooned with beribboned bottles of wine, dedicated by young men who had left for the front to the celebration they would have on their return. Two weeks later all this country had been overrun by the Germans. On our return, we took a light plane to Amiens and transferred there to a larger passenger plane. As we sped down the runway for

[222] Benjamin Chapman Browne (1901–1968) worked on measuring the acceleration of submarines. He became Head of the Geology Department in Cambridge.

[223] T. G. Hodgkinson worked in the sound ranging section during the First World War.

the take-off, mechanics ran out to stop us, and we were able to pull up in time. When we got out, we saw that the rear wheel (planes then had two heavy front wheels and a light rear wheel) had crumpled up. The luggage store was opened and the reason immediately became apparent. An infantry colonel returning from India had, against orders and while no one was looking, piled into the store his cases, his sporting gear, tent, bedding and bedstead, camp table, his batman's pack and his arms, a load which threw out all calculations and broke the rear wheel. We were fortunate that it was noticed before we left the ground; the camp commandant's language was lurid.

Soon after our return, Hitler marched in, France collapsed and the evacuation from Dunkirk took place. We could hear the guns in Cambridge. Carrie, a Canadian who had been very kind when I was in Ottawa, was then in England with his unit and he offered hospitality to our children if we wished to send them to Canada.[224] Alice and I were faced with a very difficult decision: so many of our friends were sending their children to Canada and to the United States because an invasion of England seemed imminent. We decided not to send them; Alice felt that we would largely lose them as our own children if they left us and that in her position as head of the WVS in Cambridge it would be destructive of moral if she took advantage of a possibility denied to all but the very few. She was right on both counts. I visited many of the children in Canada in 1941, and many were sad people; though treated with such boundless kindness, they did not belong. The saddest were the mothers and children, who were so completely dependent on charity and felt it deeply. And Will Spens, the Commissioner for the East Anglian area, told Alice that the fact that the children of so many in the know had remained in England had a very big effect.[225]

In 1941 Appleton asked me to go to Canada as Scientific Liaison Officer between Canada and the UK. Darwin, then head of the NPL, was sent at the same time to take charge of the scientific mission in Washington. America was of course not at war at that time—Pearl Harbour had still to happen—but the American scientists were hard at work at

[224] Carrie is probably George Milroy Carrie (1892–1970), an engineer and a military officer in the Canadian Army. He worked on accuracy of artillery fire.

[225] William Spens (1882–1962) was an eminent educationalist in the mid-twentieth century, an academic and Master of Corpus Christi College, Cambridge. During the Second World War, he became Regional Commissioner for Civil Defence for the Eastern Region.

war problems and we had shared much information with them. It was Darwin's task to act as a link and to keep our country informed about America's advances. Canada had a number of groups working under Mackenzie at the National Research Council's laboratory in Ottawa, and I was attached to his team.[226]

Katharine and two of the boys were going to America with Charles Darwin, and we travelled together.[227] I spent the previous night in Liverpool with Alban Caroe, who was doing war work there, and next morning went aboard with the Darwins.[228] Our steamer, the *Baltrover*, had been a trader which had plied in the Baltic in peacetime. She was about 4,800 tons, with a maximum speed of ten knots, hardly enough bunker space to take her across the Atlantic, and half a dozen cabins for passengers. A number of other cabins had been roughly built in the hold, a most uncomfortable 'black hole of Calcutta' because they were quite cut off from the outer air. We had some sixty passengers in all: a few on an official mission like Charles and myself, airmen who had flown bombers over and were returning to Canada, civil servants with their families who were to work in the US, and one very interesting party of seventeen mannequins in the charge of a woman civil servant. They were being sent out to South America by our Board of Trade to get customers for our clothes and so earn money to pay for the war. Their clothes were all being sent in another ship, which seemed to us

Figure 28 The SS *Baltrover*, convoy to Canada, 1941. (Sketch by WLB; Courtesy: the Royal Institution London)

226 Chalmers Jack Mackenzie (1888–1984) was an important person in the development of science in Canada.

227 Katharine is Katharine Darwin, the wife of Charles Galton Darwin.

228 Alban Douglas Rendell Caroe (1904–1991) was a British architect who married WLB's sister, Gwendy.

odd because this was clearly a case where it was desirable to have all the eggs in the same basket. Anyhow, they were a great addition to the party. They all wore different pastel shades of trousers, and when our escorting destroyers came near (they seemed to do so very often) the mannequins formed a colourful sight as they lined the rail and waved their hands. There were only two passengers' bathrooms on board, one of them being attached to the grand deck cabin which Charles and I occupied, and we were in great demand for loaning the bath to mannequins. They slept on palliasses in the lounge. The only room in which we could all forgather was the bar; and at night, in order that no gleam could be seen from outside, the lights automatically went out every time anyone entered or left the room, a very disconcerting business. We had no submarine incidents, due, we were told, to the clever tactics of our commandant, but we were bombed several times. I wrote a diary of the voyage out which I am adding as an appendix; it may be of interest as an example of what an Atlantic convoy was like in the Second World War.[229] We arrived at Halifax after fourteen days at sea, welcomed by one of the most brilliant auroras I have seen.

I took over from Fowler in Ottawa and, when I left eight months later, George Thomson took over from me. The Canadians had various schemes on hand. They had, for instance, a mobile set for getting radar echoes from planes and directing the fire of anti-aircraft guns. They had a range at Halifax for checking the demagnetisation of ships by a current in a cable run around the deck. They were doing work on the airman's reaction to high accelerations, to low temperatures and to low atmospheric pressures. An industrial undertaking linked with the National Research Council was producing large numbers of optical instruments. Much energy was being put into these developments, but I was not sure whether it was all spent in the most paying way. Such developments as the building of an anti-aircraft radar set was a duplication of a similar development in Great Britain, with the handicap of a smaller body of scientists and less contact with the realities of the front line. The men would probably have been more effective if they had formed part of the English team.

I stayed in a hotel in Ottawa which I believe was called the Rochester.[230] It was much favoured by politicians, and the doorman confided

[229] Unfortunately, this appendix has not been located.

[230] More likely it was the Hotel Roxborough.

to me that there was hardly a well-known one whom at one time or another he had not carried up to bed. I used to walk each morning to the National Research Council headquarters, about a mile and a half away, along the Ottawa River. Behind the CPR hotel and near where the Rideau Canal joins the river and the bridge goes across to Windsor, there was a steep bluff with a public garden at its top, and leaning over the parapet one could look down on the tops of the trees that were clustered on the steep side of the bluff.[231] It was a grand place for birdwatching, because they were not shy of an observer looking down on them. I got quite a good knowledge of the North American birds. Another pleasant haunt was up the Gatineau River opposite Ottawa. My colleague in the liaison work was Shenstone, a Canadian who was Professor of Physics at Princeton in the States.[232] He and I had many pleasant expeditions together, and he and his wife, Molly Shenstone, often invited me to their flat in the evenings.

The Canadians had a great admiration and affection for my predecessor Ralph Fowler. He was a very outspoken man who said and did quite outrageous things sometimes and the next moment apologised in the most disarming way. Though a theoretical physicist, he had considerable practical engineering insight and was a born leader. He was the moving spirit behind many of the ventures started in Canada. He often went far beyond his authority in his eagerness to get things started but was always forgiven because he was so right. I was a very pale person in comparison.

The High Commissioner at that time was Malcolm Macdonald; I came under his office for pay and allowances.[233] I hardly saw him while I was there, for which I am sorry now, because he was an ardent birdwatcher and I am sure we could have had excursions together. He wrote a book later about the birds in the neighbourhood.

I made one trip to Vancouver and back, going out by Winnipeg and Edmonton and returning by Calgary and Alberta. I went by train as I was stopping at so many places and so had an opportunity to appreciate the vast distances. The trip was made in June when all was green and flowery. The big towns, each with its large railway hotel, are like

[231] CPR, Canadian Pacific Railway.

[232] Allen Goodrich Shenstone (1893–1980) worked in Princeton on atomic spectroscopy.

[233] Malcolm John Macdonald (1901–1981) was a British politician and diplomat.

liners in the ocean, compact settled places with miles of open country around them. In winter, we were told, one has to be quite careful in going into the country because it would be really dangerous to be stuck in the car in that bitter cold. Many farmers just 'summer farm', leaving their places empty in the winter and going south to better climes. After the long stretch of prairie, one comes to the foothills and the Rockies, which I think are very disappointing. The mountains in the distance are grand, but there is no exciting foreground, just pine trees growing out of gravel. The stretch of cowboy country beyond the Rockies is exciting, real Wild-West scenery with sagebrush etc. Then one comes to the coastal range and it is like a transformation scene. The country is wet and lush, the woods full of flowers and ferns—a kind of Devonshire with a backdrop of snow mountains. In the park at Vancouver, and on hills behind the original forest, cover has been left untouched. There seems to be as much wood as air as one walks through the forest threading a way through gigantic trunks of hemlock and Douglas fir. I sailed over to Vancouver Island during my Vancouver stay and gave a talk in Victoria. The bishop had been at school with me and called to take me on a very pleasant run in the neighbouring country. On our way back, we stopped at a school and were met by the headmaster. He led me up to a small door which I innocently stepped through, and there was the whole school waiting to be addressed by me! I wished I could remember a greater number of their bishop's more regrettable pranks at school, but I did my best. Vancouver Island is drier than the mainland, but the whole place has a warm, wet, misty climate much like that of England and is a paradise for birds and flowers. It seemed quite strange seeing the brilliant emerald lawns after the dryness of Eastern America.

I had gone to Vancouver by the national railway and returned by CPR, staying a night at Lake Louise and doing the stretch between there and Banff by charabanc. I travelled with a hotel manager who dilated on the beauties of the CPR hotel at Banff, which looks down over the Bow River. He spoke with admiration about the castles in England and Scotland, the schlosses in Germany and the chateaux in France but, he said, the best features of all these are combined in the Banff Hotel and when I saw it I realised that it was all he had described. The ride between Louise and Banff was interesting as a specimen of the Rockies. We had of course a guide with loudspeaker, who described things of interest en route: 'Round the next corner we may see some mountain sheep'—and there, sure enough, were the sheep, organised for the tourist trade by putting lumps of salt on the roadside; and similarly for bear and moose. The hotels were highly organised: cowboys to take parties out riding, golf courses carved out of the wilderness, 'trails', swimming baths, etc.—fun to see just once.

I managed to cadge a lift back to England by bomber. Two-engined Hudsons were being made in Canada and flown over to England. I have mentioned that in going out by boat a number of our fellow passengers were pilots returning to bring back the next plane. There were also a certain number of four-engined bombers flying to and fro, and I was given a place in one of these; it had been roughly fitted out for half a dozen passengers. Two of us, and I was one of the lucky ones, had a berth in the tail; the others managed as best they could in the bomb rack. There was no heating, and in fact the air at minus forty blew freely through our compartment. We were advised to buy a kind of padded cocoon of flannel, with just a slit near the face for air. This was a good protection, but we had a thin time because the crew had to go over 21,000 feet to avoid icing and, although there were arrangements for giving the passengers oxygen, they forgot about us in the excitement of rising above the cold front. We passed out and were lucky we did not get frostbite. Just previously, a passenger coming to Canada, we were told, had lost both hands through exposing them for some reason and then fainting. I remember the extraordinary satisfaction of getting gulps of solid air, as it felt when we descended again and I regained consciousness. The arrangements altogether were a bit casual. We flew first to Gander and lived there in tents for a night or two awaiting good weather conditions. There was a notice in the camp commandant's hut to say, 'Pilots taking Hudsons to England, who have not flown this type of aircraft before, are advised to call at the Camp Commandant's office and acquaint themselves with the position of the controls'. A transatlantic flight was something new and exciting in those days, and those who had made it could join a special club and, I think, wear a badge. I was none the worse for our night's adventure, except for a splitting headache. I believe that as a result of our mishaps the regulations for bringing over passengers were tightened up considerably. In the afternoon we were flown to Hendon and I caught a train at Liverpool Street to Cambridge. One of my fellow travellers commented to me on how rainy it had been the day before and, when I said I had not experienced it because 'I was in Canada yesterday', the effect was terrific. Another passenger, getting off at an intermediate station, promised to telephone Alice to say I was back; so Alice met me on the Cambridge Station platform.

I returned to the strangely unreal wartime life in Cambridge. We had a number of air raid warnings and some bombing, but most of this was incidental, though quite a number of houses, I believe about 1,500, were at one time or another rendered temporarily uninhabitable. Food rationing pressed on everyone heavily. I remember poor Patience, when we were tactlessly talking about food we had in peacetime, saying, 'Oh, how I wish it were pre-war again!' The Cavendish was busy with large numbers of students taking an abbreviated two-year honours course, and with special classes in electronics for radar personnel. I went regularly to Fairlie for three-day visits to go over the ASDIC station, and to Salisbury Plain for sound ranging, and I was on some government committees. Naturally, very little research was going on. Poor Bradley was then in a bad way; the mental illness which ended his research career had started and his brilliant and imaginative brain had lost touch with reality. A very secret unit was working in the Cavendish under Halban and Kovarsky, who had escaped from France with a supply of heavy water and were using it to explore the possibility of using nuclear energy.[234] I was not in any way involved in this project. It was Chadwick who paid regular visits and supervised their work.[235] Ex-Cavendish staff and research workers were doing great things at the various radar establishments.

My father died in 1942. He worked to the very last: he was giving a BBC broadcast on Monday and died the following Thursday. His heart had been weak for some years. I remember he felt the walk up St James's Street from the Athenaeum, which is up a slight hill, very trying, and he generally had to take a taxi back to the Royal Institution. Gwendy and I were with him when he seemed to get much better and be out of immediate danger. I returned to Cambridge, and on my arrival home found a telegram awaiting me to say he had died while I was on my way. Gwendy and I were invited by the Royal Society to write our personal memories of him in 1962, the hundredth anniversary of his birth, and I put into that all my dearest and most

[234] Lew Kovarski (1907–1979), a French physicist, fled to Britain in 1940 with Hans von Halban (1908–1964) to work at the Cavendish Laboratory.

[235] James Chadwick (1891–1974) was a British physicist awarded the Nobel Prize in Physics in 1932 for the discovery of the neutron. He worked on the Manhattan Project during the Second World War.

vivid impressions of him.[236] He had been at the Royal Institution for nineteen years when he died (he succeeded Dewar in 1923[237]) and he had built the place up again. The staff at the Royal Institution were terrified of Dewar; as one of them said, 'He was a tyrant, but a gentlemanly tyrant.' Dewar had been at the Royal Institution for forty years and had carried out his magnificent work on the liquefaction of gases in its basement laboratories. When my father took over, however, there was practically no research. Cuthbertson worked regularly in the Mond Laboratory on the refractive indices of gases and there were one or two others who came in sporadically, but otherwise the laboratories were empty.[238] My father started his research school there, with a few faithfuls he brought with him from University College and with his brilliant instrument maker, Jenkinson.[239] A fine galaxy of well-known X-ray crystallographers passed through this school—Müller,[240] Shearer,[241] Kathleen Lonsdale, Bernal, Astbury and others.[242]

[236] See *Notes and Records of the Royal Society of London* (1962), **17**, 169–182; doi: 10.1098/rsnr.1962.0016.

[237] James Dewar (1842–1923) was a Scottish chemist and physicist. He carried out research into liquefaction of gases and many other areas and is best known for the invention of the Dewar vacuum flask. In chemistry he was known for his model of the benzene molecule, which he described as being flat; the model was later proved by Kathleen Lonsdale, student of WHB, in her structure determination of hexamethyl benzene.

[238] Clive Cuthbertson (1863–1943) was a British chemist and physicist who worked on the optical properties of gases, often with his wife, Maude.

[239] C. H. Jenkinson (d. 1939) was a superb instrument maker who had been trained in the works of the Cambridge Scientific Instrument Co. He worked for WHB in Leeds, at University College London and at the Royal Institution. He built the famous ionisation spectrometers that enabled WHB and WLB to make precise diffraction measurements from crystals.

[240] Alex Müller(1889–1958) worked under WHB, initially at University College London and then in the Davy-Faraday Laboratory at the Royal Institution on the crystallography of paraffins and related substances.

[241] George Shearer worked with Alex Müller on the crystallography of paraffins.

[242] WHB carried out research mainly into the structures of organic compounds until his death in 1942 while at the Royal Institution. During his career he was responsible for the education of many graduate students, and he is credited (along with his ex-student, John Desmond Bernal) for encouraging many female students into crystallography, including the Nobel Prize winner Dorothy Hodgkin, as well as Dame Kathleen Lonsdale and Helen D. Megaw. It is remarkable to note that, out of eighteen of his students, eleven were women! WLB too employed several female students and researchers, so that we see that both Braggs played a major role in bringing women into science at a time when this was discouraged by society (see Maureen Julian in 'Suggestions for Further Reading').

Two great rebuilding schemes were undertaken during his time. The first involved the library and flat above it, it having been found that the walls behind the book stacks were cracked and unsafe. The trouble was laid at the door of the wicked architect Vulliamy, who had added the Corinthian columns to the outside of the building early in the nineteenth century and, in order to space them regularly, had moved windows of the old building without taking precautions to strengthen the main walls.[243] The building had to be gutted from the basement upwards and the main rooms of the flat were involved. The family lived in Mulberry Walk while the work was going on, and it was there that my mother died. The old study in the Royal Institution was not involved and its panelling is said to be the original in the house as it was before the Royal Institution bought it. The main bedroom was also left intact. Guthrie was the architect, and he redecorated the dining room and drawing room with the old study as model.[244] His work all through the Royal Institution, in matching the old and new, was quite wonderful. The other rebuilding scheme involved the lecture room. Attention was drawn to its very dangerous condition by an explosion in the mains in Albemarle Street, which might well have caused a serious fire. The seating in the lecture room, it was discovered, was supported by a wooden framework, and the only way of escape from the room was via the doors on either side of the lecture bench, just the place where a fire might well start. The whole lecture room was rebuilt from the bottom, and separate exits to the street from the rear were arranged.

My father started a successful scheme of inviting London University students to the Royal Institution for afternoon lectures, and he and my mother also started parties for the members after the 'Friday Evening Discourses'. Another great event during his time was the celebration of the one-hundredth anniversary of Faraday's discovery of electromagnetic induction in 1831.[245] In all these schemes Martin, whom my father had appointed the assistant secretary in the Royal

[243] Lewis Vulliamy (1791–1871) was a British architect and a member of the Vulliamy family of clockmakers.

[244] Leonard Rome Guthrie (1880–1958) specialised in domestic architecture.

[245] Michael Faraday (1791–1867), a British scientist famed for his work on electromagnetism and electrochemistry. He was one of the most influential scientists in history. He made many discoveries, including the principle of the electric motor.

Institution, was my father's right-hand man.[246] From 1938 onwards I was frequently in the Royal Institution because I was appointed its Professor of Natural Philosophy when Rutherford died and I went to the Cavendish. This post involved giving regular discourses and afternoon lectures.

In the spring of 1943 I was asked by the British Council to visit Sweden for some weeks and give talks to Anglo-Swedish societies. We flew over from Leuchars near St Andrews, in a small plane piloted by a Norwegian who had been a whaler and who had escaped the German occupation because he was then in the southern seas. There were three passengers: a secretary-typist in the embassy, a Norwegian lady going to help with British Council in Stockholm, and me. We flew over without mishap by night and drove from the airport to our hotel through the brilliantly lit city—such a contrast to the blackout we had left. At that time there was a great shortage of petrol in Sweden, and the cars were running on wood, transverse sections of birch trunks quartered like hot cross buns. The taxis and private cars carried sacks of them on their roofs, and I was told that on a cross-country trip it was usual to carry an axe in case one ran out of fuel. The gas produced by semi-combustion in an affair the size of a dustbin attached behind the car was fed into the carburettor, which ran all night though with much reduced power. I remember that I could not understand why, when one took a taxi at the station, the driver waited a minute before starting off. He had to get his blower going and start the fuel burning well. After the growing asperities in Great Britain, Sweden seemed exceedingly luxurious, with fascinating things to buy in the shops, such as two dainty wristwatches which I got for Margaret and Patience. I had a long list of 'wants', including hairpins for Alice!

At that time the Swedes were just beginning to think that perhaps the Germans were not going to win, and Anglophiles felt more able to extend a welcome. I gave talks in Stockholm, Uppsala, Malmö, Lund, Helsingfors and Göteborg; these talks had a friendly reception, and I made some excellent friends. In Stockholm there was of course Westgren, and the Professor of Mineralogy, with whom I stayed for a

[246] Thomas Martin was Senior Administrative Officer at the Royal Institution from 1928 to 1950. He oversaw the publication of Faraday's laboratory notebooks in seven volumes between 1931 and 1936 (information from Frank James at the Royal Institution).

few days. I also had a stay at the embassy; the ambassador, Sir Victor Mallett, left the day after I arrived but Lady Mallett was very kind and I much enjoyed some jaunts with their son and daughter, then aged about twelve and fifteen.[247] On one occasion Westgren offered to take myself and the children on an excursion to the old capital, on an island some twenty miles from Stockholm; Lady Mallett, probably used to the ways of the country, suggested she should pack us some lunch, but I assured her that the tremendous hospitality of the Westgrens would make that quite unnecessary. We walked miles and miles, getting hungrier and hungrier, and it finally became clear that Westgren had not included lunch in the programme! The poor children—some bread and milk from a farm was the best I could do for them. Somehow the Westgrens have always been associated, as far as we are concerned, with gargantuan feasting or extreme hunger. Alice and I went one day exploring the country with Westgren, quite famished till we finally got lunch about four o'clock, and on another occasion we went sailing all day with a picnic box in full view on the boat which was never opened!

I also met on this trip the Eckermanns, who became such friends of the family. Eckermann is a distinguished mineralogist who has a private laboratory at his vast country estate near Sparreholm.[248] They invited me to stay over Easter; their home was a fascinating place dating from the sixteenth century, with a large central building and two detached wings all on one storey. He had, if I remember rightly, about 17,000 acres of woodland and lake. There were elk and roe deer in the woods (an elk chased the cook on her bicycle when I was there), ospreys on many of the islets in the lake, and kites and buzzards. Easter was celebrated with traditional festivities by a large family gathering and it was an extraordinary change from the grimness of life in our country. There is something especially lovely about that Swedish country. It is very like the corresponding country on the other side of the Atlantic, like the woods and lakes of the Laurentians in Canada, but it is blander, there are more flowers and there are such attractive towns, villages and farms.

[247] Victor Alexander Louis Mallett (1893–1969) was a British diplomat and an author. He was married to Christiana Jean Andreae.

[248] Claes Walther Harry von Eckermann (1886–1969) was a Swedish mineralogist and engineer.

Eckermann was head of the Swedish observer corps and told me they had followed the arrival of our plane. He gave me a most interesting account of the evacuation of Hanno at the end of the Russo-Finnish war.[249] The Swedish Automobile Association had offered to help the Finns evacuate their property from this town, and Eckermann headed a party of trucks sent for the purpose by the Swedes. It was a very cold winter and they were able to drive the trucks straight across the frozen Gulf of Finland. He said how strange it was to see a signpost 'To Finland' pointing out to sea on the beach. A Finnish officer had told Eckermann of his adventures just before the war ended; he with his platoon had been captured by the Russians, and they were preparing a fire to scorch their feet, as they did with all their prisoners to prevent escape, when the news of the armistice came through. The Russians embraced their prisoners and threw a marvellous party for them, at which everyone got very drunk, and their hosts swopped rifles and machine guns for souvenir buttons from the Finnish uniforms. He said it was an extraordinary sight to see his men staggering back home over the ice floes bristling with arms of all kinds.

It was strange at Helsingfors to see the Germans crossing in the ferry to and from Denmark; Sweden was giving free transit to the Germans who occupied Norway. A Swedish friend said to me, 'Would you not like to cross over on the ferry? It is only 80 öre.' From Malmö we could see the fires in Copenhagen due to our bombing. I travelled from Helsingfors to Göteborg with a Dane who was enthusiastic about our airmen's accuracy in hitting German posts and missing Danish property. He said the Germans were very correct as regards the Danes; 'even a murderer may have a canary' was the way he put it. At Göteborg I was entertained royally by the shipping magnates who spoke English with a good Tyneside accent.

It was not so easy to return to England as it was to fly to Sweden. The Germans had no knowledge of a departure from Great Britain, so it was comparatively easy to slip across the sea on a dark night till the Swedish coast was reached and the plane was safe from attack. Coming back, the Germans 'knew all about it' and it was a much more risky business. I was warned that I might have to remain in Sweden all the summer. There had been a Swedish service, but it was withdrawn because the Germans fired at one of the planes and winged the wife of a Swedish

[249] WLB is probably referring here to Hanko in Finland.

VIP. We had a Mosquito as a stand-by, so swift that it could not be intercepted, but by bad luck it had been shot down. It was not possible to fly except on moonless nights, and the nights were getting too short anyhow for a comfortable trip. However, some very thick weather, with thunderstorms, provided a chance and I got a trip in a large bomber with a number of British pilots who had come down in Denmark and had been smuggled across to Sweden by the Danes. We took nine hours to reach Scotland—we must have made a tremendous detour.

Stephen and David had both been at Harden House, a local preparatory school, while we were in Alderley Edge. Stephen went from there to Rugby with a scholarship. I do not think he lost by being at Harden House; it was not a good school but one of my old students gave him a good training in maths and he had an interesting home life. A master at Rugby told me that on the whole he preferred boys who had been to a day school at the preparatory stage because they had a fuller life. When we went to Teddington there was no convenient preparatory school in the neighbourhood and so we sent David to The Downs, near Malvern, where the headmaster at that time was Geoffrey Hoyland. Stephen was at Rugby when the war started. He got a scholarship at Trinity and got full state support for his two-year course in Honours Engineering in which he got a first. It was a strenuous time for him, with Home Guard training as well as the concentrated engineering course, and a hungry time too. The food at Trinity was very meagre, and a famished Stephen often appeared at about half-past one to have a second lunch with us at West Road. In the interval between leaving Rugby and coming to Cambridge, he worked in the British Thomson-Houston factory in Rugby. This, I think, was a very valuable experience. Coming straight from school, he was treated as a mate by the men and not as someone of another class from the university. He worked with a fitter and got a very varied experience. After getting his degree, he was reserved for technical war work and applied for a job on jet engines. C. P. Snow, then in charge of scientific appointments at the Ministry of Labour, wanted him to work with Whittle.[250, 251] I was strongly advised that it would be better

[250] Charles Percy Snow (1905–1980), a British chemist and novelist, was famous for his lecture entitled 'The Two Cultures', where he was concerned about the gulf between scientists and literary intellectuals. His novel *The Search* tells of a crystallographer called Constantine, who is evidently a portrait of J. D. Bernal.

[251] Frank Whittle (1907–1996) was the British inventor of the turbojet engine.

120 · CLIFFORDS INN · LONDON · E·C·4A 1BX

01-405 1444

14A/81

23rd March, 1971.

Dear Sir Lawrence Bragg,

I don't know whether you are aware that your name was on the Gestapo black list of those to be arrested in the event of the invasion of this country in 1940.

Some of us on the list are planning a party to celebrate the thirtieth anniversary of our fortunate escape when in the spring of 1941 Hitler decided to invade Russia instead of invading this country. Of course we have no means of knowing the exact date, but we assume it must have been early in May. So we intend to have a private dinner party in the private dining room of the Reform Club.

Those who will take part include Lord Chandos, Lord Crowther, Mr. Duncan Sandys, Mr. Richard Crossman, Mr. George Strauss, Dame Rebecca West, etc. We are anxious that the party should not be confined to politicians or politically-minded authors and have invited a number of scientists and representatives of various arts.

I do hope you will be able to attend. Even a man of your distinction has reason to be proud of the fact that you were one of 3,000 people whom the Nazis hated or feared the most.

The date will be probably the 10th or 11th May, and if you are interested I shall let you have full particulars as soon as arrangements have been completed.

Yours sincerely,

Paul Einzig

PAUL EINZIG

✓ May 5.

Sir Lawrence Bragg, F.R.S.,
6a, The Boltons,
London SW10.

Figure 29 Letter from Paul Einzig to WLB, informing him of a party in honour of Einzig's being listed by the Gestapo for arrest. (Courtesy: the Royal Institution London)

for him to start with a branch of Rolls-Royce at Barnoldswick in Yorkshire, near Skipton, where the jet was being developed by Hooker.[252] I was told that the inventor is rarely the good producer; it is much better to get experience with a big firm. This was the start of Stephen's connection with Rolls-Royce, where he is now (1964) the chief scientific adviser. Alice and I paid a visit to Skipton while Stephen was there. We were somewhat aghast when we found that he had booked rooms for us in the Railway Hotel, opposite the station and cattle market and between the gasworks and Dewhurst's mill, a somewhat dreary late Victorian hostelry with frosted glass windows to the jug and bottle departments. We found that he had chosen excellently, because the farmers attending the market came there to eat and brought offerings of game, eggs, cream, meat and poultry off the ration. The food was amazing compared with the dull rations we had at Cambridge. I remember too buying some cakes for a picnic and remarking to Alice that the paper bag actually had grease marks on it. Incredible. I had a chance to take Alice to see Bolton Abbey, and Deerstones, where we had our country cottage when my father was in Leeds.

David left The Downs in 1940 and went to Rugby like his brother. When the time came for him to be called up, he volunteered to go into the mines rather than into the armed forces. He had a grim experience in the coalmine, and this precipitated a severe nervous breakdown which gave us great concern for many years. Margaret went to the Perse in Cambridge and then to Downe House near Newbury.

2.8 Another War Ended

I had a very difficult time indeed at the Cavendish when the war ended. The two key posts in the laboratory were the Jacksonian Chair, which was held by Cockcroft, and the Plummer Chair of Mathematical Physics, held by Fowler. Cockcroft, who had been the head of army radar research during the war, was clearly booked for the directorship of the new Atomic Energy Establishment, but for a very long time this post was not officially created and he remained an absent Jacksonian Professor. I lacked his help in the laboratory, which would have been invaluable in those critical years, and yet we could not appoint

[252] Stanley George Hooker (1907–1984) worked at Rolls-Royce on the earliest designs of the jet engine.

to the chair because Cockcroft had not resigned. So I had the mortification of seeing the 'possibles' snapped up by other appointments. In the end we were fortunate in getting Frisch, from the Atomic Energy Research Establishment (Harwell).[253] There was the same quandary, for a different reason, in the case of Fowler's post, because Fowler had a long and sad illness at the end of the war. Again, possible successors went to other posts while we were powerless to make offers, and ill fate seemed to dog all our plans. Mott at one time seemed keen to come, if certain conditions could be fulfilled, and I set about making the arrangements he wanted, but in the end he felt he could not leave Bristol.[254] Peierls practically accepted while away in America, but on his return to England decided he could not leave Birmingham! In the end we were fortunate to find that Douglas Hartree, who had strangely been invited by Manchester to run the Engineering School, wanted to remain a theoretical physicist and accepted our offer to come to Cambridge. My great helper in running the laboratory was Ratcliffe, who returned some months after the war ended. He was magnificent and did the main work of organising all the classes again. He was clear-headed in matters when I was often so vague; I remember his tact when he thought I was not putting a point in quite the right way to the General Board or some other body and Ratcliffe would chip in with 'We think, don't we, Professor' and state everything clearly. He rapidly built up the Appleton heritage of ionosphere research again, and it was he who foresaw the immense importance of the radio waves coming from space which had just been discovered. He infected me with his enthusiasm.

I had many a heartache over the disintegration of nuclear research at the Cavendish. The finest men in the country had congregated there under Rutherford and I felt very responsible for trying to keep the leading position which Cambridge had attained, but it was of

[253] Otto Robert Frisch (1904–1979) was an Austrian-born physicist who worked with Rudolf Peierls to design the first theoretical mechanism for the detonation of an atomic bomb in 1940. He came to Britain in 1933 after the accession of Adolf Hitler. His aunt was the physicist Lisa Meitner who with Otto Hahn discovered nuclear fission.

[254] Neville Francis Mott (1905–1996) was a British physicist who worked on the electronic structure and magnetic properties of disordered solids. He shared the 1977 Nobel Prize in Physics with Philip Warren Anderson (1923–) and John Hasbrouk Van Vleck (1899–1980). He became Cavendish Professor, and Head of the Cavendish Laboratory, following WLB, in 1954.

course a line in which I had no knowledge or experience whatever and I could be of no help. The good men mostly drifted away, and schemes for vast nuclear machines were started at Glasgow, Liverpool and Birmingham. Ought Cambridge to have one too? Chadwick pressed strongly for it; Frisch was very doubtful and I think one can say now that Frisch was right. The vast sums spent on the machines and, more importantly, the absorption of the flower of the research men in a number of schools have produced little in the way of first-rate work. The hero in the nuclear field has been Powell, who has worked with cosmic rays, not with a machine for producing particles of high energy.[255] Cambridge had at one time a scheme for developing a linear accelerator. But such projects are no good unless they are done on a gigantic scale with the right type of man in charge. The American competition, backed by all the money available in the US, was too fierce.

I had of course no chance to do serious research during the war. Bradley and Lipson had come to Cambridge with me from the NPL, and Bradley continued his investigations into alloys, thus providing the nucleus of a laboratory for the X-ray analysis of metals. But poor Bradley was already suffering grievously from the illness which eventually put an end to his research career. His appointment as a Director of Research at Cambridge could not be renewed when the initial five years elapsed. It was a great tragedy because Bradley was brilliant. He developed methods for solving binary, tertiary and even quaternary phase diagrams which were unsurpassed; he had a wonderful flair for mastering their complexities. He created a whole new branch of metal physics. I remember that when Sir Charles Goodeve became director of the Iron and Steel Research Association's laboratory and set himself to find out what had been done in this branch during the last decade, he told me 'at the end of almost every trail I follow I come to Bradley.' There were still traces of the old fire and genius when Bradley worked at Cambridge, but his illness prevented a real advance. When he left Cambridge just after the war he was offered a position by Oliver at Jessops in Sheffield, so as to provide a centre for a man who had made such brilliant contributions to British science, and later the Royal Society helped.

[255] Cecil Frank Powell (1903–1969) was a British physicist who developed photographic methods for studying subatomic particles. He was awarded the 1950 Nobel Prize in Physics.

Metal physics was also kept going at Cambridge by the volatile and irrepressible Orowan, who had come to the Cavendish from Birmingham. Orowan was one of the original proposers of the theory of metal slip by the movement of dislocations. He had a most fertile imagination and ingenious mind, both as regards the theoretical side and the devising of striking experimental illustrations. Again and again Orowan would come to me and announce with the deepest gravity that he had, during the last fortnight, hit on a new theory which showed that all the fundamental work in metal physics during the last thirty years had been quite on the wrong track. He interested me in dislocations, and I hit on a device which has become quite well known. When adding the oil to the petrol for our motor lawnmower, I had noticed that the rafts of tiny bubbles formed by pouring in the oil showed interesting movements of adjustments. I therefore asked Lipson to make an apparatus for producing such rafts so that we could study their behaviour. Lipson reported back that nothing exciting seemed to happen, but when I saw his gear, I realised that the bubbles he had produced were far too large. We then blew them from a very fine capillary so that they were from one-half to one and one-half millimetres in diameter. The tube was placed below the surface of a soapy solution and, by using a constant pressure, the bubbles which were formed were astonishingly uniform in size. Such rafts, when distorted, show 'dislocations' running along 'slip-lines' in the most charming way. If 'cold-worked' by stirring them up with a rod, they show recrystallisation. 'Defects' can be studied when bubbles are missing or bubbles of the wrong size are introduced. Three-dimensional bubble crystals can be formed. The forces between the bubbles are in fact closely analogous to those holding the atoms together in a metal; the analogy only breaks down because the bubbles have practically no mass. First Nye and then Lomer helped me to do experiments with the bubbles and to give the theoretical explanation of their behaviour.[256, 257] Later films were made by Kodak and by the

[256] John Frederick Nye (1923–), a British physicist in Bristol, worked on plastic flow of glaciers. He was the author of a classic book on the physical properties of crystals.

[257] William Michael (Mick) Lomer (1926–2013) was originally a student of Orowan but was 'lent' to WLB. He worked later at the Atomic Energy Research Establishment, Harwell, and was the United Kingdom Associate Director at the Institut Laue-Langevin in Grenoble, France.

Macqueen organisation.[258] The bubble model is of course of secondary importance but it has had quite a vogue as a simple way of illustrating the behaviour of metals.

Another sideline of research was the development, with Crowe's help, of an optical way of producing the transform of a crystal structure or of forming a map of the crystal structure from its X-ray transform.[259] Lipson later developed the optical method into a fine piece of apparatus. A mask is made with holes which represent the atoms in the projection of the crystal structure upon a given plane. Monochromatic parallel light passes through the mask and then through a long focus lens, which forms the 'transform' of the crystal as a diffraction pattern. By cutting different masks, one can seek for an arrangement of holes which diffracts light in the same way that the crystal diffracts X-rays. It is really a kind of analogue computer using light. Such methods are no longer of much practical importance, since electronic computers have become so widely available.

A great acquisition to the laboratory during the war years was Cosslett, who came from Townsend's Electrical Laboratory in Oxford, when that laboratory was amalgamated with the Clarendon Laboratory and some posts became redundant.[260, 261] DSIR had given us one of the first batches of electron microscopes to reach England from America, and Crowe had set it up. Cosslett took it over and started a very fine school for electron microscopy; he became a leader in this country. He played a large part in all the earlier work of this kind and was much in demand by the biologists. The Cavendish was really very fortunate to acquire him, because his laboratory was one of the best of its kind, and students from other parts of the world came to work with him.

[258] One of these films can be seen at <http://www.amg122.com/twobraggs/videos.html>. The bubbles were blown by WLB's son, Stephen, who was positioned underneath the apparatus.

[259] George Crowe had previously been Rutherford's personal research assistant. He joined the Cavendish workshop in 1907. He was the best known laboratory assistant of his time. He built C. T. R. Wilson's cloud chamber.

[260] Vernon Ellis Cosslett (1908–1990) was a British physicist who worked on electron microscopy.

[261] John Sealy Townsend (1868–1957) was a British physicist at Oxford who worked on electrical conduction in gases.

Another worker in the Cavendish who was later to build up a magnificent school of his own was Martin Ryle.[262] Ratcliffe encouraged him to investigate the possibilities of radio-astronomy. Ryle at first was doubtful, feeling that nuclear physics was the line of the future. But I added my persuasion to Ratcliffe's, as I felt that too many people were going in for nuclear physics and that we were outclassed in apparatus-building by the Americans with their immense resources. It was indeed fortunate that Ryle turned his energies to radio-astronomy and created the school which, with Lovell's at Jodrell Bank, has placed Britain in the forefront of this subject.[263] At the start Ryle had a layout which, in contrast to Lovell's vast radio telescope, was relatively very inexpensive. He worked with interference fringes produced by two aerials receiving the signals from the radio-stars and got his resolution through the wide spacing of the aerials. Each aerial was like a wire mattress standing a foot or two above the ground and covering a space some fifty yards by twenty feet. Once when the Trinity Beagles were out, an intelligent hare took refuge beneath its shelter, and the beagles did some hundreds of pounds worth of damage in trying to rout it out. A stiff letter to the Master of Beagles elicited a most disarming reply, distinguished both for its politeness and its somewhat rocky spelling, and for a long time afterwards whenever the beagles met in this neighbourhood I had a note to warn us, so that the guards could be posted on the rifle range where the aerials were situated. Ratcliffe was a grand team-leader, who gave his people excellent ideas and never took the credit for them himself. First-rate ionospheric research was carried out in his group.

It was a strange time at the Cavendish because Rutherford's influence had been removed. Hitherto practically the whole emphasis in the laboratory had been on radioactivity, and the best students had naturally turned towards it when they started to research. It was as if some mighty forest tree had fallen and saplings hitherto starved of light and nourishment were beginning a more normal development. The radio-astronomy was one such young growth; the electron

[262] Martin Ryle (1918–1984); British radio-astronomer who developed radio-telescope systems at the Cavendish Laboratory. He was awarded the 1974 Nobel Prize in Physics with Anthony Hewish (b. 1924), who was also at the Cavendish Laboratory.

[263] Alfred Charles Bernard Lovell (1913–2012) was a British radio-astronomer and Director of the Jodrell Bank Observatory.

microscopy was first-rate of its kind; and good pioneer work was being done by Pippard and Schoenberg at the Mond.

But probably the work which in future years will be regarded as the outstanding contribution of the Cavendish Laboratory in these after-war years was the start of the investigation of biological molecules by X-rays. When Perutz came back to the Cavendish, he resumed his work on the diffraction of X-rays by haemoglobin, and the Medical Research Council agreed to support it. I remember well going, at Keilin's suggestion,[264] to see Sir Edward Mellanby, who was then secretary of the MRC.[265, 266] I told him that the chances of solving the diffraction pictures were practically nil, because they were so complicated, but that on the other hand, if success were attained, their importance in casting light on the functioning of proteins in the living body would be almost infinite. The product of zero by infinity is anyone's guess, but it just might come off. Mellanby sportingly took the gamble, and this was the start of the MRC Unit for Molecular Biology at Cambridge, which has become world famous. Last year (1963) its annual budget was £375,000, as compared with the modest £375 per annum with which it all started. Last year also, four Nobel Prizes were awarded to workers in the unit, a unique event in the history of these prizes.[267] Kendrew joined Perutz soon after the war ended.

For a long time, the results of this research into proteins were very meagre indeed. Perutz at times became quite discouraged, very naturally, and wished to change to the other line of research in which he was interested, the flow of glaciers. Why I continued to be optimistic I shall never understand. Certainly our X-ray crystal colleagues thought we were on a wild goose chase, and I think Perutz might have dropped the project had I not spurred him on. I remember Perutz sending me a long memorandum to say that it really was not worthwhile pushing on with measurements unless some quite new line of attack was discovered (this was just before the attached heavy atom

[264] David Keilin (1887–1963) was a British biologist, born in Moscow, who studied parasitology. He was head of the Molteno Institute of Parasitology in Cambridge.

[265] Edward Mellanby (1884–1955) was a British medical doctor who discovered vitamin D and its role in preventing rickets in 1919.

[266] MRC, Medical Research Council.

[267] These were for Perutz and Kendrew for haemoglobin and myoglobin, and for Crick, Watson and Wilkins for the structure of DNA.

method was developed[268]). In all it was twenty-five years after the experiments began that Kendrew solved myoglobin and then Perutz solved haemoglobin.[269]

In the autumn of 1945, I was invited to visit the Portuguese universities, which were celebrating the fiftieth anniversary of Röntgen's discovery of X-rays. I went under the auspices of the British Council, and the treasury agreed to pay Alice's expenses as well. Alas, Alice could not go, as she was invited to be Mayor of Cambridge, and the date of our proposed visit fell in the first month of her taking office. I was thrilled when the mayorship was offered to Alice. She was doubtful about accepting, but I urged her to take it on. Alice was an independent member of the council, and she had steadfastly refused to be classified as Tory or Labour. I think the council was glad to have an independent Mayor because there was much strong feeling between the parties just after the war; it also helped to have a woman Mayor because fractious councillors tended to behave better. Anyhow, Alice was a great success, beloved by all in Cambridge. I must not try to tell the story of her adventures as Mayor since only she can do it justice. It was very hard work, because ordinarily a Mayor has a Mayoress who shares in many of the official duties, and Alice had to be both. We realised later that she could have invited a friend to act as Mayoress during her term of office, which would have done a great deal to lighten the burden. Often she made two speeches in a day, attending this or opening that. We engaged a kind of companion to help Alice in the house and with the children, but unfortunately the companion turned out to be a broken reed, so altogether it was a time of stress. But it was a wonderful experience. There had only been two women Mayors previously in Cambridge.

My visit to Portugal was most interesting, and it was a shame that Alice could not come too. My friend there was Ferreira, who had

[268] In solving crystal structures using X-ray diffraction, it is necessary to know both the amplitudes and the relative phases of the diffracted waves. Because no lenses exist for X-rays, only the amplitudes were measured, leaving the 'phase problem' to be resolved. The heavy atom method involved chemically including an atom such as mercury which has a large number of electrons to dominate the intensities of the reflections and provide much of the phase information.

[269] The structure of myoglobin contains a single molecular unit, whereas haemoglobin has four units. Therefore it was easier first to solve the myoglobin structure and then apply it to haemoglobin.

worked for a time with my father in the Davy Faraday laboratory.[270] The British Council, which sponsored me, was represented in Lisbon by S. G. West.[271] I gave a series of lectures in Lisbon, and I was given an honorary degree by the university. This was a grand occasion. The British Ambassador, O'Malley, acted as my 'godfather' and testified to the assembled gathering that my character was good (taking it completely on trust).[272] West had a tailor make the garb for me. The foundation was a tube-like garment of ecclesiastical form, a kind of cassock, tightly buttoned from head to foot. A magnificent black cloak which could be wrapped twice round me was fastened at the neck. A large blue silk collar of ruff-like shape, with silken bows, went round my neck, and on my head was a blue silk affair of charlotte-russe shape surmounted by a tasselled spike. After getting the degree I went along the doctors on the rostrum, kissing each on both cheeks. Lisbon had one magnificent main street leading down to the harbour. But only a hundred yards on either side it was a fascinating medieval town, with little wheeled traffic. Donkeys and mules served as transport, and heavy objects were carried by a guild of porters on their heads. It was fascinating to see a removal going on, with the family dining tables, wardrobes, chests and so forth going on men's heads through the narrow streets. The women carried vast burdens in a similar way. At the end of my Lisbon stay I went to Oporto and was fortunate to share a carriage with one of the Gilbeys in the port trade who, with her brother, directed a 'lodge' in Oporto.[273] Her brother made me a temporary member of 'The Factory', which is the club of the English colony. This is a fascinating place, with lovely old rooms and furniture. I was shown the visitors' book, which quite a long way from the start had the signatures of officers in Wellington's army and who had been made honorary members of the club when Wellington drove the French out of Portugal. Gilbey took me to his lodge on the opposite bank of the Douro and I was initiated into some of the mysteries of blending and tasting port. Each lodge has its own farmers

[270] Herculano Ferreira Amorim (1895–1974) was a Portuguese physicist, meteorologist and politician.

[271] S. G. West was Inspector of the Overseas Division of the British Council.

[272] Owen St Clair O'Malley (1887–1974) had been British Ambassador to the Polish Government in exile during the war and was Ambassador to Portugal from 1945 to 1947.

[273] The Gilbey family were famous for importing wine to sell to the public.

up the Douro valley; these farmers are the ones who press the wine and fortify it with brandy provided by the firm. It is a great crime, like enticing away a cook, to steal someone else's farmer. The vines grow on the barren stony hills; I was told that a hole had to be made with a crowbar when planting them, and each only carries two or three bunches of grapes. The new wine comes down the river in strange old-fashioned boats. The art consists in blending the raw wine so that in a few years' time the mixture will mature into what the market wants. The lodge centres round the taster, who advises on this process. The Gilbeys took me into the tasters' room, with heavily padded doors to cut it off from the outside world, a range of sample bottles, glasses and a capacious sink. They soon forgot me as they tasted and inspected, and they talked a language which an ordinary mortal could not understand. I was told of prodigious tasting feats. It is the custom that when a firm sends a consignment out, the heads of the rival firms are politely invited to sample it. At one of these parties a famous don said, 'An excellent wine, but you cannot export it, it tastes of corpse.' A man was sent down into the vast port barrel, which was made of staves of Baltic oak and was as large as a room, and reported that nothing could be seen. The don insisted, and when the man made a more rigorous search, he came up with—a dead mouse! The experts deplore the British taste for red port; it can only retain this colour if the vessel is lined with glass. They consider that the best port is that left in oak barrels and which has a colour like sherry. I was taken down into the holy of holies, a cellar with just one barrel in the centre, from which a sample of nectar was drawn for me with a silver pipette.

The reception by the university was very formal; I was met at the station by a deputation in evening dress, though it was only early afternoon. I made a great mistake in my lecture, because I tried to make it simple. I gave it in French and made jokes. I fear my Portuguese colleagues thought they were not getting their money's worth because it was not recondite and unintelligible.

Suggia, one of the world's greatest players of the cello, was a Portuguese heroine at that time, and I treasure the honour of having once been dug in the ribs by her.[274] When I was in Oporto there was a drive to collect money for Portuguese hospitals, and Oporto had an exhibition

[274] Guilhermina Augusta Xavier de Medim Suggia, known as Guilhermina Suggia (1885–1950) was a Portuguese cellist who studied under Pablo Casals.

of British surgical apparatus in one of the hospital wards. Incidentally, the hospital was an extraordinary building. In order to avoid the expense of special architects' plans, I was informed, they had bought the plans of a projected building in the centre of London which never materialised, and it was strange to see the typical city architecture with grooves in the cement to simulate the stones of arches and other details so reminiscent of London in the middle of Oporto. Inside it had an air of 'The Lady of the Lamp'. I was later reminded of the scene by a remark Santos made to me in Coimbra: 'Salazar has done wonders for the country's economic advancement, but I wish it was not still necessary so often to have two patients in each hospital bed.' [275, 276] As part of the appeal for money, there were addresses and a concert in the Oporto theatre. Suggia, as the most famous woman in Portugal, and I, representing Great Britain in the celebration of Röntgen's discovery of X-rays, were incongruously placed together in the royal box. At one point when a gentleman in evening dress was making a fluent speech, Suggia gave me a violent dig and whispered, 'Get up and bow, he is talking about you.' As one of my friends remarked to me, it was extraordinary that someone with a face like an old boot could so radiate personality and be so fascinating.

I returned from Oporto to Lisbon by car, stopping at Coimbra en route. I had been billed to give a lecture there on the Wednesday returning to England on the Friday plane, but West discovered that there was no Friday plane and that I must leave on Thursday if I wanted to get home in time for the honorary degree ceremony at Cambridge. I was keen to attend this as Montgomery and Eisenhower were going to be honorary graduates. [277, 278] So the lecture was cancelled, and I just paid a short visit. This was a great brick to drop, since Coimbra is

[275] João Rodrigues de Almeida Santos (1906–1975) was a Portuguese physicist from the University of Coimbra. He studied for a Ph.D. in WLB's research group in Manchester between 1930 and 1934.

[276] António de Oliveira Salazar (1889–1970) was a Portuguese professor and ultra-right politician who served as Prime Minister of Portugal from 1932 to 1968. He was widely considered to be a dictator.

[277] Field Marshal Bernard Law Montgomery, First Viscount Montgomery of Alamein (1887–1986), nicknamed 'Monty', was Chief of the General Imperial Staff during the Second World War.

[278] Dwight David 'Ike' Eisenhower (1890–1969) was First Supreme Allied Commander Europe during the Second World War. He was President of the United States from 1953 to 1961.

the senior university. What mistakes one makes through not appreciating situations. At Coimbra I met again Santos, who had worked in our Manchester laboratory and was now Professor of Physics. He was rather disillusioned. I remember his saying, 'Ah that paradise, Manchester. There I was a scientist, I did research, I read *Nature*. Here we have examinations every six months which last two months. If I fail the son of the Director of the Coimbra Tramways, I receive a note to say that this must not happen again. Recently, the government gave us a sum to improve our laboratories. We could not decide which laboratories to improve, so in the end we built an avenue with a statue at the end of it. Now all I can do is to read *The Illustrated London News* and have children, of whom there now are nine' (if I remember rightly).

The old apparatus in the physics laboratory which Santos showed me was extraordinary. Apparently, in the heyday of Portugal's wealth and importance, the Marquis de Pombal decided that its university should be worthy of so great a nation.[279] Everything was to be of the best. The apparatus was made of rare woods and gilt where possible. The steel yard was beautifully engraved. In order to demonstrate that sieves with a range of holes let through particles of different sizes, a rather simple observation, the sieves were held aloft by gilt angels on a walnut stand with lion's feet. The library was an amazing place. Enormously heavy tables with legs as thick as a man's thigh were made of precious woods brought from Portugal's overseas possessions. The ladders to reach the bookshelves were rococo and gilt. It was a wonderful place. I enjoyed Portugal and its people. I was told that it was a very poor country and run on a shoestring, but the people seemed to have a gay time in spite of their poverty.

The year 1946 was Alice's last year as Mayor and it was a very busy one which left little time for holidays. Patience, Alice and I had a stay with Kenneth and Giulia Lee in March, and this included an excursion to their summer villa at Ferring.[280] Patience characteristically appreciated to the full being lapped in luxury at the Lee's. It was on this occasion that we first met the charming Davises who afterwards

[279] Sebastião José de Carvalho e Melo, First Marquis of Pombal (1699–1782), was a Portuguese statesman.

[280] Kenneth Lee, First Baronet (1879–1967), was the director of the firm Tootal Broadhurst and played a significant part in the 1951 Festival of Britain. WLB wrote Lee's obituary in *Nature* (see <http://www.nature.com/nature/journal/v216/n5118/pdf/216945a0.pdf>).

Figure 30 The Pleasure Boat Inn, Hickling Broad, May 1948. (Sketch by WLB; Courtesy: Patience Thomson)

became such friends.[281] Their daughter Georgina was about Patience's age (eleven) and looked as if butter would not melt in her mouth. We heard from Patience afterwards that Georgina had enticed her into collecting cigarette ends and smoking them behind bushes in the garden. In June Alice and I went for the first time to the Pleasure Boat Inn on Hickling Broad. I'd had a bout of flu and was still running a temperature when our family doctor told Alice to bundle me there and risk it. He was quite justified as I was as fit as could be after a day or two in the open. We hired a swift small sailing boat and explored reaches of the Broads which were inaccessible to the cabin yachts in which we had previously had Broads holidays. The Pleasure Boat Inn only had three small bedrooms and the dining room was an attached tin shed, but the food was good and it had a pleasant situation at the end of a staithe leading to the great expanse of the Broad.

[281] The Davises are the family of Admiral Sir William Wellclose Davis (1901–1987). After the war he was made Director of Underwater Weapons at the Admiralty. His daughter was Georgina Elizabeth Laetitia Davis (b. 1935).

Figure 31 Our boat on Hickling Broad. (Sketch by WLB; Courtesy: Patience Thomson)

In the summer we had our first holiday at Blakeney, staying for three weeks in the Manor House Hotel. We had later holidays there and the children came to love the place. The main summer preoccupation, dinghy racing, centres on the tides, because the harbour only fills up for a few hours at each high tide. The racing boats have to make their way somehow against the strong current in the narrow cut leading to the harbour and then sail their course in a large sheet of open water some two or three miles in extent. Returning, they have a fight up the cut against the tide which is now flowing out strongly. We hired a boat from the fish-and-winkle shop and had great fun. At low tide the stretches of sand and mud are great places for watching the waders—dunlins, godwits, ringed plovers, grey plovers, turnstones, sanderlings, whimbrels, redshanks, greenshanks, sandpipers, oyster-catchers and curlew, without number. At nearby Cley there is a sanctuary where one can see rarities such as stints, spotted redshanks, ruffs, reeves, wood sandpipers, curlew sandpipers, bitterns and the various ducks. It was here that I got to know the shorebirds and waders. There are also colonies of common terns, little terms, sandwich terns and roseate terns, though I never had the good fortune to identify one of these last.

This was Margaret's first year at Downe House near Newbury, and we paid our parental visit to the school that year. We were grateful for all that Downe provided for our girls. It was very definitely the antithesis of a cramming school, in the sense that the girls were taught in a way to arouse their interest rather than to prepare them as examinees. In some curious way, Downe seemed to give them poise and stability and make them socially competent.

During this busy year I could not be the help to Alice which a Mayoress would have been able to give a Mayor. Nevertheless, I was drawn into some of the town activities and got to know and respect those who ran them. Town and gown are strangely apart in Cambridge, neither knowing much of the other. There is always a link in that a number of members of the council are university nominees but, generally speaking, the university is not so much in any way opposed to the town as forgetful that it has an independent existence. Alice gained from running the WVS during the war, because this led to her having many friends in the town and a wide understanding of its problems. My part was a very minor one, and I could not but be touched by the warmth with which the little I did was welcomed and appreciated; I wish I had done more.

All through the war Patience had begged for a dog, and we had weakly put her off by saying that we would 'think of it' after the war was over. Poor Patience had so often been told that we could not have this or that, or do this or that, because of restrictions in wartime and for some years afterwards. She was only four when the war started and so could barely remember a period of plenty. Anyhow, at a time when Alice was away for a week, and Patience and I were keeping house together, we heard of a litter of puppies at Wimpole. Off we went by car with her nurse Hilda and came back with a puppy a few weeks old, Scrap, whom we had bought for a pound. We did not dare to write to her mother and confess. When Alice arrived home and as she entered the front door, Scrap was promptly sick on the hall carpet. He was rather a cocktail dog—half fox terrier, one-quarter Irish terrier and one-quarter foxhound, as far as we could make out—but it was a mixture of amazing vitality and intelligence. He would have made an excellent poacher's dog, as he could have been trained to do almost anything. We could not keep him in the house and garden; he roamed over the town and one would sometimes meet him in the street a mile or two from home. In due course we often saw dogs of the town bearing an unmistakeable stamp of Scrap paternity. He would fight any dog and chase any form of game, especially cats. As the people next door were cat friends and had a score or so which overran our garden we did not discourage this last propensity. We were very fond of Scrap.

About this time we also acquired half a pony. Margaret's great friend was Christina Bowen, and her father proposed to me that we should go shares in a pony for the girls.[282] We settled on an attractive white Welsh pony, Joey, who had been pulling a cart belonging to a vendor of winkles and cockles. Joey was an excellent investment. Christina and Margaret got endless fun from riding him in the field in which he was kept and taking him to the gymkhanas in the neighbourhood. He was strong, a good goer and quite a jumper. Like so many ponies, he was very tough and could live in a rough field getting his pickings for nearly the whole year; he only had to be given shelter and receive supplementary rations when the winter was very inclement. The main drawback to Joey was a mouth like iron. Christina and Margaret, being both girls of some weight, could manage Joey, but he was too wilful for Patience to control, and sometimes he ran away with her.

January of 1947 was very cold and snowy. We tobogganed and skied on the Gogs, and Patience even skied in our garden with a frantically excited Scrap.[283] The garden was as flat as a pancake, but at the end of the lawn there was a large pit, by tradition an old sandpit, which had been smoothed off to form a 'feature'. This pit was the centre of endless games. There was one particularly mean form of local croquet, in which the hoops were set at the very verge of the slopes round the

[282] Edward George Bowen (1911–1991) was a British physicist who developed the first airborne radar. He was an early radio-astronomer.

[283] Cambridge is very flat apart from a few hills to the south known as the Gog Magog Hills.

Figure 32 Crown Imperials in the West Road garden; the lilies were great favourites of Alice. (Sketch by WLB; Courtesy: Patience Thomson)

pit. Only the family knew the tactics of avoiding a calamitous descent of the ball to the bottom of the pit; it was a refined form of gamesmanship played against friends. In this cold weather, the pit became a kind of inverted Matterhorn and centre for winter sports. When the snow melted there were widespread floods, because the ground beneath was frozen. The Backs were all under water, and it came quite a distance up West Road though not as far as our house.

I had a lecture at the Royal Institution about the time the thaw started and had gone there with the faithful Crowe to help with the demonstrations. We only got to about Bishops Stortford on our return, as the railway had been washed away beyond that point. We had to sleep the night on the floor of an inn and come on by bus next day to a place where we could again board the train to Cambridge.

We took Margaret and Patience with us to the Pleasure Boat in April and taught them to sail on Hickling Broad, a great holiday for them. Scrap caused various diversions; if he saw a coot even hundreds of yards away he would leap overboard, paddle furiously towards it and have to be rescued. I shall always remember the service in Hickling Church because the parson was such a rebel: 'You have just had

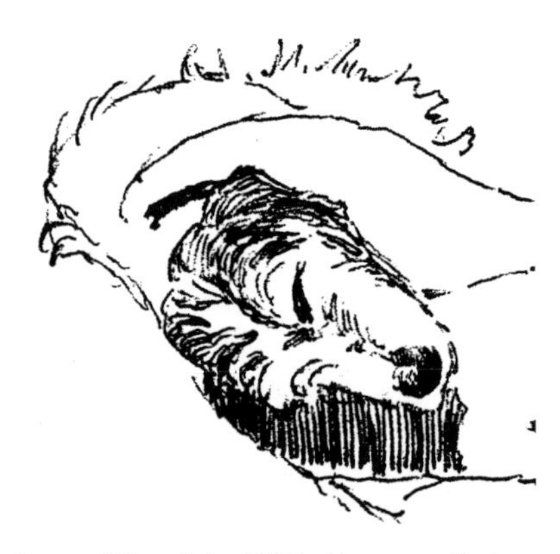

Figure 33 Scrap. (Sketch by WLB; Courtesy: Patience Thomson)

a letter from your bishop, but take no notice of it, it's a lot of rubbish,' in broad Lancashire, and also, 'What people care about nowadays is not G, O, D, God, but D, O, G, Dog.' It was little wonder that Alice and I were most of the congregation apart from the choir and sidesman.

That June we had a wonderful holiday with Stephen. We took him with us to Sweden, for his first holiday abroad. It is curious how one remembers small things. As we came into Göteborg harbour there was an advertisement of a Swedish version of a movie then much in vogue in England, *Alltid Amber*, and Stephen was deeply impressed by this realisation that he was entering a foreign land.[284] We stayed in the Siegbahn's flat in Stockholm, where the kind Mrs Siegbahn warmly adopted Stephen.[285] Siegbahn laid on for Stephen a 'smorgas' of truly oriental splendour at Skansen, and I shall never forget Stephen's delight. Plates of great lobsters, smoked salmon, raw fish of various kinds, bear-meat, 'sandwiches' in endless variety, dainties with lashings of cream, cheese and cakes; it was just too marvellous after the severe and dull rationing which was still in full force in England. He had great fun too in stocking up his wardrobe—and of course there was the Swedish railway system to study. We also enjoyed the boundless hospitality of Edeby, the country home of the Eckermanns with whom I had stayed in the war.

[284] *Alltid Amber* is translated as *Forever Amber*.

[285] Karl Manne Georg Siegbahn (1886–1978) was a Swedish physicist who was awarded the Nobel Prize in Physics in 1924 for research in the field of X-ray spectroscopy.

Figure 34 The Eckermann's house at Edeby in Sweden, 1947. (Sketch by WLB; Courtesy: the Royal Institution London)

After Edeby, Stephen had to return to England. Alice and I went to Göteborg, and then to stay with the Hedvalls in their country place on the island of Orust on the Swedish west coast.[286] Hedvall was Professor of Chemistry at Göteborg Technical College. We were driven across Sweden from Edeby to Göteborg by a friend of the Hedvalls and so had a chance to see a good cross-section of the Swedish countryside. The weather was marvellous that summer, cloudless and warm for nearly all our visit. The west coast of Sweden is a mass of islands and islets, well wooded near the mainland but getting barer as one goes seawards, until the final islets as one reaches the open sea, which is a score of miles from the mainland, are just bare rocks smoothed into whalebacks by ice in the past. Orust, on which the Hedvalls had their place, is a more sophisticated large island some twenty miles across with villages and shops. There is such a countless number of small islets all round, however, that the ownership of a patch on the main island carries some kind of proprietary right for camping, picnicking and

[286] John Arvid Hedvall (1888–1974) was one of the first to recognise that chemical reactions are possible in and between solids.

bathing on two or three islands in the neighbourhood. Clothes are discarded for bathing in the intimacy of family and guests. I worked out that one must not bathe without clothes within a hundred yards of a lady to whom one has not been introduced (there is a rigid social code for everything in Sweden). We ate our meals in the open air under the pines, waited on by a little maid from a family of fishermen and whom the Hedvalls had practically adopted as a daughter, with bearded tits flitting round in the branches overhead. In Stockholm the host made a speech of welcome at every meal. I had got hardened to replying as best as I could, but I thought that here in the woods, eating our lunch clad in bathing suits, formalities would be discarded. Not a bit of it. I saw the only too familiar boiled look come into our host's eyes, and with a severe face he made a speech welcoming us to Stillingson while we looked very grave and the little maid stood rigidly to attention. I hope my reply was adequate. The Hedvalls had a motorboat in which we cruised among the archipelagos. We had one visit to a fishing village on the ocean verge; it was strange to see the cottages, the fish-racks and the boats all on an island which was just smooth round rock without a trace of earth or vegetation.

Figure 35 Hedvall's house in Stillingson, June 1947. (Sketch by WLB; Courtesy: the Royal Institution London)

Figure 36 Stillingson, June 1947. (Sketch by WLB; Courtesy: the Royal Institution London)

Figure 37 Orust, June 1947. (Sketch by WLB; Courtesy: the Royal Institution London)

After the Hedvalls, we went back by boat further north up the coast to Fjällbacka, not far from the frontier into Norway. The scenery here was much more bare and rocky. The main street of the tiny town ran under an almost perpendicular cliff, and at either end there was a notice saying, 'Caution, beware of falling boulders', though what one could do if they chose to fall we could never decide. Our hosts were the Westgrens: he was the secretary of the Swedish Academy, and they had already entertained us at a sumptuous dinner in Stockholm. We sailed one day in their boat to visit the famous Svedberg of centrifuge fame and who summered on an island, and we were very pleased to see a provision hamper of reassuring dimensions placed aboard.[287] Alas, although we had an early morning start and did not get back till late afternoon, the hamper was never broached! All we had during the day was strong drink with Svedberg; we were ravenous.

We also sailed to a bare outer island where there was a small base for coastguards and pilots. It was so windswept that there was a covered trench from the cottages to the look out on top of the island. Alice and I were fascinated by the clefts a foot or two wide which traversed the bare rock. They were packed with wild rose, campion, bramble and other plants just like miniature gardens, which could only grow where shielded from the wind. We flushed an eider in one of these cracks; the nest with its many eggs smelt abominably.

In August we went to stay with the Balfours at Balbirnie near Markinch in Fife.[288] They had a very lovely Georgian mansion in a large 'policy', an especial feature of which was an immense shrubbery of rare rhododendrons. The occasion of our going there was a meeting of the British Association in Dundee. We then went on to stay in Killiecrankie, at a pleasant hotel just above the River Garry. The Scottish hills and moors inspired Patience to write poetry; there is a photograph of her in the throes of composition on the slopes of Ben Vrackie. Patience made a firm friend of a fellow guest, I think a commercial traveller, who took us on a number of trips in his car including one to the wild country around Loch Rannoch. He gave Patience a book on all the tartans, which made her a great authority for a time.

[287] Theodor Svedberg (1884–1971) was a Swedish chemist distinguished for his work in physical and colloid chemistry and the development of the ultracentrifuge. He was awarded the 1926 Nobel Prize in Chemistry.

[288] In the late 1960s, the Balfour family had to quit the 700 acres of Balbirnie, where they had lived since 1642, to make way for the new town of Glenrothes, providing housing for miners at two huge coal mines (which proved to be a geological disaster).

Figure 38 Balbirnie, the Balfours' house, August 1947. (Sketch by WLB; Courtesy: the Royal Institution London)

Figure 39 Lady Balfour sketching at Balbirnie, August 1947. (Sketch by WLB; Courtesy: the Royal Institution London)

Figure 40 Beeches in the Balbirnie 'Policy', August 1947. (Sketch by WLB; Courtesy: the Royal Institution London)

Figure 41 Threshing at Balbirnie, August 1947. (Sketch by WLB; Courtesy: the Royal Institution London)

Figure 42 The River Garry at Killiekrankie. (Sketch by WLB; Courtesy: the Royal Institution London)

Perutz had returned to work in the Cavendish Laboratory and published in this year the first of a long series of papers on haemoglobin, on some of which I collaborated. This paper was entitled 'The X-ray Study of Horse-Methaemoglobin'. We were trying to find some way of reading significance in the wealth of X-ray diffraction data provided by the crystal, and we conceived the ambitious plan of measuring all the data and turning them into what is technically called a 'three-dimensional Patterson projection'. If there were any regularity of structure in the atomic pattern of the molecule, it ought to show up in the Patterson projection.[289] The plan involved an immense amount of measuring by the patient Perutz, and the reduction of the results was a formidable task. Feeling greatly daring, a large sum was set aside

[289] The Patterson projection was devised by Arthur Lindo Patterson (1902–1966) as a means to solve the so-called phase problem in X-ray crystallography. He spent two years with WHB at the Royal Institution, where he learnt about X-ray crystallography. To solve a crystal structure one needs to have both amplitudes and relative phases of the diffracted X-ray waves, but the phase information is lost in the diffraction process. Patterson devised a method to get round this while at MIT in 1934. The Patterson method remains today one of the most powerful techniques used by crystallographers.

to get one of the big electronic computers to do the work. I think this was the first occasion for the full-scale use of a digital computer. We read some significance into the results, but as it turned out most of the conclusions we drew were illusory. When the structure was finally worked out by Perutz ten years later it turned out to be so complex that our attempt was bound to fail. Nevertheless our illusions had the useful effect of stimulating us to carry on.

The year 1948 was packed with incident and travel. Alice and I went with the daughters to Freshwater Bay in April, in order to see something of Harry and Greta Hopkinson and afterwards had another holiday in the Pleasure Boat on Hickling Broad over Easter. The girls became quite expert in sailing. In June Patience went to France to live with the Chaisemartin family and learn French. Alice had had a letter from Mme Chaisemartin to enquire if we knew of an English girl who would be a companion to her Anne, who was then about thirteen, with an agreement that Anne should reciprocate the visit in England. Patience heard of this invitation and to our astonishment proposed that she should go! It was a great adventure for her, because she had always been a bit shy of staying away. Alice took her over to

Figure 43 Freshwater Bay in 1948. (Sketch by WLB; Courtesy: Patience Thomson)

Paris, wishing to meet the Chaisemartins, and left Patience in a family of small boys and girls, the children ranging in age from Anne to the current baby. At first they behaved like angels, but naturally this did not last very long and gang warfare directed against the English stranger became the order of the day. Madame was a strict disciplinarian; altogether Patience had a toughening time but a very interesting one too. They all went to the Riviera for the summer, and Patience had to get accustomed to the sea urchins and beer for breakfast. She became thoroughly French; for instance she demanded knife-rests on our table, as at a French meal, when she came back.

Alice and I were much intrigued at the manner of her return. She was put in the charge of Cook's and crossed the Channel in the cabin reserved for couriers; on arrival at Dover she avoided any tiresome customs and passport red tape by going straight off the boat with her courier friends.[290] The Chaisemartins had kindly sent me a bottle of brandy and thoughtfully provided a bottle of white wine for Patience; the latter arrived in her mackintosh pocket, with very little left in it. Our link with the Chaisemartins was formed owing to our having received Charles Chaisemartin—Anne's young uncle—into our family earlier in the year. He had been most miserably billeted with quite the wrong kind of family in Newmarket, and a mutual friend asked us to have pity on him and rescue him.

This year was also notable for Margaret's first May-week dance; she went to the Clare ball with Richard Adrian.[291] At the end of June Alice and I motored to Downe House, near Newbury, for Margaret's school festivities. Then in July I was invited over to Holland to give lectures at a number of universities, on a round of visits which included one to Phillips in Eindhoven. The firm hospitably invites its visitors to spend a weekend holiday of any kind they choose as guests of Phillips. I imagine a whirl of gaiety in Amsterdam is the usual choice of the visiting industrialists; I must have made history by saying that I should like best a weekend bird watching. It was laid on! The charming young president of the local natural history society accompanied me to the island of Schouwen.

[290] Cooks is the Thomas Cook & Son travel agency.

[291] Richard Hume Adrian, Second Baron Adrian (1927–1995), was a British peer and physiologist. He was the son of Edgar Adrian, who won the Physiology Nobel Prize in 1932 and was President of the Royal Society from 1950 to 1955. In 1967 Richard married Lucy Caroe (now the Lady Adrian), daughter of Alban Caroe and Gwendy Bragg. He was Vice-Chancellor of the University of Cambridge from 1985 to1987.

Figure 44 Zieriksee, on the island of Schouwen, Holland, where I spent a weekend birdwatching in 1948. (Sketch by WLB; Courtesy: Patience Thomson)

Figure 45 Bridge in Amsterdam. (Sketch by WLB; Courtesy: Patience Thomson)

We crossed with our car in a ferry and had time to explore the fascinating old walled town of Zierikzee. It used to be an important port, where cargoes were trans-shipped from seagoing vessels to the barges which navigated the inland waterways, and vice versa; but the channels became silted up and Zierikzee remains as a perfectly preserved specimen of a medieval town. We went on to stay at a hotel near the outer coast of the island. The island is of course below sea level, except for a strip of vast sand dunes, a mile or so in depth and up to a hundred feet high, on the North Sea side. In order to simplify the pumping problem, the dykes are double. Between the sea dyke and the inner dyke there is a belt of marsh and mud flats, with gates in the outer dyke to allow the water to flow away at low tide and bar the entry of the high tide. The main body of the land is drained into this belt, not straight into the open sea, and in this way the pumps have to lift the water a lesser height. The belt is the haunt of numerous waders. Avocets were there in great numbers, scything in the shallow pools with their curved bills. Kentish plover were also common, and I had the pleasure of studying a bird which is so rare in England and of getting its appearance fixed in my head. The dune country was fascinating. The marshy valleys were jewelled with flowers, including wintergreens. I was particularly pleased at being shown the nest of a Montagu's harrier—the nest had two fierce young in it, and they tried to fasten their bills on our hands—and also to see black-tailed godwit in summer plumage. An oriole whistled from the trees around our hotel. Altogether, it was a delightful interlude in a strenuous lecturing trip.

At the beginning of August, we had a dance for the children at 3 West Road—I think it was the only full-scale dance we held there. Later in August, Alice, Patience and I had a stay with Sylvia Fletcher-Moulton at Barcombe in Sussex, while Stephen and Margaret went off for a holiday at Polzeath in Cornwall with the Salters.[292, 293]

[292] Frank Reyner Salter (1887–1967) was a British historian and the president of Magdalene College, Cambridge, from 1951–1957.

[293] Quotation from Lady Heath: 'As Stephen was going off to MIT, our mother thought it would be a good idea for Stephen and me to go away together. The Salters frequently had reading parties and I think that is why our mother thought we could go to them'.

I had been elected President of the Conseil de Physique Solvay V, and on 25 September we went to Brussels for the first conference at which I presided. The subject was 'Elementary Particles'. Two conferences, one for physics and one for chemistry, were endowed by the famous Solvay, who invented the 'process' and founded the firm, and the family continued to support them. They are held triennially; the first physics conference was held in 1911. The number of participants is limited to about twenty; their expenses are met and they are most hospitably entertained in Brussels. The first conference I attended was that on 'Atoms and Electrons' in 1921, and I can remember the great Solvay coming to address us at our opening meeting and expressing the hope that our labours would be fruitful. Lorentz was wonderfully well suited for the presidency of an international conference. He spoke English, French, German and I think Italian, fluently. After each speaker, he would briefly translate what he had said into one or more other languages, and it seemed to me that he often expressed the points much better than the original speaker. With a somewhat dry and precise manner, he combined a wonderful sense of humour. It was not an easy conference to organise, because there were political undercurrents and antipathies in Brussels University. Our friend, Ernest Solvay, the son of the founder, was strongly pro-royalist and right wing, whereas the secretary was as far to the left as it was possible to be, a strong anti-royalist and not favourably inclined to any Solvay activity. I found it very hard, for instance, to get answers to letters, and nearly all the organising of selecting guests and arranging speakers fell to me. I tried to help in every way I could to keep the peace. More than once I asked Solvay to be relieved of the responsibility, but he pressed me to stay on until the atmosphere became less strained. I presided in 1948 ('Elementary Particles'), 1951 ('The Solid State'), 1954 ('Electrons in Metals'), 1958 ('The Structure and Evolution of the Universe') and 1961 ('Field Theory'). This last, the twelfth conference, fell on the fiftieth anniversary of the first conference and was a very special occasion. I fear I found the subject completely unintelligible! I enjoyed the conferences greatly; if I had to choose the one I liked best it would be that on 'The Structure of the Universe', a magnificent subject to cover in a week. We started with furthest space and, by the time we had worked inwards to our own galaxy, it seemed almost parochial.

Then, on 18 October, Alice and I started on a tour of the States. We travelled on the *Media*, a very pleasant Cunarder. It was on this trip that we first met the Allens.[294] He was in the civil service and was going to the west coast of America as a Commonwealth Fellow. He has now (1964) risen to great heights in the Home Office and then the Treasury, and it has been a great pleasure to have met him again in our recent negotiations about lectures for Civil Servants at the Royal Institution.[295] I remember one incident on the voyage. I befriended a small boy of six or seven whom I was sorry for because his parents were laid low with seasickness for the whole voyage. He was always asking me to show him a whale. I could point out puffs of steam in the distance but he naturally found these unconvincing. Then on almost the last morning he came up with his usual request, and I said, 'Let's look over the side', when to my astonishment there was a whale within twenty feet of the ship and which had rolled over on its embarrassment at being nearly run down and was showing its fins and tail and all. My small friend said, 'Please can I stay with you always'.

Figure 46 The Kellys's house at Short Hills, near New York.[296] (Sketch by WLB; Courtesy: the Royal Institution London)

[294] Philip Allen, Baron Allen of Abbeydale of the city of Sheffield (1912 –2007).

[295] In 1954 WLB moved to the Royal Institution to become Director of the Davy-Faraday Research Laboratory and Superintendent of the House. He instigated special lectures for Civil Servants and also for schoolchildren.

[296] Mervyn Joe Kelly (1895–1971), American physicist, and director of Bell Labs from 1951–1959. He shared his love of gardening with WLB.

We first stayed in Boston with the Karl Comptons and had a very good time because Stephen was there with his Commonwealth Fund Fellowship.[297] I had been on the election committee for these Fellowships for many years and had found the work to be intensely interesting. We first sorted out the applicants into a short list of about twice as many as the number of places. The New York office then took great pains in making provisional arrangements with American universities for all on their short list. Two representatives came over to England to attend our meetings for the final selection, so that each successful candidate could then be interviewed by them and all his plans put into final shape. As well as the year in America, with a possible extension to a second year, each Fellow was encouraged to make a tour of the States in the vacation. Members of the committee, in turn, were sent out to America to see and report on the Fellows. Alice and I were to have gone in 1939 and had booked our passages and even bought all our clothes when, alas, the war started and our trip was cancelled. I had left the committee before these visits were organised again so we had never had our turn. Stephen had been successful in getting a Fellowship and was working under Hawthorne in MIT.[298] Later he made his tour of the States with Douglas MacLellan, always known in the family as 'Small but', this being an abbreviation for 'Small but Sound'.[299] Stephen thoroughly enjoyed his time in MIT.

We next went to Toronto, where we had a stay with Bullard and his family; he had recently gone there as professor.[300] From there we went to stay with Huggins and his family in Rochester.[301] Huggins

[297] Karl Taylor Compton (1887–1954) was an American physicist and president of Massachusetts Institute of Technology from 1930 to 1948. He carried out research into electronics and spectroscopy, especially in solid materials.

[298] William Rede Hawthorne (1913–2011) was a British professor of engineering who worked on the development of the jet engine. He was Head of the Department of Engineering in Cambridge in 1968 and was appointed Master of Churchill College, Cambridge in the same year (1968–1983).

[299] Douglas MacLellan is possibly George Douglas Stephen MacLellan (1922–1999), Rankin Professor of Mechanical Engineering from 1959 to 1965.

[300] Edward 'Teddy' Crisp Bullard (1907–1980) was a British geophysicist and one of the founders of marine geophysics and the theory of continental drift.

[301] Maurice Loyal Huggins (1897–1981) was an American chemist who was one of those who conceived the idea of hydrogen bonds holding organic molecules together in crystals. He produced a model of the α-helix of proteins in 1943, roughly eight years ahead of the modern model of Linus Pauling, Robert Corey and Herman Branson.

was in the research department of Kodak, and I met a number of old friends. I remember one incident when I was lecturing in the university. I could not think why the audience was looking a little anxious, until with one final gesticulation I swept my glasses off the table and they smashed on the floor—the audience had watched them getting nearer and nearer the edge of the table. But next morning I received a brand new pair, designed by a local firm of opticians from fragments they had collected, and sent with their compliments!

From there we went on to Ann Arbor, where we met again our old friend Dean Kraus, a mineralogist who had been Dean of the science faculty when we went to Ann Arbor in the summer of 1924. The university had grown very greatly. The law school alone had 4,000 students, not much less than the whole of Cambridge before the war. There was a very fine Graduate House 'donated' by Ford's lawyer, with rest rooms, a library, concert rooms and lecture rooms for postgraduate students, as well as staff to look after them. One name on a door intrigued me very much: 'Dr —, Religious Adjuster'. From Ann Arbor we went to Chicago, where we stayed with Mrs Schaffner, who had been so very kind in sending us parcels during the war. We were embarrassed on our arrival, at about 2:30 in the afternoon. We had decided that this was too late for lunch, so we had eaten on the train rather heartily. Then, when we arrived at the Schaffner home, there was an immense turkey waiting for us! It was a great effort to seem to do justice to it. Mrs Schaffner still had the same faithful chauffeur that we had originally met in 1924, when Alice and I first stayed in Chicago. He was a great character and friend of the family; Joe Schaffner had built a small bookcase under the dashboard so that the chauffeur could catch up with his reading while waiting for his master, and I remember that one day when we were discussing the arrangements for our departure (in the back of the vast car) the chauffeur overheard us and asked Joe if he could keep us for a few days longer because he (the chauffeur) was making a study of our accents.[302]

[302] Joseph Halle Schaffner (1897–1972) spent most of his life in his native Chicago as Director of the clothing-manufacturing firm Hart, Schaffner, and Marx. He formed several collections reflecting both literary and historical interests. His greatest collecting passion was the history of science, and he bequeathed to the University of Chicago an extraordinary collection of several hundred scientific and medical rare books and manuscripts.

The main part of our stay was in Pittsburgh, and this was really hard work. Silverman, Professor of Chemistry, was responsible for inviting us there.[303] We stayed in a hotel opposite the university, which is aptly termed the 'Cathedral of Learning'. Instead of spreading departments horizontally, they are piled vertically one on top of the other in an immense skyscraper. I gave my lecture in the Mellon Institute. It was very interesting to study the structure of this unique body. The idea is that firms who wish to do research and have not a laboratory of their own can use the institute. It sets aside a small team of experts to tackle the firm's problem. Quite often, the firm is so satisfied with the results that it later sets up its own organisation, with the members of the Mellon team as a nucleus. My lectures, I was told, drew record numbers to the Mellon Theatre, but they were very strenuous occasions. Silverman worked me very hard, in that he would remorselessly call for me at the hotel at 6 p.m., and it was followed by an invariable dinner with turkey and cranberry sauce, etc., until the lecture started at 8:30. A very pleasant feature of the visit was that Mr and Mrs Chapman (Storm Jameson[304]) were staying at the same hotel; both had lecturing assignments in the university. Storm Jameson had been asked to give a course on how to write novels. After three lectures she said she had told the class everything she knew about writing novels and threw in her hand. Instead, she proposed to the class that each member should write a novel and she would then discuss it with them. One elderly lady in the class brought a story based on some family letters dating from early days in that part of the States—and I remember Storm Jameson saying that when she told her, 'You know, this is the real thing,' the lady burst into tears. Alice and I have always enjoyed our visits to America enormously, and the hospitality we have experienced has made us feel how hard it is ever to approach its standards when Americans visit us. I am afraid though that we greatly enjoyed getting into a huddle with the Chapmans and having a real heart-to-heart about American ways, so as to let off steam. I remember noting at Pittsburgh a phenomenon then new to me, though it has become so familiar since. This was the sight of suburban streets lined with cars,

[303] Alexander Silverman (1881–1962) was an American chemist who carried out research into ceramics and glass.

[304] Margaret Storm Chapman, née Jameson (1891–1986) was a British journalist and author known for her novels and reviews.

parked there because the houses had no garages. They remind me of aphids on a rose bud; I suppose that soon no one will remember how pretty the streets looked in the old days, with trees and gardens on either side, and no cars.

Alice went to Duke University to give a talk while I finished my assignment in Pittsburgh, and we reunited in New York. We were to have sailed from there, but a dock strike prevented it, so we were put on a special train and despatched to Halifax. It was a strange journey, because we never stopped at a station the whole way; we only pulled into sidings, so we lived for thirty-six hours in a packed huddle of grown-ups, children and baggage. No sooner had we pulled into Halifax than Henderson, Professor of Physics in the university, descended on me and demanded a lecture.[305] There was just time to hunt out some slides and give it before the *Media* sailed.

We had an alarming incident in the night. I heard a strange swishing sound in the small hours and putting my hand to the floor of the cabin I found that it was covered with some inches of water. I woke Alice in great haste and tried to remember were the lifebelts were. What had happened was this. The salt water supply to the baths had been cut off in harbour, for obvious reasons and, when the woman next door tried to fill hers, nothing came and she inadvertently left the taps open. She also hung up some stockings to dry over the bath, and these subsequently fell into it. In the night, the water supply came on, the stockings blocked the plughole, and the result was a flood which in lessening degree filled some half-dozen cabins next to hers. Alice had bought a great many clothes at Lord and Taylor's for herself and the girls just before we sailed, because everything was still in such short supply in England. Luckily as a last afterthought when we went to bed, I removed the boxes containing the clothes from the floor under our bunks and put them on top of the wardrobe. So the clothes were saved. Our carpet was wet for the whole trip.

The *Media* had no cargo because of the dock strike and rolled abominably. Nearly everyone was ill. Alice stayed in her bunk till Scotland was in sight. Her stewardess while attending to her was thrown down and broke her arm. An American lady missionary, working in India, was one of the few occupants of the captain's table well enough to

[305] George Hugh Henderson (1893–1949), physicist at Dalhousie University, Canada, was famed for work on radioactive haloes in rocks.

appear at meals. She used to give me daily reports of her roommate, who was very sick. Finally, at lunch she said, 'Now you can only see the whites of her eyes', and I was not surprised to hear a little later that the sufferer had been taken to the sick bay. I took a great liking to the missionary; her stories of trying to run a hospital in India were memorable.

So we arrived in England on 4 December after a record year of stays in different places—Isle of Wight, Hickling Broad, Newbury (Downe House), a number of places in Holland, Paris, Barcombe, Brussels, Boston, Toronto, Rochester, Ann Arbor, Chicago, New York (where I ought to have recorded that I received the Roebling Medal of the American Mineralogical Society[306]), Pittsburgh, Duke and then home.

One of the most charming incidents of my life came about in the spring of 1949. Patience, then thirteen, conceived the idea of taking me on the Broads as her guest for a week. She had, I believe, been saving her money for two years with this end in view and had not told anyone about it. She confided the plan to Alice when she felt that enough money had accumulated and suggested that we start the following Saturday, indicating that she could cope with everything except 'Daddy's beer'. Alice, while thrilled, tactfully suggested that rather more notice might be necessary. We got the boat however for the Easter holiday—and we had such fun, just the two of us with Scrap. The *Wild Rose* was a delightful cabin yacht for two. I had not been on the Broads for many years and admired the improvements of electric light, Calor gas and a mast in a housing above the deck so that when lowered it did not foul anything in the forepeak. The weather was very cold when we started, with frost in the morning, but it turned into a heat wave at the end of the week. We explored all the rivers above Acle when we got the boat. Scrap slept at the end of the bunk in a recess which also received Patience's feet and, when in the middle of the night he had a call, he used to jump overboard onto the mud and walk back over Patience's face. We had endless adventures. There are always a good many inexpert navigators on the Broads, and I used to be proud to see Patience cunningly outsailing them. It was a very windy Easter, and we saw several dismasted boats. Once, we had just gone through

[306] The Roebling Medal is the highest award of the Mineralogical Society of America. It is named after Colonel Washington A. Roebling (1837–1926), who was an engineer, bridge builder, mineral collector, and a significant friend of the Society.

a bridge and I was engaged in the tricky task of raising the mast again. Patience supplied intelligence in clearing fouled stays and halliards but not brute strength, and I could only just manage it. At the critical moment, when I had got it past the sticky point, a boat with four young men in bobble caps rammed our stern, and down came the mast on top of me. Patience said afterwards, 'Daddy, I didn't know you knew those words.' Rationing was still in force and, as a very special treat, Alice had given us a small packet of rashers of bacon. We decided we would have them for our final supper, and as we were rather late in the evening, I laid out the bacon in the frying pan and made other preparations for supper, so as to start cooking directly after we tied up. One of the many sailing distractions occurred—and when we had coped with it we saw the rashers hanging out of Scrap's mouth! They were rescued, and then Patience looked at me and I looked at Patience and without a word between us the rashers when back into the frying pan. The Broads have always fascinated me. I think it is the way in which one moors at night right away from everything, with the vast expanse of sky, the aromatic marsh vegetation and the night noises of bird and beast, and yet one has all the comforts of a snug home which one carries with one like a snail its shell. Alas, the Broads are becoming too crowded, and the despicable motor cruisers, which take away all the adventure, are replacing the sporting yachts. The Broads seemed to be a vast expanse on a yacht—one could happily explore them for a whole fortnight—but they shrink to a day's trip from end to end in the motor boat. Wherries are no more. I am glad I remember when they were common with their great sails coloured, as Alice used to say, like the inside of mushrooms. They could outsail and outpoint the yachts. When sailing close-hauled, they used to hug the lee shore, piling up water between boat and shore in a way that seemed to serve as a kind of lee board. It was fascinating to see the two men on board sail up to a bridge, lower sail and mast in the nick of time, shoot the arch and raise mast and sail again before the boat had lost way.

Patience and her friends played a trick on us when we paid our June visit to her school. Each child told its parents that the first event would be a picnic of four families in the woods, and each lot of parents was asked to bring food for the whole party, which they did on a lavish scale, so that their particular offspring should not lose face. The result was a gargantuan spread. I believe the perpetrators made a cache of much of it in a secret place in the woods and fed on it for weeks.

Alice and I went for a June holiday in the Shetlands. At Cambridge it is common to take a holiday between the end of the May term and the start of the Long Vacation. We had often said we would like to go to the Shetlands. We flew from Aberdeen and had a chancy landing, because the cloud cover over the Orkneys was so low. The pilot had several attempts at a run in and at last broke through the clouds in a possible place. The little plane on to the Shetlands only just managed it too. A car met us and drove us through mist across the dreariest moors, with hideous houses at long intervals on the road, to a one-storey whitewashed inn at Spiggie. I wondered why I had brought Alice so far at such expense just for this—and it turned out to be one of the best holidays we ever had. The others in the hotel were all fishing enthusiasts and spent the day on the loch in rowboats with, it seemed to us, very little encouragement in the way of returns. We explored. The cliffs are marvellous, and running inlands for a short way is a strip of vivid green turf, I suppose encouraged by the innumerable sea birds, on which the island sheep and cattle are grazed, and this was ideal for walks. We got boatmen to row us across to islands where we picnicked and were called for again in the evening. It was quite light, and the larks were singing all night. Trees, and most broadleaved plants, cannot survive in the salt spray which often drifts right across the island, and the ling and heather are dwarfed to a few inches. But as if to compensate, the sheltered clefts in which the little streams ran were brilliant with flowers, which seemed to be more brightly coloured because they got so much light. The marshy bits were a mess of mimulus. In places where the sand drifted over the land (machair), the turf was a carpet of thyme, buttercups, orchids, pansies and other dwarf flowers. All available sites on the cliffs were taken by nesting birds—fulmars, kittiwakes, razorbills, guillemots and gannets. The skuas flew at our heads if we went near their nests and, when we were on the beach, the seals used to swim towards us alive with curiosity. It was a holiday quite of its own kind.

Later Gwendy and I had one of our all-too-rare holidays together. It was my idea to go to Portisham near Weymouth for a sketching holiday. It was not a very good place to choose. It is strange how some localities just lend themselves to sketching, while others, though in charming country, are quite barren of stimulating scenery. It was a great delight, though, to have her all to myself. Gwendy and I were robbed of many of the times we might have had together. She was

Figure 47 Loch Spiggie, Shetland, July 1949. (Sketch by WLB; Courtesy: the Royal Institution London)

only a few years old when I was at Cambridge. Then came the war, and I could only have fun with her on the rare occasions of my leave—and then I was overwhelmed by my difficulties in finding my feet at Manchester University. After two years came my engagement and marriage. We never had much of the usual family brother-and-sister time, which we have always enjoyed so greatly when it has been possible.

In August we went to Polzeath, to a house lent to us by the Salters. Polzeath is an attractive place, well known for its surfing, and there were pleasant walks in the neighbourhood. We had Anne Chaisemartin, Patience's opposite number, to stay with us and also Idolette, a daughter of the Swiss family with which Margaret had been billeted in Vevey. But it was rather a sad holiday. Alice was not really well; the strain of the war and her year as Mayor had taken a great deal out of her. Then, in the middle of the holiday, David turned up with the absolutely appalling 'Nigel'.

David had been staying with the Hoylands at Painswick in Gloucestershire (Hoyland was the retired head of his prep school, The Downs) and working in a local pottery. He had made the acquaintance of this Nigel, who persuaded him to go off on a walking tour, finding hospitality where they could and earning their living by selling sketches and drawings. Nigel had a real gift as an artist. We had heard of them from time to time in various parts of the country, generally when money had run short and parental help was needed. They appeared one afternoon at our house, with their immense rucksacks, and stayed with us for the rest of our holiday. Nigel spun us a number of stories; he was a close relation of a well-known man, Sir Oliver Leese, he had been a welfare officer and he had orders for a number of Christmas card drawings with which David could help.[307] David had clearly enjoyed the adventure of tramping over the south-west of England and, as we were grateful for the benefit it was doing him, we never thought of doubting any of these stories, especially as Nigel really could draw and design with talent. The boys planned to stay on at the local inn, with a contract to do murals for its dining room, after we left Polzeath. By degrees the truth came out. I think Alice first suspected something was wrong when she met them in London and Nigel was wearing one of David's suits. He explained that it did not fit David well and that he was arranging for another one for David with 'his tailor'. I had an angry letter from the manager of the hotel to say that, after starting some daubs on the walls, Nigel had decamped; then one came from the owner of a caravan they had occupied claiming rent which Nigel had never paid. David was of course quite unsuspecting and innocent in all this. So I wrote to Nigel's supposed relation and found that he had no such nephew. Then we discovered that Nigel had wormed out of David the addresses of our various relations and friends in that part of the world, gone to stay with them and borrowed quite large sums of money on the ground of his acquaintance. By this time, they were settled in a flat Nigel had hired in London, where the Christmas card scheme was supposed to be going forwards. I descended upon Nigel, taking Malcolm with me for both moral and physical support. The place was a shambles, with empty bottles everywhere. No rent had been paid for weeks, and I had to stump up to the proprietor, who was very decent and sympathetic. Nigel was wearing David's suit and other clothing. I went through the place, rescuing David's belongings and incidentally coming across Nigel's identity card which showed that his name was not the one he

[307] Sir Oliver William Hargreaves Leese, Third Baronet (1894–1978), was a British general during the Second World War.

had given us. We had to repay everyone and it was a costly business. The only scrap of pleasure I got was that Nigel was throwing a party the evening of the day of our raid and, by the time I had insisted on his handing over all David's belongings, he had hardly a stitch on him. Sometime afterwards I had lawyers' letters asking for the address of Nigel—who had falsely represented himself as Sir Oliver Leese's nephew—and letters from doctors who wanted their bills paid. Nigel was most specious; he deceived us, and it is no wonder that David, to whom he had given a great adventure, was completely taken in by him. He was, I suppose, a typical psychopath who just could not go straight in spite of his very real gifts. It was a sad experience.

Perutz this year produced his second paper on the structure of haemoglobin, in which he proposed a model of the protein-chain configuration. Later research showed that it was much too simplified a version, but much interesting information was obtained. These false starts were part of the long search for the key to the haemoglobin structure, a search which was finally so triumphantly successful.

All this time I was doing my best to cope with the Cavendish, often feeling my difficulties very keenly indeed. I felt especially my failure to keep up the Cavendish tradition of nuclear research. Frisch was of the highest quality, but an individualist, not taking kindly to the leadership of a team or to large-scale research. The cyclotron was kept going. A high-voltage machine of the Van der Graf type was installed in a wartime tower shelter. The Cockcroft–Walton accelerator was going.

Ratcliffe had an inspiring laboratory, and Ryle was producing very fine results indeed. Ratcliffe's team concentrated on upper atmosphere structure and movements, and I was fascinated by the elegance of the optical principles they developed. Ryle got his first brilliant results with relatively simple equipment. This side of the Cavendish work was really inspiring. So was Cosslett's unit, which was doing electron microscopy and developing high-resolution X-radiography. The Mond had a good team headed by Pippard and Schoenberg. Orowan's metal physics was always interesting. But the nuclear physics, apart from Frisch's small team, was slowly dying and I do not think anything could have saved it, though I was so worried about it at the time.

In April 1950 we had a short stay at Portisham near Weymouth. I continued my links with ASDIC, the detector of enemy submarines, for some years after the war and visited the experimental station on

Portland at regular intervals. These visits always had a pleasant flavour. Weymouth is attractive, and the station is on the edge of the harbour. Anderson, the director, was a warm friend, and after the day's work he would take me on a tour in the Dorset countryside. The work of the station was of high quality. So much research for defence purposes between wars suffers in becoming detached from reality, but the naval work seemed to escape this deterioration, I suppose, because of the demands made on any gear which has to work reliably at sea. A pleasant hotel on the front, the swish of the waves all night, the drive along the narrow neck to Portland, good scientific talks, lunch in the mess when the pressure to drink 'the other half' made me very sleepy afterwards, the evening drives in the pretty country, all created an impression which made me say, 'I must bring Alice here.' Hence, our stay at Portisham, at first in a rather primitive hotel and then kindly put up for a night or two by Fisher, whose family was away from home. In the hotel we shared the bathroom with the proprietor and his wife. I could not bear the sight of his hair brushes and, to the horror of the family, I washed them (this is a Pepys-like confession); what he made of it was never revealed. Fisher had a wonderful collection of jade, much of which must have been priceless. He had acquired it at sales through his expert knowledge and judgement. I remember his telling us that some of the pieces were of such complexity that they could not have been carved by one man in his lifetime—they were the product of a long-established guild of craftsmen.

In June Alice and I went to Geneva at the invitation of Paul M. Haenni, who directed an International School for Training in Management.[308] We stayed in the guest flat of the school, where I gave a series of scientific talks to a class of about twenty. They were young people sent there by their companies, which supported the school; I believe a preponderance of the companies dealt with aluminium. Alice thoroughly approved of the flat. There was a very completely equipped cocktail cabinet, tickets which we could exchange for restaurant meals in the town, rolls of tram-tickets, letters and postcards already stamped, maps, guides and a wide choice of cigarettes and cigars; in fact, everything one could think of had been provided by our kind hosts. After this stay we went to Grindelwald. This was our

[308] Paul M. Haenni (b. 1900) was Director of the Centre d'Etudes Industrielles, at the University of Geneva.

first taste of a summer holiday in Switzerland. Alice adored sailing up the mountain side, over the pastures bright with flowers, in the chair-lift. The flowers were a special treat for both of us; we have had great enjoyments throughout our married life from our mutual interests in exploring pretty country, flowers, anything to do with water and architecture.

The great event of this year was Stephen's engagement to Maureen Roberts, the daughter of Dorothy Roberts (Amos) who had been a great friend of Alice's and Enid's as girls, especially of Enid's. Maureen had visited us when at Cheltenham Ladies College, and earlier in that year she had stayed with us for a dance which her parents gave jointly with the Henns (Rosamund Henn was her crony) in St Catherine's College.[309] In August, when we had a family holiday at Blakeney, Maureen spent part of it with us. We had got a delightful house for the summer through the kindness of Buxton, who owned it.[310] We had heard of it on an earlier visit when exploring for a summer place, but found, as was only to be expected, that it was far beyond our means. To our surprise, we had a letter after our visit offering us the house and gardener at an almost nominal rent, provided we promised never to write him a letter during our tenancy. He must have been pestered by some former tenant, I imagine. His kindness went further. David, in the fervour of painting still life in the garden house, knocked over and broke an enormous and impressive vase. When we confessed this, Buxton's comment was 'I always hated that vase'. We took the opportunity to have various friends such as Sylvia Fletcher-Moulton and Violet Cunliffe-Owen to stay in such a delightful place.[311] Sylvia and I put in good birdwatching in the estuary.

We again hired the boat from the fish-and-winkle man and, to my intense surprise, we nearly won a race. Patience and Alice persuaded me to enter for an open event, and we got twenty minutes start from the smart Fireflies and Nationals. We had to go twice round the harbour. There was a very stiff breeze, and the old winkle-boat roared

[309] Thomas Rice Henn (1901–1974) was a literary critic. He was President of St Catherine's College, Cambridge, from1951 to 1961.

[310] Buxton is presumably Aubrey Leland Oakes Buxton, Baron Buxton of Alsa (1918–2009), who was a British soldier, politician, television executive and writer. He was one of the founders of the World Wildlife Fund.

[311] Violet Cunliffe-Owen was born in 1880. She was the daughter of Frederick Philip Lewis Cunliffe-Owen and Emma Pauline de Couvreu de Deckersber.

along to such good purpose that we entered the cut in the final lap while all the other competitors were still only half way around the second lap. The narrow cut was, alas, our Waterloo. Beating against a head wind and strong current was too much for our gallant craft, and a Firefly just beat us to the finishing post. However, we had a gun all to ourselves as second in, and it was hard to say whether Patience or the winkle man beamed most.

In early September we had a stay with the Trevelyans at Wallington in Northumberland.[312] It was by the grace of God that we ever got there. Our car had been bought second-hand in 1935, when Patience was born, and had been worked very hard in the war. I should have had an overhaul before we set out, Alice, Margaret, Patience and I, with a large amount of baggage, on a journey right up to the Border. Every tyre went in turn, and tyres were in very short supply. We just managed somehow to get a replacement in one town after another; in more than one case, it was the last in the place. Patience (who was fifteen) was at that time writing a novel, which she used to read to Trevelyan and Patrick Duff: Trevelyan took it very seriously and evinced great interest.[313] He read us Border poetry in the evenings and took us on long tramps over the hills.

In October I had an invitation which made a useful after-dinner story at the time, though its force is lost now. I was invited by the General Electric Company at Schenectady to a dinner to celebrate the fiftieth anniversary of their starting a research laboratory under Steinmetz and asked to reply for the guests.[314] I said that, alas, I would not be in the States at that time and then got a cable back to say, 'We will fly you over on 7 October and fly you back next day'. I felt very important at being ferried over the Atlantic just for an after-dinner speech and hope I rose to the occasion.

Another fiftieth anniversary this year was that of the foundation of the Nobel Prizes. The normal awards on 8 December were made the occasion of a special celebration to which former Nobel Laureates were

[312] The Trevelyan family owned the country house Wallington Hall.

[313] Patrick William Duff (1901–1991) was Vice-Master of Trinity College, Cambridge, and Regius Professor of Civil Law.

[314] Charles Proteus Steinmetz (1865–1923) was a mathematician and electrical engineer. Born Karl August Rudolph Steinmetz in Breslau (now Wrocław, Poland), he went to the United States to work on electricity. He was called the 'forger of thunderbolts', being the first to create artificial lightning in his football field-sized laboratory.

invited. A great event was a dinner in the town hall, with 1,400 guests. Each course was brought by a regiment of waiters who marched down the stairs at the end of the hall to the accompaniment of music, holding the courses and drinks aloft in great salvers—a wonderful sight—especially the ices on vast blocks of ice into which electric lamps had been frozen.

I remember well on this occasion that George Thomson had a heart-to-heart with me about the Royal Institution. It was becoming apparent that it was headed for disaster. Andrade, with all his great gifts, was by temperament entirely unsuited to direct a place where tact and patience are the first requisition, and although he had only been appointed recently, he was already at loggerheads with the officers and managers.[315, 316] Lord Brabazon, the president, had, I learnt afterwards, warned the managers against making the appointment because it was a dangerous thing to do, but the secretary Rankine pressed on with it in a headstrong way against all opposition. [317, 318] George Thomson was a manager at the time and could not forgive himself for agreeing to the appointment. It was known to be Andrade's great ambition to go there, and it was hoped that the realisation of his wishes would soften his asperity. This was perhaps not unreasonable. I hoped myself that perhaps all might go well, but the actual debacle was worse than anything we had pictured. Things went from bad to worse, until finally the members refused to pass a vote of confidence in Andrade and he resigned his post. It was tragic, but it was clear that if he had gone on the Royal Institution would have foundered in a year or two.

[315] Edward Neville da Costa Andrade (1887–1971) was an English physicist, writer and poet. He had carried out important work on γ-rays.

[316] There is some early history here. In a letter dated 4 June 1917 from WLB to his father during the First World War he wrote: 'Andrade got jolly well kicked out of the show here as he became absolutely the limit. He is a hopeless chap. I am sorry for him too sometimes but he has got a bad kink in him somewhere. There was an awful to do about it all, the officers in his section refused to work with him any longer and told their colonel so, and as they were A1 chaps and about eight others had one by one begged to leave his section, he departed. He used to tell them off in front of the men!' (RI 37B/2/22).

[317] John Theodore Cuthbert Moore-Brabazon, First Baron Brabazon of Tara (1884–1964), was an English aviation pioneer and a Conservative politician. He was the first person to make an official aircraft flight in England in 1909.

[318] Alexander Oliver Rankine (1881–1956) was, from 1919 until 1937, Professor of Physics at Imperial College.

It was far from being altogether Andrade's fault. The real clash was between him and Rankine, who had at first supported him with such fervour. The strange and difficult organisation of the Royal Institution provided the fuel for the blaze, because it does not lay down any clearly defined chain of authority. A normal organisation would have a board of trustees (the managers), who lay down the broad lines of policy, and a director, who is responsible for administration and gives all orders to the staff. Instead of this, as it is a private society, administration is technically in the hands of the treasurer and the secretary, who have first call on the services of the office staff, and the director is an employee appointed by the managers on whom certain duties are devolved. The honorary secretary had an office in the institution. The greatest tact and patience are required on both sides to avoid friction, and Rankine lacked these qualities just as much as Andrade. Towards the end he became quite distraught and committed indiscretions of which he would normally have been incapable. George Thomson saw the situation developing, and we had anxious but fruitless discussions as to what could be done to save the place.

In 1951 the First Conference of the International Union of Crystallography was held in Stockholm under my presidency. The Swedes were kind hosts and the king came to our opening meeting. All went successfully. I consider the International Union as one of the highly successful ones which has amply justified its formation. The creation of new unions is rightly watched jealously by the Royal Society and other bodies, because specialised varieties can so easily multiply. But X-ray crystallography has always been a coherent branch of science, with clearly defined aims and its own disciplines and techniques. The union has been responsible for excellent international compilations such as its tables and review of structures. Its journal, *Acta Crystallographica*, covers practically all the new exciting work in this field. Wood was one of the first to press for an international body, and Ewald and Bernal were largely instrumental in getting it going.[319, 320] I was not present at the

[319] Elizabeth (Betty) Armstrong Wood (1912–2006) was an American crystallographer who became the first female scientist to work at the Bell Telephone Laboratories in 1942.

[320] Paul Peter Ewald (1888–1985) had been a student of Sommerfeld in Munich and is said to have provided the clue to Max von Laue that crystals might be able to diffract X-rays.

first meeting in America when plans for the new union were made; I much appreciated being asked to be president at the first formal meeting. My main contribution, I suppose, was that of collecting the money from firms in this country to start the new journal on its way.

A landmark of this year was Alice's appointment as a member of the Royal Commission on Marriage and Divorce. The Commission met regularly over four years and gave Alice a most interesting time. She made her mark in the final report, which embodied some of her contributions. I was very proud of her.

We had our summer holiday at Buttermere that year. Stephen joined us for part of the time, and Maureen drove over with her parents from Darlington to spend the day with us. I had the pleasure of giving Patience a really stout pair of nailed boots for fell-walking. I tried for the first time on this holiday to sketch in pen and ink but was not successful. I felt much happier with biro pen or with watercolours. Alice and the young climbed a lot over the hills, but I found I now had to go a bit slow on uphill work. In the autumn I presided again at the Solvay Physics Conference, which was on the metallic state this year. Stephen and Maureen were married on 15 September.

WLB's account ends here. It remained unfinished, probably because when he subsequently moved to the Royal Institution in 1954 he walked into a maelstrom left by the departure of Andrade following the vote of no confidence. WLB had to work hard to sort out the problems at the Royal Institution, which he did to great effect. However, because he was seen (incorrectly, in fact) by certain members of the Royal Society to have been responsible in some way for Andrade's ousting, he sometimes found himself snubbed by the Royal Society, and this caused him a great deal of upset. Nonetheless, his time at the Royal Institution was certainly highly successful, both from the point of view of bringing science to the public and in his involvement in the research being carried out there. One of the greatest scientific triumphs during his tenure was the solution under his guidance of the first enzyme structure (lysozyme, an enzyme found in egg white, saliva and tears and responsible for destroying bacteria) by David Chilton Phillips together with Louise Johnson.[321, 322, 323] *WLB finally retired in 1966 and died in 1971.*

[321] David Chilton Phillips, Baron Phillips of Ellesmere (1924–1999), was a pioneering structural biologist and an influential figure in science and government.

[322] Louise Napier Johnson (1940–2012) was a British biochemist and protein crystallographer.

[323] The original publications were Blake, C. C. F., Koenig, D. F., Mair, G. A., North, A. C. T., Phillips, D. C. and Sarma, V. R.(1965). *Nature (London)*, **206**, 757–761, and Johnson L. N. and Phillips D. C. (1965). *Nature (London)*, **206**. 761–763.

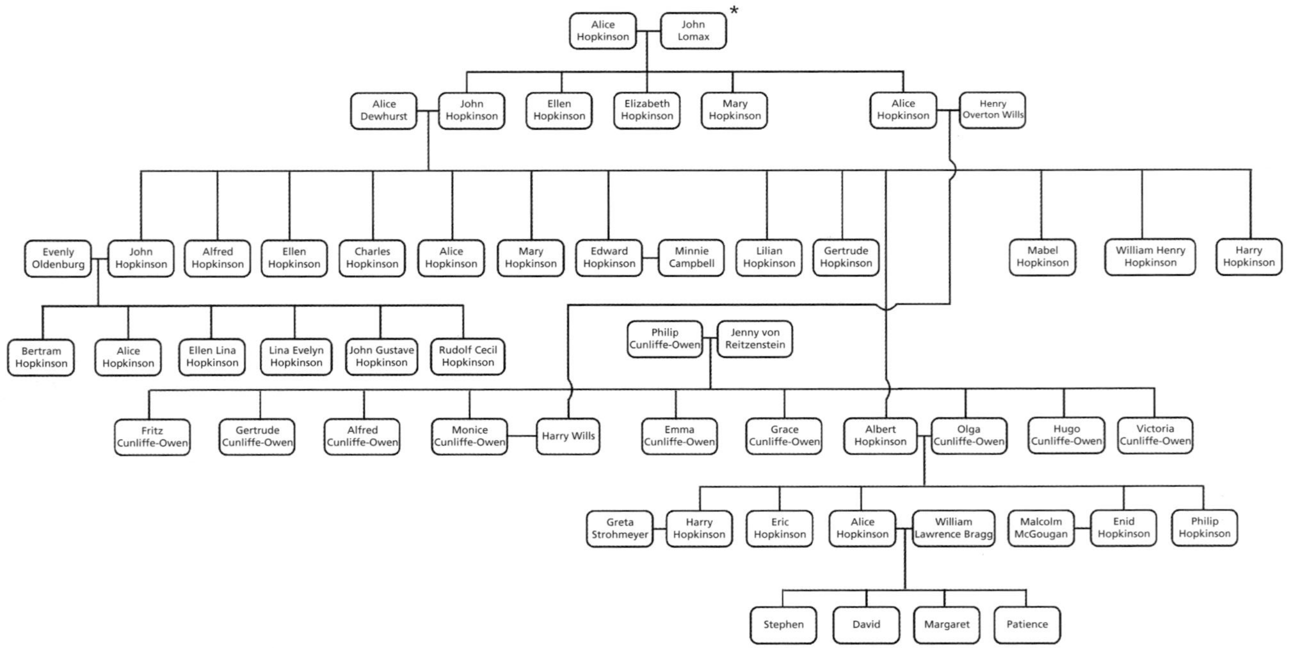

Alice Hopkinson's Family Tree Constructed with the Help of Ian Butson

*John Lomax and Alice Hopkinson never married and the family tradition had it that his parents so disapproved that they threatened to 'cut him off'

3

Alice Grace Jenny Bragg: 'The Half Was Not Told'

3.1 My Hopkinson Grandparents

My grandfather was illegitimate.[1] It was a pity, as this was to have a profound and disturbing effect not only upon him and his sisters but on his children. It was also inconvenient in the conventional Victorian age, when members of the family were to be high sheriffs, mayors and the like, and blanks in lineage had to be left. My grandfather was known as John Hopkinson but, had he been the child of a lawful marriage, he would have been John Lomax. Neither name seems particularly attractive.

John Lomax, my great-grandfather, came from Harwood, a village near Bolton. His family can be traced back to the seventeenth century—they were described as clothiers, trading in Bolton and Manchester, and were certainly people of means. The family house, once Lomax Fold, now Harwood Lodge, is a charming Georgian building with a lovely view down the valley with the mill at the bottom.

A Lomax built the village church in 1820. In this church there is a plaque to our great-grandfather; it reads:

> A tribute of respect
> from his affectionate Nephew to
> John Lomax Esq.
> of Manchester.
> Youngest son of Michael and Ellen Lomax
> of Harwood
> who departed this life June 5th 1827
> aged 63 years.

[1] An interesting account of the history of the Hopkinson family can be found on the internet at <http://www.johncassidy.org.uk/hopkinson.html>.

Figure 48 Harwood Lodge; originally built by John Lomax during the reign of James I and then extended in the mid-1700s; once owned by the Lever family and latterly the Porritts; believed to have once been visited by Sir Winston Churchill. (Courtesy: Fine and Country Homes)

We also know that he was closely connected with a Sunday school at Stockport and that a medal with his name was struck to commemorate the opening of the school.

Why then should this respectable man have a liaison, resulting in five children, with a stonemason's daughter from Bury—Alice Hopkinson (my great grandmother)? We shall never know. The family found it all so shameful and distasteful that two of my aunts burned all the letters and papers that would have given us the information. Why did he not marry his Alice? Had he a wife already, perhaps in South America, where he often went on business? Perhaps she was a deserted wife? Perhaps he felt a class barrier? At any rate, he appears to have been faithful to Alice, setting her up in a house in Manchester, where he had his business, providing for her amply and finally, in his will, leaving everything to her except his wine. There were five children: Alice, Mary, Ellen, Elizabeth and John, my grandfather. John Lomax died when his son was three years old.

One wonders how they were received in Manchester in their irregular set-up. We know that they became pillars of the local Congregational chapel and that all taught in Sunday school. They must have had some social opportunities, for the four girls married comfortably.

According to my mother, 'They looked like duchesses, with very good bone structure of the face, but their hands and feet revealed the cloven hoof' (she was always dry). One, Elizabeth, married a solicitor called Rooker, who, for public services, had a statue of himself in Plymouth; the statue exists to this day.[2] Another, Ellen, married Mr Tubbs, a clergyman,[3] and Alice married H. O. Wills, the famous tobacconist.[4] Mary married Charles Tubbs, a dentist in Plymouth.[5]

My grandfather, John, seems at an early age to have taken on responsibility for his mother and sisters.[6] He was extremely religious, with a hard-working sense of duty that excluded much humour or joie de vivre. He had to live down the circumstances of his birth and make good, and he did. His childhood was dominated by his mother, who was a kind and intelligent woman, and there was money for first a nanny and then a governess to look after the five children before he went to school. When he was sixteen, he became a 'gentleman apprentice', bound for five years to a Manchester firm of engineers. He had obviously a strong bent towards engineering; he passed this trait on to his sons, grandsons and great-grandsons (my eldest son is an engineer), and to great effect. After his apprenticeship he had enough capital to buy himself a partnership in the firm which came to be Wren and Hopkinson. The work included inspections, investigations, the erection of large mills and works and the designing of machinery.

Soon after this he married Alice Dewhurst of Skipton.[7] The Dewhursts made cotton, and to this day reels of cotton that one buys bear the name Dewhurst. They were married for over fifty years and had thirteen children, of whom my father was one of the youngest. My own picture of grandmother is that of a very old lady with white

[2] Alfred Rooker (born in Tavistock in 1814). He studied law and was articled to Mr Bridgeman of Tavistock, head of the best-known law firm in the town in the middle of the nineteenth century. He then moved to Plymouth, where he established a practice. He became much involved in the local government of his adopted city, becoming an alderman in 1848 and the Mayor in 1851.

[3] George Ibberson Tubbs (1813–1893), born in Mildenhall, Suffolk, married Ellen Lomax Hopkinson (1817–1900).

[4] Henry Overton Wills III (1828–1911) was the first Chancellor of the University of Bristol.

[5] Charles Foulger Tubbs (1816–1914), born in Mildenhall, Suffolk, married Mary Lomax Hopkinson (1822–1866).

[6] This is John Lomax Hopkinson (1824–1902).

[7] Alice Dewhurst (1824–1910) married John Hopkinson in 1848.

curls and a cap, lying in a huge four-poster bed, the light subdued in the room by pale-green venetian blinds. As children, my sister Enid and I were pushed towards the bed and given one end of her ear trumpet, which was rather like a bicycle pump. Into this in turn we recited a hymn at breakneck speed—and never to her satisfaction. Either we spoke too softly or we hurt her ears by shouting. Then we were given sixpence each and retired thankfully to eat boiled mutton and caper sauce downstairs. She was the only one of my four grandparents alive when I was a child.

But in her heyday grandmother was a character and ruled her family by the most authoritarian and possessive methods, by which she interpreted Christian principles. She gave her children unbounded affection and admiration, especially her sons, so that they came to depend on this assurance to such an extent that later on they looked for it unstintingly from their wives. In return she demanded constant attention, obedience to her wishes, regular letters when away and general mother-worship, which made it hard for future daughters-in-law. The poor daughters of the family were sacrificed to her indifferent health, brought about by incessant childbearing. They had to help with the younger children, sew, mend and spring clean. There is little record of any fun, parties, music or theatre, or of interesting and exciting people coming to the house. The carrying-out of an almost fundamentalist form of Christian life was paramount. Both grandparents were completely consistent in their piety.

They were devoted to and dependent on each other. At the same time they seem often to have gone away from home independently, grandfather on his engineering business, and grandmother, with or without a child, to spend several weeks 'resting' with friends or relations. Then, letters passed almost daily, full of love, affections and piety but often expressing on both sides great depression and poor spirits. This depression has been inherited by many of us and is even spoken of by their grandchildren as 'Hopkinson depression'. They were extremely detailed on both sides, those letters. If, for instance, one of the thirteen children was unwell, being weaned or cutting teeth, my grandmother would generally record it. There were many religious reflections and many incidents of how her responsibilities weighed upon her. Annie, the nurse, had staggered her by asking for a rise in wages from £11 to £12 a year. 'Ignorance, impudence and self-conceit pointed every sentence'. But grandmother prayed that she might repent, and poor Annie said no more. My grandfather's letters after some paragraphs devoted to cheering his wife's spirits were devoted to telling her about his work, and he worked very hard. It seems to me that in living down the stain of illegitimacy, grandfather for his part had to strive too hard and attempt too much.

As grandfather's career became established, he turned to civic matters. For about forty years, he was a member of the Manchester City Council. He served on gas, street lighting, rivers (of which there were several in the city) and the watch committee. This last was also responsible for the fire service, and there was no notable fire in the city at which grandfather was not present to see the workings of the service for himself. In fact, he invented some special fire-lighting appliance which he called the Grinnell sprinkler.[8] Later he was Mayor though he accepted this with some anxiety, as his illegitimate birth was raised and his acceptability queried in some quarters at first. He says himself that he had 'never been disposed to prominent public positions', on account of his distrust of his own powers and of his wife's sensitive, shrinking nature. However, my grandmother said loyally that she was 'prepared to stand by her husband anywhere and at any cost'. He took counsel from his friend the Reverend James Griffin, who told him that he had 'not a shadow of doubt that he would be able, with Divine assistance, to fill the office of Mayor with dignity and effect' and ended by writing 'you cannot decline to be Mayor'.[9]

As Mayor he interested himself in the art gallery and was the first chairman of that committee; he instigated various structural improvements there. I think that it was while he was Mayor and certainly when he was the chairman of the art gallery, that he made the acquaintance of Sir Philip Cunliffe-Owen, my mother's father, who, as the director of the South Kensington Museum, came up to advise Grandfather Hopkinson about pictures, thus making the first link between the two families.[10]

By this time my grandparents lived in a spacious house, Grove House, where the present Whitworth Gallery stands, with a lovely

[8] It is unlikely that John Hopkinson actually invented this himself. Frederick Grinnell (1836–1905) was a pioneer in fire safety and was the creator of the first practical automatic fire sprinkler.

[9] Rev. James Griffin was the first pastor of Rusholme Road Independent Chapel.

[10] Francis Philip Cunliffe-Owen (1828–1894) was an exhibition organiser and a director of the Victoria and Albert Museum, London.

garden and paddock for an Alderney cow (Norah) to keep this large family in milk. At table, nothing less than a baron of beef, my father used to say, fed the family. In summer they made a mass exodus to the seaside, with two nursemaids, tin baths, perambulators and the general equipment for thirteen children.

I imagine that my grandmother had little time or energy to play with the children, although she did read aloud to them and certainly instructed them in Bible knowledge. Grandfather fostered their hobbies, such as stamp collecting, and gave them a love of nature, showing them butterflies, birds and plants when they went to the country. At home he had a small store room at the top of the house, filled with simple apparatuses, so that his children could take an interest in natural science. There was a glass cylinder and rubber for generating electricity, the Leyden jar which was charged from it, a stool with insulating glass legs, on which they sometimes stood, so that their hair was uplifted as the cylinder revolved. There were magnets, retorts, test tubes, blowpipes and cog-wheels, which the boys fitted to a crane and thus lifted their toys from the basement to the attic. No doubt the eldest son, my Uncle John, who was to become such a brilliant engineer, took the lead in these scientific ploys and was the one who, becoming more ambitious, decided that a galvanic battery was needed.

Alfred (the second son) reports that grandfather provided the cells required; they were cast in his works and, shortly afterwards, the two boys went there to wind coils for galvanic machines so that they could use the whole apparatus.[11] Alfred says: 'One of the early uses of the batteries and coils so made was in the cause of charity, to provide entertainment at bazaars for Church or Sunday School. Someone clutched each of the handles of the machine and then a circuit was formed of young men and maidens, sometimes old people too, taking hands which were vigorously and affectionately squeezed under the influence of the galvanic current. This current was made stronger and stronger as long as the arms and fingers could bear it. The apparatus was then a scientific novelty and adapted in this special way to human needs'. No wonder all the sons were scientifically orientated.

Grandfather's finances at some point failed, and the family were for a time in reduced circumstances; he suffered for a year such mental depression that a male nurse had to be installed. However, he

[11] Alfred is Alfred Hopkinson (1851–1939).

recovered and, when he retired from business, he became a consultant engineer and the family moved to a fine house in Bowden outside Manchester and which was their home for the rest of their lives.

All the Hopkinson aunts and uncles had the same physical characteristics. They had fine regular features, piercing but watery blue eyes, and skin that tended also to take on a blueish colour in cold weather. They had poor circulation resulting in a hereditary drop on the nose, which unfortunate trait many of the descendants exhibit in cold weather. All were remarkably thin and, like the Biblical cattle, however much they ate, they never grew fatter. Their legs were remarkably long and they had a peculiar, lilting walk, inherited even by some of their grandchildren. This made them adept at walking, climbing and cross-country running. Though all to the good in the male of the clan, it was less acceptable in the female members and definitely unattractive in evening dress. My parents sometimes played a curious competitive game about their respective families' appearance. My father would say provocatively, 'The women in the Cunliffe-Owen family of course incline to embonpoint, my dear', to which my mother would reply with spirit, 'Well, your sisters are so thin it wouldn't matter if you turned their heads back to front.' 'Also,' teased my father, 'your family has no features (gently stroking his own patrician nose), though I would agree they have good skin, eyes and hair.' My mother then closed the game by saying, 'The Cunliffe-Owens may not have features, but the Hopkinsons have skin, especially their necks, like boiled fowls.'

Their character and personality bore the stamp of their upbringing. They had a tremendous sense of duty, a strong evangelical religious feeling, a passion for hard work, and self-discipline (they usually took cold baths till the age of eighty). They had their parents' emphasis on service to the public, so much so that members of my own generation have been asked, 'Are you a civic Hopkinson?' Personal relationships were not their strong suit, for they were reserved, inclined even to be withdrawn except within their own domestic circle, though they were not without a considerable rugged charm. Their sense of clanship and loyalty was very strong and, should remote cousins be reported in their vicinity, they would make every effort to make contact. Of the arts, painting, theatre and music, they knew little. Occasionally, they went to the Hallé Concerts in Manchester but one does not hear of music in their home and, in the chapel that they attended, 'no organ was allowed as long as any member of the congregation objected'. As to books, they were not allowed to read novels when they were young. *Robinson Crusoe* and *Swiss Family Robinson* passed. My Uncle Alfred records that there was no Shakespeare in the house, but poetry, read aloud and learnt by heart, was a lifelong pleasure to the whole family. Milton, Longfellow, Tennyson, Wordsworth, Coleridge, Browning and many others found great favour. I might add here that my father was reading Tennyson aloud while my mother was expecting my sister, who thus came to be called Enid. I fear had she been a boy she would have been Geraint, which does not go well with Hopkinson somehow. It could not be said that they were in the least cosmopolitan; they had that strong Lancashire characteristic: a distrust of too much charm and of the graces of life. The uncles admired pretty women in a rather nervous way, as if they feared too much might be asked by the women of them. When I remember my Uncle Alfred saying that the best-looking girls in the world could be seen in Oxford Road between the hospital and Manchester University, I feel inclined to query their standards. From heredity and upbringing, sex, I should guess, was sublimated, and had I heard of the mildest affair connected with any aunt or uncle, I should have been astounded. (How different my mother's family!)

My uncles had to achieve whatever they set their minds to and, if possible, they had to set records. If one climbed a mountain with one of the uncles, one had to, however tired, reach the top and, if possible, set a time record for one's age-group. Stopping to pick bilberries or heather, or even to look at a view, was not encouraged. Rock climbing was my uncles' passion. At one time, five of the uncles were members of the Alpine Club, and all over the English Lake District are chimneys and stacks called after them because no one had climbed them before. 'Abroad' to them was Switzerland, and Switzerland was the Alps. In these recreations they showed a glorious sense of adventure, that, though one must admire, could and did lead to tragedy.

But as Professor James Greig has said in his biography of my eldest uncle, John, 'the Hopkinson family rather ran to memoirs'.[12, 13] The

[12] James Grieg was William Siemens Professor of Electrical Engineering at King's College London.

[13] The biography is entitled *John Hopkinson: Electrical Engineer* (Science Museum booklet, 1970).

Hopkinsons have written at least four books about each other. (The Cunliffe-Owens wrote nothing; they enjoyed living too much.) For this reason I shall not give an equally detailed description of all my thirteen aunts and uncles but only spotlight a few of the most colourful in the group. The spotlight will be uncertain. I can only pass on my own adolescent impressions and what I have heard from my parents and from others who knew the family. These aunts and uncles made a considerable impact on my personal family life, and I was very fond of most of them, and their children—much fonder indeed (with certain notable exceptions) than of the Cunliffe-Owen relations, no doubt because the Hopkinsons were much more interested in me and my brothers and sisters than the Cunliffe-Owens were.

3.2 Aunt Mary and Others

John was the eldest of the Hopkinson uncles, and twenty-five years older than my father, who came near the end of that large family. I never knew Uncle John, as he died the year before I was born, at the early age of forty-nine; but since his widow, my Aunt Evelyn, his daughter, Nellie Ewing, as she became, and his granddaughter, another Alice, have been warm friends throughout my life, and his youngest son, Cecil, was to become my husband's greatest friend, I must record something about my Uncle John from what I have been told by my father.

Uncle John had a brilliant career as an engineer, carrying all before him at Trinity College, Cambridge, where in his finals he was Senior Wrangler. He also had a fine record there for both rowing and running. His college made him a Fellow, not without difficulty as Dissenters were not eligible. And it needed a change in college law, with the sanction of the Queen's Council to enable him to hold a Fellowship. He was called to the Bar and was subsequently much in demand as an expert witness, for which he received very large fees. But it was as an engineer that he became famous. Indeed his biographer, Professor Greig, describes him as 'one of the greatest electrical engineers of his day'. Years later in Russia I bathed in reflected glory from Uncle John, when during a visit to a factory my husband happened to say that I was a niece of the man who invented the Hopkinson dynamo, which the Russians were using there.

Personally, I gather he was a somewhat austere and withdrawn man. He himself wrote to his mother, 'I think there must be in my nature something that makes people not at home with me.' Certainly my mother, when she entered the family, found it difficult to get on with him, for he had no small talk.

Like all my Hopkinson uncles, he was an expert rock climber and went to Switzerland every summer. In 1898 this led to a tragedy which shocked the country. He and two daughters and a son were killed climbing at Arolla. Half the family were wiped out; there remained the widow, Aunt Evelyn, Bertram (to become Professor of Engineering at Cambridge and father of my friend Alice), Nellie and Cecil, all of whom come into my story.

Alfred, the next uncle, was never very popular with my family; though he lived near us throughout my childhood in Manchester, we hardly ever saw him. He, like his brother John, had a brilliant university career, but at Oxford instead of Cambridge, and he read law not mathematics. Thus, he contributed to the family passion for achievement and reaching the top of appropriate trees. A member of Lincoln's Inn, of which he became a bencher and subsequently treasurer, he practised at the Chancery Bar, was twice elected to Parliament and was Vice-Chancellor of Manchester University.

While holding that office he once did something for us that gave us great pleasure, and that was to send us tickets to enable us to watch the arrival in Manchester of King Edward and Queen Alexandra, from a window in the university museum. My sister Enid and I had new pink smocks embroidered with honeysuckle. We greatly enjoyed seeing the royal landau stop outside the university buildings and watching the King knight our Uncle Alfred.[14] The royal party went on to open the new infirmary close by, and Enid and I were put out that if the King was knighting uncles he missed out our beloved Uncle Charlie (of whom more presently). Charles Hopkinson had been engineer to the Royal Infirmary.[15]

Alfred's wife, our Aunt Esther, we seldom saw.[16] Actually, she had been governess to the Hopkinson family, but I understood that she was rather better born and later, as a result of some legacy, had

[14] This was in 1910.

[15] Charles Hopkinson (1854–1920).

[16] Esther Wells (1846–1931) was born in Nottingham.

some independent means. Occasionally, she paid a formal call on my mother, arriving in a 'victoria'. I have the impression of an elegant woman with a feather boa, pale veilings, a black velvet band to support her neck, and a lace parasol. She spoke in a very soft refined voice and smelt seductively.

Now we have Charles Hopkinson, a great favourite with all nephews and nieces, and particularly of my own family. He was a consulting engineer in Manchester, a partner in fact with my grandfather. Academically he was less dazzling than his three brothers and did not go to the university; whether because he was not up to scholarship standard or because at that particular time his parents could not afford it has never been clear. My mother always insisted that not having been to Oxford or Cambridge 'made him so nice'. As an engineer he had a great deal to do with the Manchester Royal Infirmary, and I remember his telling me that he with his brother Edward had laid the tram lines in Belfast. I was always amused that, when the Great Wheel at Blackpool stuck, with the couples in the air in their little cars, Uncle Charlie was called in to advise.

He found his wife, our Aunt Mabel, in Belfast.[17] Their home, the Limes, was only a mile or so from my family in Manchester, and we often went there. Although they had no children, they were very fond of them and had 'young people's parties' at which we played paper games and shu-vette, where on a large circular board one flipped draught-like discs.[18] I have never seen this game anywhere else, and to this day it is popular with my grandchildren. At dinner we invariably had jugged hare, and hot chocolate sponge pudding with whipped cream, using tablecloth and napkins of the finest, as Aunt Mabel's family in Belfast made linen. She was meticulous in the ordering of her house: breakfast at 8 o'clock, no one could stay in bed unless ill, and one could never have coffee. My mother caused consternation by washing her hairbrushes after the housemaid had polished the bathroom taps for the day. In her outlook on life too she was rigid, and my mother sometimes called her (to her face) 'my Ulster mule', but strangely enough Aunt Mabel took this as a compliment. In appearance she was forbidding, but this concealed a kind heart. Her great interests in Manchester were the Gentlewomen's Association and the

[17] Aunt Mabel is Mabel Campbell (1863–1942).

[18] Shu-vette is otherwise known as crockinole.

Princess Christian College for Childrens' Nurses. As the nurses could not enter for training unless 'ladies by birth and education', Aunt Mabel spent much committee time considering this question and was outraged by a newcomer's suggestion that there might be merit in taking ladies by birth or education.

This uncle and aunt were great gardeners; but, as everything in those days was very dirty around Manchester, nieces and nephews were employed on Sunday afternoons to scrub the rhododendron bushes with soap water. Uncle Charlie, like all the Hopkinsons, was irritatingly accurate and used to tease my mother. Once in the garden she told him there were hundreds of sparrows in a tree. 'Count them, Olga,' he said, chuckling.[19] My poor mother only got up to ten. 'Cunliffe-Owen exaggeration' said my uncle, but they were great friends. He had a gentleness and sensitivity unusual among the Hopkinson brothers. When I was born he apparently bought a bouquet of roses and, when invited to admire me, went straight to my mother, saying, 'Old friends first!'

Charlie danced, played golf, was the best rock climber in the family and sketched. True, he painted almost entirely mountains, or scenes from a bedroom window, but he had some feeling for colour. Unfortunately, he was exceedingly deaf, a legacy from scarlet fever; but the family were so fond of him that everything was always repeated to him in a loud voice. As a girl, I used to hate being left in a silence, shouting some joke or worse some platitudinous remark several times to Uncle Charlie, but it gave me a great sympathy for deaf people. He did not share the Hopkinson passion for writing their memoirs or autobiographies, but I believe he had a happier life than most of them and certainly he was much loved.

I remember my Uncle Edward, who came next in order, with affection and gratitude.[20] He was an electrical engineer and, after a scholarship and splendid record at Cambridge, he joined the firm of Mather and Platt and later became the vice-chairman of the Chloride Electrical Storage Company. He lived outside Manchester in a large Victorian house, 'Ferns', where we were always made welcome by him and Aunt Minnie, 'the little Aunt'. She was very small; my mother spoke of her as piquant, and she was Irish—in fact, she was the sister

[19] Olga is Olga Hopkinson, née Cunliffe-Owen (1866–1940).

[20] Uncle Edward is Edward Hopkinson (1859–1922).

of Uncle Charles' Mabel. Life at Ferns has been described by their only daughter Katharine (now Lady Chorley) in her book *Manchester Made Them.*

I said that I remembered them with gratitude and affection. Uncle Edward several times sent generous cheques for our best holidays and, at a lower level but equally welcome, boxes of his hothouse peaches and other garden delicacies would be dropped in upon us. My relationship with him was particularly friendly. He liked a bit of dash, and I suppose I had that, and I was sorry when I lost ground with him in my college days after reports reached him that I was very wild. However, I was reinstated when I became engaged to someone who was so completely acceptable to the Hopkinson clan.

I suppose before I consider my six Hopkinson aunts I ought to say that there were two more uncles who died in early childhood. Several daughters died young also; the family was scourged by scarlet fever which, when it did not kill, left several of them deaf. I will resist the temptation of describing all six and will mention only the two who made an impact on my childhood.

First, my Aunt Mary, who was a great character. In appearance she was masculine, big-boned, big feet and hands and watery blue eyes. She had no clothes sense and wore shapeless coats and skirts and sensible shoes. Had she lived in more modern times, she would certainly have been to a university and had an interesting career. As it was, she dedicated her life to her parents, especially her mother. She was the purveyor of news within the clan and had to be told everything first. My father stood slightly in awe of her. My mother found she had a sense of humour and, as she used to say, 'stood on a broad base'. I was very fond of her and as a teenager enjoyed trying to shock her. I did not succeed. When I shouted a tale of one of my escapades down her ear trumpet (she had the Hopkinson deafness), Aunt Mary would say, eyes twinkling, 'but that would have been a somewhat injudicious course of action, would it not?' The account of any sort of love affair she enjoyed greatly. I used, as a girl, to spend the night with her at her house in Bowden, near Manchester, where she welcomed me with outstretched arms and reminded me of one of those stuffed bears on its hind legs, holding a salver. The only thing I did not like was her beloved deceased collie dog, Bell, which had been skinned and made into a rug in front of the spare-room dressing table. Aunt Mary, a strong nonconformist and inevitably a Sunday school teacher, held prayers before breakfast for Annie and Julia, the two retainers and any guest present. Julia occasionally burst into uncontrollable laughter at certain parts of the Bible reading and had to be suppressed with an authoritative 'that will do, Julia'. Often my aunt explained or paraphrased the 'portion of scripture' as she went along.

Police were my aunt's hobby or, to put it more accurately, her life's work. It would not have been so remarkable today but, a hundred years ago, it was a curious form of welfare work for a young woman to adopt. My aunt started her activities in her early twenties by holding a Sunday school for young policemen in Manchester. She then organised a police orphanage and a police benevolent society and embarked on an endless round of visiting members of the force at their homes, in hospital or wherever help was needed. Her conversation had to be interpreted in terms of police. Reference to a station, a superintendent, a sergeant or headquarters, when made by my aunt, could only mean police. All her relatives knew of this obsession; she was, it was often said, 'wedded to the police'. In fact she never married. She could have married her first cousin H. H. Wills, of the tobacco firm (in each generation a Wills wanted to marry his Hopkinson cousin) but she refused him.[21] We were very proud of her and I remember, as a child, my father telling us at breakfast one day that Aunt Mary had been co-opted on the Manchester Watch Committee and that a woman had never served on it before. Everyone seemed so impressed that we children did not like to ask what my aunt would watch but we felt sure it could only be the police. From then on it was fun to go about with her in the city and be saluted by policemen and handed across the road, with my aunt holding up her long braided skirt and showing a flannel petticoat.

Towards Christmas time she would start on one of her major tasks for the police, the sending out of about three thousand personal cards. I can see her now, settled in her sitting room, which was all sage green and dark red, and the table with a velvet cloth fringed with bobbles and covered with cards, texts and copies of her own specially printed Christmas 'message'. All three items had to go into each envelope.

[21] Henry Herbert Wills (1856–1922) was a tobacco manufacturer and philanthropist in Bristol. He funded the H. H. Wills Physics Laboratory at the University of Bristol. He was married to Mary Monica Cunliffe-Owen (1861–1931), an aunt of Lady Bragg.

She was assisted by the cook, who in cap and apron sat sorting and folding but was forbidden to address the envelopes. Annie was with my aunt for fifty years and was allowed certain liberties, such as an occasional joke about police, which my aunt would not have tolerated from any other quarter. When my aunt went to the railway station to carry out her various missions in Manchester, Annie would hold her bicycle while she mounted, hand her an opened umbrella in wet weather and, with a gentle push, send her on her way.

A short time after the start of my marriage, when we were living in Manchester, I most mistakenly decided to take advantage of my aunt's unique position with the police. One Sunday morning when we were away, the wall of our garden was breached, apparently by some heavy vehicle. I wrote to the Chief Constable, casually mentioning whose niece I was, and asked his help in the matter. Within twenty-four hours, the act had been pinned to the driver of an ice-cream van. Armed with his address, we confronted him in his shop, with the result that he and his mate came over and mended the wall. Delighted, I made a good story of this to my aunt but she was not amused; indeed, she was very angry. I was only restored to favour by giving a tea party in our garden to a selected group of sergeants and wives; they arrived, with my aunt at the head of the group, in two busloads. Smoking was not permitted in her presence but I think she knew quite well that my husband escorted some of the guests to the bottom of the garden for a quiet smoke behind the bushes.

There was no question that the police respected, admired and, I think, loved my aunt. On one occasion, when she was crossing the road, she was knocked down by a tram, and it was difficult to persuade the police that this accident was in no way the fault of the driver. The force was shocked. I was driving my father to the hospital to visit her one day when I was stopped by a policeman on point duty. I had that instant feeling of guilt that grips one on these occasions and waited. Slowly the man walked to the window, his face grave. 'How's your auntie?' he asked anxiously.

After fifty years of devoted service, the police decided to mark the occasion by giving my aunt a party and a presentation. This, it was generally understood, would be the moment for her retirement. A watch bracelet and radio, suitably inscribed, were given to her and there was an air of farewell about the occasion. But the police had reckoned without my Aunt Mary. She rose and, thanking them all heartily, said that this tribute had put new life into her and she now felt ready for another ten years work with them.

Time passed, and my aunt was approaching ninety when she confessed to us that she wished that she might witness her own funeral. There would undoubtedly be a police band, police singing her favourite hymns, and the Chief Constable reading the lesson. Since her active participation in the event was out of the question, she went over the ceremony in detail with the Chief Constable and derived great pleasure from discussing arrangements which were, in due course, carried out according to plan. She was indeed a Hopkinson of great stature.

Now we must look at Aunt May, the youngest, as we saw a good deal of her and her four children, who were more or less the same age as we were, and I am now very fond of one or two of her grandchildren.[22] She was always referred to in our family as 'the little playmate', as she was my father's constant companion in that large family. She and her husband, Uncle George Anson, had, near Keswick, a fine house where we all used to stay. Uncle George was a quiet, inoffensive business man but my sister Enid and I bore him a deep grudge. In the first war he made clothing of some sort, equipping the troops. As we were the poor relations of the clan, Uncle George sent a roll of flannel, presumably for soldiers' underwear, to make school nightdresses for Enid and myself. It was a pale duck-egg blue, thick and very scratchy, and no amount of lace or feather-stitching on my mother's part redeemed them. We hated to be different but managed to turn them into a joke; they were known at St Leonard's School as 'Uncle George's nighties' and are still remembered by contemporaries.

Aunt May was sweet and gentle natured, if a little prim. I fell from grace with her for having a mild affair with her youngest son, Ellis, during the war. He and I had fun dancing and doing things together when he was on leave, but she regarded me as a frivolous teenager and wrote me a letter, warning me against 'playing with fire'. In those days 'fire' would not have the same definition as nowadays by a long way.

Albert Hopkinson, my father, came tenth in the family, but I have left him till last, because with his marriage to my mother this story runs on naturally into that of the Cunliffe-Owens, her family.[23] He

[22] Mabel (May) Hopkinson (b. 1864) married George Anson in 1890.

[23] Albert Hopkinson (1863–1949).

had the Hopkinson appearance with fine bone structure and regular features and he had his share of the family characteristics, though with all his edges, as it were, softened. His mother wrote of him, as a young man: 'In Albert there is no guile'. He was simple and humble-minded, subdued no doubt by being the youngest son and brought up not only by his parents but by his formidable elder brothers. They threw him into deep water to make him swim, chased him with frightening pictures and generally disciplined him. They must have tempered this teasing with kindness, for he had an admiration and reverence for them and indeed, for authority generally, that remained with him throughout his life. This sometimes goaded my mother into assuring him that in the sight of God he was worth the whole family put together. (Typical of a Cunliffe-Owen to speak for the Almighty.)

He went to Cambridge, to Emmanuel College, but did not try for a scholarship, the family being under less financial strain than at the time when his elder brothers were being educated. Like them he was a fine sprinter and jumper and rowed in his college boat. He was also a competent rock climber and a member of the Alpine Club. He had always wanted to be a doctor and would have liked to specialise in diseases of children but when the time came there was not enough money to support him as a specialist while he waited to establish himself as such. There was a choice of practices for him as a GP either in Weymouth or in Manchester, and no doubt his parents steered him towards his home town. He took a partnership in Withington, South Manchester, and there he started and then worked for thirty years. The story of his life will have its place later, interwoven with that of the family.

He had his full share of Hopkinson recklessness, oblivious to danger. I remember once two dogs were locked in deadly combat, to the agony of their owners. My father, arriving on his bicycle, shouted to the gathered crowd to stand clear and rode full tilt into the dogs. Frightened, they parted and slunk off to their respective owners.

Although a good and even saintly character, he had flashes of temper occasionally when the family blue eye became glassy, and the effect would last for hours. Once when a drunken patient, whom he was trying to help, called him a liar, he hurled him down the front door steps. Again, and perhaps a more curious example, his anger was roused when a patient's husband refused to let her breastfeed her baby, as it would keep her from the hunting field. My father and the baby won that round.

Like the rest of the family, he strove towards achieving a certain goal but it was a different one from theirs. It was not wealth, getting to the top of any tree, or power; it was to be a good doctor, alleviating human suffering and being his patients' counsellor and friend. When he retired after thirty years of general practice, he was offered the post of demonstrator by his old professor of anatomy at Cambridge University; and budding doctors, many of whom gained great distinction in their profession, learnt their anatomy from 'Hoppy', as they affectionately nicknamed my father. When he died, the Master of a College wrote to me, 'He was a fountain of love', not a phrase lightly employed by a Cambridge don.

3.3 My Mother's Family

My maternal grandfather (later Sir Philip Cunliffe-Owen) was destined to follow his father's career and joined the navy at the age of twelve, in 1840. However, the life was too strenuous for his health, and after five years he had to retire, but not before he had seen quite a bit of the world. It then appeared that he did nothing for a time, spending 'several years in idleness travelling with his father' (I quote from my mother's diaries). It was while doing so that, when staying at Vevey in Switzerland, he met his future wife, my German grandmother Jenny von Reitzenstein.[24] She was staying with her uncle Baron Henry von Roeder (the German minister to Switzerland). There is a legend that my grandfather walked through the snow in bare feet to court her. One wonders why! They were married and then returned to London.

My grandfather, and indeed my grandmother, died before I was born but I have a vivid picture of him; he was rather stout, with a fair beard and twinkling blue eyes. He exuded geniality and charm. He made friends with everyone and had that gift of never forgetting a face, name or date. Extremely kind, he would help anyone, as is apparent from his correspondence. Never very well off himself, he was always generous with what he had, and his pockets always held sweets for children, whom he loved. He did not drink, smoke or gamble and he was very religious. Although he left school at twelve to join the navy, he must have been a cultivated man and spoke several European languages. Later he was to attract to his home the painters Alma

[24] Jenny von Reitzenstein (1830–1894) was born in Potsdam, Prussia, and was the daughter of Baron Fritz von Reitzenstein.

Figure 49 Baron Henry von Roeder. (Courtesy: Patience Thomson)

Tadema and Leighton; the musicians Strauss and Stainer; the writers Tennyson, Meredith, Watson and George Gissing; and generally interesting people from every walk of life.

The exact nature of his work immediately after the start of his marriage is not known, though my grandmother later on refers to their life in Ealing as the time when 'he was a little clerk'; but he must have worked in the newly formed Science and Art Department of State. This would be in about 1853, shortly after the Great Exhibition in 1851. Henry Cole (head of the department) had his eye upon the young clerk and, eventually after serving in various subordinate positions, grandfather succeeded him as the director when the department had blossomed into the South Kensington Museum, forerunner of the Victoria and Albert.[25] One thinks of a director of a museum nowadays

[25] Henry Cole (1808–1882) was an English civil servant and inventor who facilitated many innovations in commerce and education in nineteenth-century Britain. Cole is credited with devising the concept of sending greetings cards at Christmas time, introducing the world's first commercial Christmas card in 1843.

as being an authority on china, pictures or scientific exhibits, whatever it may be, but it is clear that my grandfather was not. He knew however who was, and he seems to have had a genius for using experts and appointing the right people to the museum's different departments. He also had excellent natural taste.

His greatest contribution was to be in the field of the international exhibitions which sprouted up all over Europe in the nineteenth century. From 1855 (the date of the first foreign one, in Paris) onwards, my grandfather played an important part. Perhaps the two greatest were the Vienna Exhibition in 1873, and the second Paris one in 1878. The Prince of Wales (afterwards Edward VII) was president of the British section for both events, and grandfather was the secretary. The Vienna Exhibition was a dazzling affair with a wonderful site along the banks of the Danube. Unfortunately, the city had a bad outbreak of cholera at the time, and my grandfather received a letter from one Henry Lennox asking him to 'book a room in a good hotel for the Exhibition, but preferably not one in which a cholera patient has died'.[26] The scope of these exhibitions was limitless, to judge from the letters that my grandfather received.

The Paris Exhibition of 1878 followed the same pattern as Vienna. This time, my grandfather and grandmother had a house in Paris in the Rue de Mars and he was more or less absent from the museum for nearly a year. This was not frowned upon; grandfather was excellent at delegating and when necessary travelled back and forth. About this time, he was made KCMG (Knight Commander of the Order of St Michael and St George). As usual, he worked extremely hard and seems to have seen to everything himself. Princess Alexandra, in recognition of his services, sent him a set of studs.

There was one last big exhibition for grandfather to play a part in and that was in 1886, the Colonial and Indian Exhibition. Again, the Prince was president. Sir Arthur Sullivan was invited by grandfather to play at the opening ceremony, and there is a letter of thanks from him for the bestowal of a royal badge.[27] At the end of this exhibition, held in the International Galleries at South Kensington, he was made

[26] This is probably Henry George Charles Gordon-Lennox (1821–1886), known as Lord Henry Lennox, a British Conservative politician who sat in the House of Commons from 1846 to 1885 and was a close friend of Benjamin Disraeli.

[27] Arthur Seymour Sullivan (1842–1900) was an English composer, most famous for the Gilbert and Sullivan operas.

KCB (Knight Commander of the Order of the Bath). He received this information from the prince in the following letter:

Dear Sir Philip Cunliffe-Owen,

I have much satisfaction in acquainting you that the Queen at my instance has been pleased to approve of the Honour of a KCB being conferred upon you for the valuable services which you have rendered in connection with the Colonial and Indian Exhibition.

I trust that this announcement will be agreeable to you.

Believe me,

Truly yours

ALBERT EDWARD

It is clear from the letters that the Prince of Wales wrote to him that he and my grandfather, working so closely together, came to know each other well. All of course are written in the prince's own hand (difficult to read). Some are about buying china or furniture for a royal establishment, when grandfather's help was sought. One asks him to find 'a millionaire friend to contribute to the Restoration of Wolfenden Church on the Sandringham estate'. Others ask his opinion on the suitability of men for certain appointments. There is a very pleasing one regarding grandfather's salary:

Longleat, Westminster

December 8/81

My dear Owen,

Many thanks for your letter received this morning. I quite feel and have always felt that your salary might be increased.

I had an opportunity of speaking to Lord Spencer on the subject today, without him having the slightest idea that you had either spoken or written to me on the subject. He spoke in the kindest and most complimentary terms about you and will give the matter his fullest consideration. But there are some difficulties and the Treasury is one . . .

It gave me great pleasure to hear that your visit to Berlin went off so well. The Crown Princess wrote to me how successful the opening of the new Museum had been,

With kind regards to Lady Owen

Believe me,

Yours very sincerely,

ALBERT EDWARD

We have none of his letters to the Prince but it is clear that he must have told him that he was feeling the pinch.

The result of this friendship brought him in contact with many other members of the Royal Family and there are various letters asking him to do something or other for them. There are many from the Empress Frederick, eldest daughter of Queen Victoria.[28] Quite the most interesting is one which seeks his help and influence in stopping roulette playing on the Riviera. The Empress always refers to this as the 'Monaco Cause' in her letters. It was well known that my grandfather was strictly anti-gambling. The first letter states that the Empress has read in the newspapers that the owner of the gaming tables in Monaco is dead, so 'this may be the moment to strike a blow'. Grandfather is to use his influence wherever he can, to put an end to Riviera gambling. She adds that she herself hopes to see Bismarck at an evening party shortly and will endeavour to enlist his help in Germany. Grandfather had no sense of class; it just so happens that these letters have been saved, but there were many other people whom he helped and encouraged, although there is no record.

His home life with his wife and nine children was very happy. The family lived in one of the several houses, 'The Residences', on the museum site, guarded always apparently by a policeman in those days. On one occasion as my grandparents were driving off to some state occasion, grandfather let down the window of the carriage and called, 'Constable I've forgotten my teeth. Run up to my bedroom and get them—in a glass on the wash stand.' Each child on marriage had a present from the Metropolitan Police Force. My mother had a little silver tea set. In a simple family way, for they were never well off, they entertained the most distinguished and cosmopolitan guests and made everyone feel at home. They had a summer house at Lowestoft, then a quiet fishing village. While there grandfather always took a great interest in local affairs, especially anything to do with seamen, for example, the Mission to Seamen, the Seamen's Church and the Seamen's Church Institute. He also interested himself in the Lowestoft School of Science and Art. He was a much loved figure there, and it was at Lowestoft that, full of honours from almost every European country, he died in 1894.

[28] Empress Frederick (1840–1901) was the oldest child of Queen Victoria and Prince Albert of England. She married the future German emperor, Frederick III, in 1858, thus becoming Empress of Germany and Queen of Prussia.

Figure 50 A picture of Friedrich (Fritz) Freiherr von Reitzenstein in 1843, presented to him by his officers on his retirement. He was married to Paulina von Roeder. (Courtesy: Patience Thomson)

My grandmother, Jenny von Reitzenstein, was German. Her father commanded the Prussian Garde du Corps, and her mother was a von Roeder. Both came from very old and distinguished families. My grandmother was a forceful, rather formidable character and was deeply religious. In later years, in London, she attended Holy Trinity Church, Brompton, where she sang all the hymns (with her beautiful strong voice) in German, for she was quite unselfconscious. She was a devoted wife and perfectly adaptable, exchanging an aristocratic, indeed feudal, way of life in Germany for a small house in Ealing and very small means, when she was first married. We have many letters from her to grandfather and these were all love letters, generally signed, 'Your bird'. They grieve at his long absences abroad and yearn for his return. At first she often went home to Germany with her babies but as more arrived she had to wait at home. Later of course, as he rose in the museum world and she had an excellent German governess and a nurse for the children, she often went abroad with him. Both parents seem often to have left the children for weeks at a time.

At home it was she who managed everything, keeping meticulous accounts and ordering her household with German precision. It was a very happy household, full of life, gaiety and music. She was very musical herself, and the whole family played some instrument and sang. She was a welcoming hostess not only to a brilliant and cosmopolitan group of friends but to lonely and friendless people. Capable and resourceful, when grandfather had smallpox, for example, she never disclosed this to the household but isolated him in an attic bedroom and nursed him herself. Incidentally, when Vienna had its outbreak of cholera at the time of the Great Exhibition there, grandmother stayed on to help nurse the patients.

Physically, she was a very big woman—over six feet in height—and, though not beautiful, she had a pleasing face, with twinkling brown eyes. Her hands were so large that she had to have her gloves especially made. She was particular about her clothes but a very careful and economical buyer. For example, she had one new evening dress a year but this was made for her in Paris in the Rue de la Paix. She was quite content to wear it again and again. One was of such beautiful wine-coloured Lyons velvet that, though it is over a hundred years old, I still have it; I had it cut down as a skirt and wear it with pride. Her handkerchiefs were of the finest linen, each with a seven-point coronet hand-embroidered in the corner. In London she often shopped at Harvey Nichols and, if she had a mid-morning fitting there, she would tell the manager that she expected 'a glass of hot milk and some thin bread and butter' brought to her, and this was always done. Her children were all devoted to her and she spent as much of her time as possible with them. When she had to go off somewhere with grandfather, she endeavoured to take one at least with her.

Everything about my Cunliffe-Owen grandparents and their nine children was as different from the Hopkinson family as could be but they had one strong trait in common.[29] All four grandparents were

[29] The nine children were Philip Lewis Fritz (b. 1855), Gertrude Agnes (b. 1856), Alfred Mark (b. 1858), Mary Monica (b. 1861), Emma Paulina (b. 1863), Grace Elise (b.1864), Olga (b. 1866), Hugo (b. 1870) and Victoria Mary (b. 1875).

extremely religious. Unlike the evangelical Hopkinsons, the Cunliffe-Owens veered to the High Church, several of the children becoming Roman Catholic, but both families were brought up in a God-fearing atmosphere. Here similarity ends.

The Cunliffe-Owens were sure of themselves; they were happily confident that what they did and how they behaved was right. This enabled them to be quite uninhibited. Perhaps for this reason none of them I think had nervous breakdowns or anxiety complexes. They were at ease with every sort of person and inspired devoted service in anyone who worked for them. They were rather inclined to interfere in other people's affairs and tell them what to do and how to arrange their lives; but they were kind and friendly and all very generous. They were all good organisers, so that it followed that they thought it stupid to do things themselves. Besides, they liked to be comfortable and very much disliked getting tired. There is no record of any of them taking an interest in civic affairs or being Mayors and so on; I think they would have thought that 'rather middle-class'. The Hopkinson passions for climbing, roughing it in old clothes, making records and keeping fit with cold baths in the morning would have been anathema to them. The truth is that they were cosmopolitan in outlook, all spoke French and German fluently and, when they travelled about Europe, far from climbing mountains, they sat about in cafés, went to concerts and galleries and chatted to people in foreign languages. They had no clan-feeling and did not care for 'get-togethers' of any kind, but loved tête-à- têtes. Charming, accomplished, highly sexed and full of character, my Cunliffe-Owen aunts and uncles were not intellectual and quite unscientific. They all had a great love of life and would not have entertained the idea of putting their thoughts on paper or writing their memoirs. I see what a curious mixture of Hopkinson and Cunliffe-Owen I myself seem to be.

Of the children who were our Cunliffe-Owen aunts and uncles, only one, Monica, impinged on our lives; and there is a great deal to be said about her, as we saw so much of her.[30] She was a special aunt, since my parents and family spent every Easter holidays at her home until I married. She played a considerable part in my life. She was, as her future husband recorded in his diary, 'not beautiful, but bright and vivacious with a very good figure'. Attractive to men, she was in Victorian

[30] This is Mary Monica Cunliffe-Owen, later Mrs Wills (1861–1931).

language 'something of a flirt' as a girl. Typically Cunliffe-Owen, she spoke several European languages and played and sang well. Johan Strauss was among her friends, and a letter has come to us from her saying, 'Strauss himself (in 1885) not only played the Blue Danube specially for me several times at the Albert Hall (the year of the Musical Exhibition) but I have a light blue satin box with his waltzes, given me by him with his autograph, and presented with a bouquet'.

I remember as a child telling my sister it was a pity that she sang so out of tune in the village church. Really, of course, she was singing alto. She married my father's first cousin Harry Wills, whose mother Alice was one of the illegitimate Hopkinson daughters, so we were doubly related. He was well off when they married and became very rich, and they were extremely generous to us all.

Aunt Monica was a curious mixture of tastes. She was genuinely as pleased with a sixpenny present from one of us as with a piece of jewellery (which incidentally Uncle Harry never gave her till their silver wedding, when she had a string of pearls—actually two strings, because she could not decide between them). On the other hand, she felt a car was a necessity that no one should be without and dished them out to poor clergy and relations at the rate of about three a year.

She was quite unathletic but, though frightened of horses, when first married she rode around her home in Somerset. There is a delightful story of her setting off to return a call on horseback. No doubt she was helped to mount at the start, but she could not dismount when she arrived at her destination. But her friend saw her from a window and the call was paid, Aunt Monica remaining on the horse while tea was served up to her on a tray.

Her married life was reasonably happy, although husband and wife had not many tastes in common and there were no children. The suggested cure for this in those days was a trip round the world, and this was taken, but to no avail. Otherwise, they seldom went abroad and, if they did, Uncle Harry took England with him and demanded roast beef and Yorkshire pudding in Egypt in the hottest weather and only wanted to see engineering works. All his life he worked the day long and had no idea of amusing her or himself.

Left alone so much, she became deeply religious. This had a practical side, for she built churches and paid the expenses of young men entering the church. She became something of a mystic and had her own oratory, to which she would retire to meditate and say the offices of the day. Bishops and clerics often stayed at her home. Uncle Harry himself had quite a puritanical streak in him and actually played the organ in church. Clergy in the house sometimes irritated him and he one day referred to his wife as pious Monica. We did not like this and, through one of those complicated family jokes connected with her love of gardens, changed it to Pyrus Japonica; and from then on, my brothers and sister always called her Japonica. After my uncle's death, she was created a Dame of St John of Jerusalem for, I believe, her war services and her generosity to Bristol City. I have been told that she was the first person ever to be addressed as 'Dame'.

One cannot write about her without describing Uncle Harry, who was also an important figure in the lives of me and my brothers and sister. Harry Wills was one of the seven sons of Henry Overton Wills of tobacco fame. He was a very shy, withdrawn man, who found it difficult to make easy relationships with people. Always bent on engineering, he went into the tobacco works as soon as he left school, refusing to go to Cambridge, as he feared the competition of his brilliant Hopkinson cousins, so he said. He lived at home till he was married, a home strongly influenced by his puritanical Hopkinson mother (sister of my grandfather), with no theatres, parties or dances. Then, on a visit to his Uncle Frederick Wills, he met Sir Philip Cunliffe-Owen, who was also staying there.[31] My Cunliffe-Owen grandfather in his kindly way arranged for the young man to see the Inventions Exhibition in London and to come to dinner with the Cunliffe-Owens at the Residences; and there he saw Monica, his future wife. He records in his diary, 'the atmosphere at No. 4 the Residences was completely different from that of my own home. Sir Philip's work brought him into contact with interesting people in every walk of life, and diplomats and statesmen, artists and scientists were constant visitors there'.

After he married Monica, he frequently introduced her as 'my wife the daughter of Sir Philip Cunliffe-Owen'. He was very proud of her. Actually, as recorded in my account of Mary Hopkinson, before he met Monica, he wanted to marry Aunt Mary. His whole life was spent in or near Bristol, and practically every day of the week until he retired he went to the tobacco factory. It now seems acknowledged that he was the brains of the firm, for he had considerable vision, and

[31] Frederick Wills, First Baronet, of Northmoor, Somerset (1838–1909), was the son of Henry Overton Wills II.

scientific ability. He had to have his own way and, as tact was not his strong suit, there were violent quarrels with the Wills relations in the firm, as they were less clever and thoroughly conservative. It was Uncle Harry who insisted on new factories (calling in his uncle, John Hopkinson, my grandfather, as a consultant engineer), who took up new inventions, who pressed for another engineer and got Hugo Cunliffe-Owen in the face of great opposition, who was responsible for machine-rolled instead of handmade cigarettes (and sold them in one-penny packets!), and finally it was Harry who waged war on the American tobacco trade and founded the Imperial Tobacco Company, as we know it.[32] When Harry joined the business, the company's profits were £25,000 a year. After the Imperial Tobacco Company was formed, they were £750,000. As children, staying with him in the Easter holidays, we often heard him tell the story of his battle with the autocratic head of the great American Tobacco Company, J. B. Duke, who came over to try and break W. D. and H. O. Wills.[33] It is a long, complicated but fascinating story out of which the Imperial Tobacco Company and the British American Company were born. At the end, Harry would tell us that Duke said, 'All my life I have beaten everybody, but now I fear H. H. Wills is going to beat me,' and, added our uncle gleefully, 'That's what I did'. Finally, Duke asked Harry Wills to be the chairman of both companies, here and in the USA. This he did for a time, before he retired.

It is clear that my uncle had little time left for the lighter side of life; but ever since he was a young man he had fished, as his parents had part of a salmon river in Norway, where many of my Hopkinson relations went to stay. Later he shot; in fact the only holiday in the year that he took was six weeks in Scotland renting either a deer forest or a grouse moor. Once, staying with him as a girl, I said that this seemed to be the only fun he had with all his money and that I thought this a pity. But I think his pleasure was giving it away, especially to institutions in Bristol, the university, the hospital and many private charities. He used to take me sometimes to look at land such as the

[32] Hugo Cunliffe-Owen, First Baronet (1870–1947), was an English industrialist and the son of Sir Philip Cunliffe-Owen.

[33] James Buchanan Duke (1856–1925) was a US tobacco and electric power industrialist best known for the introduction of modern cigarette manufacture and marketing and his involvement with Duke University, USA.

fort which he thought of buying in Clifton for part of the university. 'It's always buildings you want to pay for, Uncle Harry,' I said once. 'Oughtn't you to pay some professors to work in them?'

'I will think about it' said my uncle, and he did. Although too shy to be really at home with any children, I think he accepted us, and we him.

Emma, another of my mother's sisters, was a character. She married her cousin Edward Cunliffe-Owen when she was seventeen.[34] He was in the civil service and a rather dull young man but the first person to ask her and she said that she did not want to be left on the shelf. This was an unsound marriage eugenically (and in other ways too for the high-spirited and accomplished Emma) since their five children all died before they were forty. My mother used to say that, Emma and Edward being first cousins, none of the children had complete insides. Be that as it may, I think that they were rather a queer lot, though we hardly knew them. Emma was proud of them and quite a devoted mother. I remember her writing about the youngest, Alexander, to say that he was successfully selling deluxe cars with a firm in Piccadilly and was able to offer with each sale the gift of a pedigree white mouse, as he had recently been made President of the All England White Mouse Society![35]

To my family's surprise, I was asked to be bridesmaid at the weddings of both her daughters. Sybil married a grandson of Charles Dickens and, as he was a Catholic, they had a wedding in Brompton Oratory.[36, 37] Dorothy married a not very intelligent young man called Lynch-Blosse, who although not well off gave the bridesmaids sable muffs, urged on by my extravagant Aunt Emma.[38] I should have loved a sable muff but was not allowed to participate at either wedding, partly because I was at school far away in Scotland and also because my father did not approve of society weddings of this kind for teenagers. I was most disappointed.

[34] Edward Cunliffe-Owen (1882–1918) was invested as a Companion, Order of St Michael and St George.

[35] This was Alexander Robert Cunliffe-Owen (b. 1884).

[36] This was Sybil Dickens, née Cunliffe-Owen (1889–1934).

[37] Mr Philip Dickens was the son of Mr Henry Dickens, KO, and the grandson of Charles Dickens.

[38] Dorothy Mary Cunliffe-Owen (1887–1926) married Sir Robert Cyril Lynch-Blosse, Thirteenth Baronet (1887–1951), in 1911.

Aunt Emma had the Cunliffe-Owen gift of attracting devoted service to her. She had a maid, Hoare, who as ladies' maid or general factotum, with or without wages, remained with her always. Emma was very extravagant and, as my mother would say, only had peaches and carnations in her flat when they were out of season. She battened on her rich brother, Hugo, and her rich brother-in-law, Harry Wills, and too often my mother was the thankless go-between. At the same time, Emma had her family's talent for organisation and hard work. Her triumph was in the First World War. When the great recruiting drive was on, she bethought herself of all her hunting, shooting and fishing friends, who though just overage were hale and hearty and thirsting to serve their country. Emma went straight to the War Office to see Lord Kitchener; after being headed off several times, she did see him and told him she was ready to raise a sportsmen's battalion, to recruit the men and to equip them personally. There was a condition; she must have a suite in the Hotel Cecil, as it was then.[39] Kitchener, though hostile, especially to the condition, was no match for Emma and finally sent a message saying, 'give this woman anything she wants as long as I never see her again'. Recruits rolled up to the Hotel Cecil, and here Emma weighed and measured and generally recorded them. She saw to their equipment, even to their toothbrushes, buying them all very economically. We had a first-hand account of all this from C. E. Montague (of the staff of *The Guardian*), who was one of her sportsmen and greatly admired her initiative and drive. Finally, at the end of the war, Emma in a hat with ospreys and an ermine tippet said goodbye to her battalion at a rally in Hyde Park. Being then somewhat arthritic, she sat in a bath chair, with the sportsmen's drum across her knee, and on her breast she wore the decoration awarded to her.

There was something of the grande dame about her. She always signed the book at Buckingham Palace, even when later she was living in Loughborough. 'Do you mean Emma, even when crossing London to change stations, you stop and sign the book?' my mother asked her incredulously.

'Certainly', replied Emma, looking very prim. 'I know our dear Pater would have wished it.' She was definitely rather eccentric. A pair of slippers made by her and embroidered on one toe with a J and the

[39] This was a grand hotel built in 1890–1896 between the Thames Embankment and the Strand in London. Only the façade remains today.

other with an S were sent by her to Stalin at the Kremlin, because she felt he had all too few well-wishers. A reply to this present is not recorded. Perhaps it should also be mentioned that she was at one and the same time, a Roman Catholic, a Spiritualist and a Christian Scientist.

Really she was rather fun but we hardly ever saw her as she considered coming to stay in Manchester with us much too dull. Years later, uninvited, she suddenly appeared at my wedding, a startling sight for my new Bragg in-laws as, being then short-sighted, she had made up her mouth and eyebrows about half an inch above their real positions and completed her outfit by a huge picture hat on her over-reddened hair. This was the last, and only about the fourth time in my life, I saw my Aunt Emma.

Victoria Mary Louise Adelaide Cunliffe-Owen was the youngest child. She was said to have been born with scarlet fever, which her mother had at the time of her confinement. Medical evidence says that this is impossible but it was made the excuse in the family for Aunt Vicky's character, which was not of the best. She was to have been called Louise but the Empress Frederick, a great friend of my grandparents, intimated that she wished to be her godmother and that she would be called Victoria, after her. The Empress gave her for her christening a lovely brooch of rose-coloured spinel and yellow diamonds. Unfortunately Aunt Vicky, when hard pressed in later life, twice pawned it, but Aunt Monica Wills somehow or other found out and rescued it, and I have it to this day.

My grandmother died when Vicky was just grown up and she went to live with 'Tutz' (Aunt Grace[40]) in Withington, close by our family. Manchester was not the place for Vicky; she was bored and soon married a Manchester publisher.[41] I doubt if marriage was for Vicky; there was a divorce, and the rest of her life was spent under a series of 'protectors'.

She was of course considered a bad influence and example to us children, so I only saw her once; but that was quite dramatic. She was visiting Manchester with someone called 'Foxy' and asked to come to Sunday tea with us. My mother and Aunt Monica, who was staying

[40] Grace Elise Cunliffe-Owen (1864–1909).

[41] Otto Kyllmann (d.1958) was a publisher and Senior Director of Constable and Co. from 1909 until 1950.

with us, allowed it but Foxy was banned and we were not to be told who she was. My sister Enid and I, in clean, embroidered pinafores, went to the drawing room at teatime; the three ladies were talking heatedly in German, but one, whom we had never seen before, darted forwards, fell on her knees and clasped us to her heart. I remember to this day the exquisite smell of her scent. She had a wonderful floating sort of chiffon dress and a huge picture hat, such as was seldom seen in Withington, and seemed to us young and lovely. In spite of the prohibition, she said, 'Who am I, darlings?'

'Violet' I remember saying doubtfully, naming the fascinating and saintly Swiss cousin whom we had never seen.

'Violet' cried Aunt Monica. 'Mein Gott'. We were immediately taken in to tea, the three talked German all the time (my father must have been out) and we were bored, until my Aunt Monica forgot and addressed the stranger as Vicky. It was not lost upon us—we could not wait to go upstairs and discuss it all.

Poor Aunt Vicky. She had generous allowances from both Uncle Hugo and Aunt Monica but never enough, and my kind mother, who would never ask anything for herself, was always the go-between, pleading her cause, especially with Hugo. Vicky was shocking indeed to the puritanical Hopkinson family. Occasionally some mild peccadillo of mine later on prompted an aunt to suggest to my mother that there might be a touch of Aunt Vicky here, and then my mother flew to my defence like a tigress.

I have left Olga, my mother, to the last. She came near the end of her family, a few years before Hugo and Vicky. She was small and shy as a child and nicknamed Mäuschen (the little mouse). The Cunliffe-Owens were not addicted to writing about themselves, but Olga wrote a simple account of her childhood for the interest of her five children, and I cannot do better that quote from it presently. Like my father in his family, she was always overpowered by older brothers and sisters, enough to give her an inferiority complex. It made her modest and humble but could not dim her personality. When she married, she obviously puzzled her new Hopkinson relations; they never really understood or appreciated her. Her first meeting with the clan was perhaps a little unfortunate. There was one of the famous family rallies in the Lake District, and my mother, just engaged to my father, Albert, arrived at Coniston in her London high-heeled shoes and a hat with feathers. But she was prepared to please and be pleased. It

Figure 51 Olga Hopkinson, née Cunliffe-Owen, with Alice. (Courtesy: Patience Thomson)

was, she used to say, 'bourgeois not to get on with one's in-laws'. She seldom referred to the gay and interesting life that she had given up; indeed, several of her nephews and nieces never knew her maiden name. Like all her family, she could adapt herself to the circumstances in which she found herself. I only twice recall remarks, made in a rather dreamy voice, that could have been considered snobbish. Once, someone in her Manchester days boasted about the expensive florist she was employing at her daughter's wedding, and my mother replied that the Duke of Abercorn had sent all the flowers from his gardens for her own wedding. And, again, a lady told my mother that she had been to great trouble to arrange to have her daughter presented at Court. Afterwards my mother said that seemed waste of effort unless you were going to be asked to Court parties; 'After all,' she added, 'my mother had the entrée, and we went to all the things,' although she did not say this to the lady. I remember this because at the time I thought the expression meant a dish served as some course at a dinner party, and I was puzzled.

My mother was not an intellectual but she was not stupid. In fact, with all my girl's public school and Newnham education, she was much more talented than I ever was. She spoke and read French and German fluently, was well read in art, literature and history, and in a more practical sphere she held a teacher's diploma in cooking and had published a cookery book for charity. Not that I ever saw her cook anything and, when once asked what she could cook, she said, 'Quails in aspic and that sort of thing.' She sang and played the piano and could accompany very well. In her modest way, she had a great admiration for university education. I remember admiring enviously some beautifully smocked frocks one of our school friends wore, made by her mother. 'Ah,' said my mother, 'she can do that sort of things because she was at Girton and has a trained mind.' For a long time afterwards, I thought Girton must have been a domestic science college.

Her character was simple and straightforward. She spoke her mind and did what she thought right. We, her children, loved her very much. In fact, I find it hard, even today, to consider her with detachment. She lived very much for her family and a very small circle of friends. Completely loyal, she defended us to the world but criticised us freely later on. If she thought we spoilt our own children or brought them up in a way she did not like, she did not hesitate to say so. None of us minded. I think we were always conscious of her fundamental goodness, for she was deeply religious. It was not what she said to us about this but the small way she lived her spiritual life that left its impression on us and those who really knew her. She wrote to each of us children twice a week. They were on the whole rather dull letters, full of short sentences. One knew they would include lines like 'My bulbs are coming up well, Gertrude is making marmalade, Mrs So and So is having another baby, Nanny's knee is better, try not to get too tired, Canon C came to lunch as we were having a joint.' Never mind, we loved them and, if she was ill and could not write, we missed them dreadfully.

Her life as a dedicated GP's wife in Manchester must have been dull. My father had seldom time to play with her in any sense but she made a life for herself within her family. When my father retired to Cambridge, he set out to amuse her, travel with her, take her to the theatre and made up for the thirty years in which he had been too occupied to do these things. My mother never, I think, really cared for Cambridge. She loved her house and garden there, but she found it a man's world, and a rather withdrawn and intellectual man's world at that. The attitude held by so many there, that there was nowhere in the world other than Cambridge, rather irritated her. Her first encounter with a professorial caller always amused me. Professor Burkitt (whose wife, my mother maintained, was born on Mount Ararat) came to call one Sunday afternoon before going to the lodge at Trinity.[42] Suddenly my mother said to him: 'But you cannot see Lady Thomson with that hole in your trousers.[43] I can soon mend it. You need not take them off.' No wonder those who did come to know my mother enjoyed her.

On the whole, I think, she had a happy life, perhaps because she demanded very little of it. Her great sorrow was the death of my brilliant brother Eric.[44] From the moment the telegram in 1915 came saying, 'seriously wounded and missing', her life was clouded. Her natural courage kept her going and she helped and inspired those who had also suffered. All her children and their spouses had a perfect relationship with her. She made a point of always siding with the in-law children in any dispute, not with her own. We loved her and she knew it. She wrote an account of her life until she married and kept it for us. Like all children, we wanted to know about our mother's childhood, especially as we gathered it was very different from our own. Thus, she was persuaded 'to write it up' for us.

3.4 My Mother's Diary

I was born on September 26th 1866 at No 4 The Residences South Kensington Museum in London. My father, Philip Cunliffe-Owen, was Director of the Museum, and our house was a Government house. My dear mother was German, Jenny von Reitzenstein, and belonged to an old and noble family. Of my parental grandparents I know little. My

[42] Francis Crawford Burkitt (1864–1935) was a British theologian and scholar.

[43] Rose Elisabeth (née Paget) Thomson (1860–1951) was the daughter of Sir George Edward Paget and the wife of the physicist Sir Joseph John Thomson.

[44] Eric Humphrey Hopkinson (1894–1915).

Figure 52 Charles Cunliffe-Owen. (Courtesy: Patience Thomson)

grandfather, Captain Charles Cunliffe-Owen, had been a sailor.[45] He was quite a character, a clever, courteous old gentleman, who spent the evening of his life in Avignon, where he died. His wife died much earlier, and we knew little of her. . .[46]

Of my mother's parents I knew much more. Her father, Baron Fritz von Reitzenstein, was Commander of the Garde du Corps in Berlin, a splendid man physically and morally.[47] His portrait in the white uniform of his regiment, riding his spirited mare, hung in our dining room. He lived a life of great simplicity and neither drank nor smoked. His wife, Pauline von Roeder,[48] was a very cultivated woman, but most of her life was spent in bed, but from her room she ruled her household,

[45] Charles Cunliffe-Owen (1786–1872) was a captain in the Royal Navy.

[46] Mary Peckwell Blossett married Charles Cunliffe-Owen in 1821.

[47] Fritz von Baron Reitzenstein was Aide-de-Camp General to Frederick Wilhelm IV of Prussia. He gained the rank of Commander in the service of the Royal Prussian Horse Guards.

[48] Pauline von Roeder (1802–1849).

and guided the lives of her children, one son and three daughters, of whom my mother Jenny was one . . .

I was the eighth child of my parents, and the fifth daughter. I was a tiny baby and in no way remarkable. When I was four years old my brother Hugo was born. Now I must tell of my elder brothers and sisters—of Fritz and Alfred I saw very little, they were mostly away at school. Of us girls Gertrude the eldest, ten years older than I, was an admirable elder sister, then came Monica, Emma and Grace. Grace or Tutz, two years my senior, was my chief playmate and friend.

Next to our mother our greatest friend was Fräulein Emma Clemens, our German governess for nearly 20 years. We also had a nurse, Mrs Pollard—she had red hair and sharp features and was a great disciplinarian and tyrant. She was splendid at teaching us hymns and telling us long stories. We were brought up very simply. At 7 am we were wakened and had to strip our beds, empty our slops. At 8 we had breakfast and prayers. 9 to 12 lessons, then a walk in Hyde Park. At 10'clock came our dinner, 2 to 3 another walk, lessons and work till 6, then schoolroom tea, and then we were with our mother till bedtime. No dainty ways were allowed at meals, our food was wholesome and plentiful, but very simple. If it could not be eaten, we had it again for tea. How I remember tapioca pudding, spinach and other childish bugbears. A great treat was treacle for tea, or penny buns. We all had music lessons and attended an excellent French class given by Monsieur Roche.

Winter had its own delights, with Christmas as its highlight. For weeks beforehand we prepared presents and made shopping expeditions to the Baker Street Bazaar, where for an outlay of 1d and 6d presents could be bought. On Christmas Eve at 5 o'clock we gathered together outside the drawing room and sang 'Hark the Herald Angels Sing', then the doors were opened and there stood a beautiful Christmas Tree with coloured candles. Our presents were displayed on tables covered with white cloths, no one was forgotten and often and few lonely friends came in to share the festivity. On Christmas Day we went to church, and after lunch old friends called. At night we had Christmas Dinner followed by round games and music.

As a general rule we were not allowed to go to big parties, though we had many friends and were usually out on Saturdays. The preparation for an occasional party was not all joy, such scrubbing of hands and necks, such stiff and starched frills, scratchy stockings, and a general feeling of 'malaise', soon forgotten. There was often a grown-up dance or dinner party at home, and then we children were very excited. How we admired our Fräulein in her party dress of black and green silk with thistles in her hair. When we had been safely tucked up in bed the fun began. Our rooms were on an upper floor, opening on to a gallery from which a good view of the floor below was obtained, where the ladies unrobed. Softly three little figures stole out into the gallery and lay on quilts, peeping down at all the smart ladies. Emma contributed refreshments, having made lemonade in the water bottle on her wash-stand. Next morning we raided the ice-pails and feasted out of great wooden spoons of pink and white ice . . .

1878 was the year of the great Paris Exhibition and our father was Secretary to the British Section and was obliged to spend the greater part of the year in Paris. We children were sent with Fräulein to Sandgate, near Folkestone, installed in a comfortable lodgings and were very happy. We three girls, Emma, Grace and I, were regular tomboys and never so happy as when we were playing rounders with other children. We had riding lessons, bathed, paddled and took long walks. One day there was an unusual stir in the streets and on the shore. It was a glorious summer day with a calm blue sea, and yet there before our eyes, a large ship, a German man of war went down. Two of these big ships, the *Grosser Kurfürst* and the *Kaiser Wilhelm,* had collided, and in a few minutes the *Grosser Kurfürst* disappeared.[49] We could see the heads of the unfortunate sailors in the water and watched the lifeboat putting out. It reached the scene of disaster too late to be of any assistance. Afterwards, when the dead bodies of the German sailors were brought ashore and buried with military honours in Folkestone Cemetery, we made wreaths and tied them up with the German colours to put on the coffins. We were present at nearly all the funerals.

I went to Paris for about three months of that summer. How keenly I enjoyed that time. First the excitement of being fitted out with new clothes and then the memorable journey. My father's house was in the Rue de Mars, just opposite the Exhibition. I went constantly to visit the Exhibition, a veritable fairy palace, which fairly dazzled me. Then there were churches and galleries to be visited. I was nearly twelve years old, and could take an intelligent interest in much that I saw. I

[49] SMS *Grosser Kurfürst* was sunk on her maiden voyage in an accidental collision with the SMS *König Wilhelm*. The two ships, along with the SMS *Preussen*, were steaming in the English Channel on 31 May 1878. The three ships encountered a group of fishing boats and, in turning to avoid them, the *Grosser Kurfürst* inadvertently crossed too close to the *König Wilhelm*. The latter rammed the *Grosser Kurfürst*. A failure to adequately seal the watertight bulkheads aboard the *Grosser Kurfürst* caused the ship to sink rapidly in about eight minutes, taking between 269 and 276 of her crew with her.

remember one day being taken to see Belleville, that part of Paris where the communists chiefly resided in 1870, a very poor part of Paris.

Interesting visitors came to our house for lunch. One day it was the great Gambetta of balloon fame,[50] another German Emperor then a rather awkward youth.[51] We drove daily in the Bois de Boulogne—I noticed the differences between Paris and London. Paris had a look of dainty cleanliness, with its green-shuttered houses, and streets bordered with shady trees, the poorest people looking neat in their blue cotton blouses. My dear mother encouraged me in my interest in all around, and I had to write a daily account of all I saw and heard.

There follow accounts of family weddings: Gertrude, Monica and Emma were married and left home. My mother and the delicate sister Grace (Tutz) made trips abroad, generally to Switzerland, with their mother. Occasionally, all the winter months were spent in the south of France, while I suppose my grandfather remained at the museum.

In 1884, we went to Cannes and stayed at the Hotel California, where there were many pleasant people staying, and we had some nice excursions. Here I had my first love-affair. My suitor is now over 80, and has never married. Whilst we were there Prince Leopold died at Cannes.[52] I remember going to see his funeral procession pass, and how strange it looked to see the street lamps lit, and veiled in crepe.

And again in 1886:

Mother, Grace and I, attended by a maid went to Cannes in October where we stayed at the Hotel Continental till April. As we were there for so long, we had to occupy ourselves, and I had gymnastic classes, music, French, and sang in the choir. We had a good deal of pleasant society, and there were concerts, picnics, and dances. My father and Hugo came out for Christmas.

In between these trips abroad, my mother pursued her education in London. She went to the school of cookery and took a diploma, also attending various 'classes' in French, English literature and history.

[50] Léon Gambetta (1838–1882) was a French statesman, prominent during and after the Franco-Prussian War. During the war he left Paris by hot-air balloon to take control as minister of the interior and of war in Tours.

[51] This was Kaiser Wilhelm II, Frederick William Victor Albert of Prussia.

[52] Prince Leopold, Duke of Albany (1853–1884), was the eighth child and fourth son of Queen Victoria and Prince Albert of Saxe-Coburg and Gotha. His early death was due to his haemophilia.

In 1868 my sister Grace and I were presented at Court. Then it was rather a dismal affair. It was an icy March day. The hairdresser arrived at 10 am to do our coiffures and arrange our plumes. Then we were dressed, and at midday we drove off with Lady Hunt who was to present us, and had given us our dresses. In those days there was no buffet, no kind evening light, and the old looked very old and wrinkled in the hard winter light. Still, I felt it a great occasion. With my long train laid down by officials, then the deep curtseys to her Majesty and the Princesses, then walking backwards to make a graceful exit. Afterwards a photographer and friends to see us at home—but it was the entrée to many pleasures, balls, royal concerts and garden parties. My father was splendid when he took us to such entertainments, for he knew everyone, and could explain who the various celebrities were.

One day still stands out in my memory. It was during the year of the great Colonial Exhibition and we were invited to Windsor with all the distinguished colonial visitors. A special train took us to Windsor where royal carriages with postillions met us. We drove up to the Castle, and after a cold luncheon, were presented to Queen Victoria. Her Majesty extended to me her hand to kiss, but in my confusion I shook it instead. Then we were shown round the chapel, the mausoleum, and finally feasted on strawberries and cream in the royal dairy.

Another day my parents and I had tea at Kensington Palace with Princess Louise and the Marquis of Lorne, and various Indian princes were invited.[53, 54] Thus do childish dreams come true, for in our nursery walks past the Palace I had often longed to go inside and see the house where Queen Victoria had lived as a girl.

In 1887 we were present at the first Jubilee, and had good seats on a stand outside the Abbey. The outstanding figure in the Royal procession was the Crown Prince Frederick of Germany, who looked like one of King Arthur's Knights in his white and silver uniform, with flowing yellow beard.[55] I had a chance to see him close to as he was a great friend of my mother, whom he had known in childhood and he came to see us, and was most charming. . .

[53] Princess Louise, Duchess of Argyll (1848–1939), was the sixth child and fourth daughter of Queen Victoria and Prince Albert.

[54] The Marquis of Lorne, John George Edward Henry Douglas Sutherland Campbell, Ninth Duke of Argyll (1845–1914), was a British nobleman and was the fourth Governor General of Canada from 1878 to 1883.

[55] Later Frederick III (1831–1888) was father of Wilhelm II and was German Emperor and King of Prussia for only 99 days.

> In the summer of 1890 the others were all abroad and I paid my first visit to Manchester, and stayed with the Hopkinsons at Bowdon close by. I visited the ship canal, and met for the first time my future husband Albert. His sister Gertrude had become my great friend and had stayed with us at the Residences.

The Hopkinsons and Cunliffe-Owens had various links. The mother of Harry Wills (who had married Monica Cunliffe-Owen) was one of the illegitimate Hopkinson daughters.

> At the end of March I went for my first visit to the Lake District as the guest of my future brother-in-law Edward Hopkinson and his wife Minnie—and there on Latrigg, I became engaged to Albert Hopkinson.
>
> My last summer at home passed quickly. In July, Albert came to our summer home in Lowestoft for a visit, then my mother and I stayed with the Hopkinson parents. On October 2nd 1891 I was married, and so ended my happy uneventful girlhood. When I said goodbye to my London home I did not realise that I should never see it again.

Within a year or two both my mother's parents died, and she was too tied domestically to go home to be with them.

> I came north to Lancashire, making a great change from my gay London life to that of a general practitioner's wife in the suburbs of a great manufacturing town. I lived there from 1891 to 1919 and in those years I learned to love the north and the warm-hearted Lancashire people.

3.5 Childhood at Parsonage Nook

My father and mother had five children: Harry and Eric close together, then a gap of five years, me and Enid, another gap of five years and then Philip. Harry and Eric did everything together although they were very different. Harry was warm-hearted and generous but quick tempered. His performance at Rugby and Cambridge was perfectly adequate but he was overshadowed by Eric, a fact which I do not think he minded at all. When we were children, he was special to me and taught me to swim and to ride a bicycle. When Eric was reported missing in 1915 he was devastated and became lost, wandering to various spheres of war, firstly joining the RAF and then later the RAMC, vainly hoping for news of him. Finally he became a doctor. Eric 'belonged' to Enid, and he taught her his skills at running and climbing. He was very tall, untidy in appearance but with a dazzling smile. He had a brilliant record at Rugby, from where he got a top scholarship to Trinity, Cambridge, and a first in the only examination he had time to take there. How well I remember my father's excitement. Enid and I were having our Saturday night bath together, when he rushed in with a telegram. 'Eric has got a £100 scholarship at Trinity,' he shouted. 'Trinity what?' we said, catching the excitement. 'Cambridge of course', he answered, running off to telephone relations. Eric was also a natural leader of men, as was shown even from his early days at Rugby, where he was Head of House.

Enid was born about twenty months after me. I was annoyed and said, 'Put her out in the garden'. Asked what would happen if it rained I replied, 'Give her an umbrella'. Whether I really had such a command of language at that age, I doubt. At any rate, I became used to her and we were inseparable throughout our childhood. She was delicate and often had to spend weeks away somewhere out of the Manchester climate, with our nanny and Philip, five years younger than Enid, in attendance. I resented this because I missed her so much, and when someone said she was like a frail snowdrop I pushed her into the coal-box and had a complex for some time after. As our nanny said, I had probably 'done her a mischief'.

Philip was considered an afterthought. It is thought that he had some minor trouble with his heart, so that he was always quite unathletic. As a small boy he was very pretty, the type that the photographer of the day took sitting on a huge mushroom, with nothing but a wisp of chiffon draped appropriately. All his life he had an amazing memory, and when very small could say the names, once he heard them, of all the birds in the standard book. Although he became a clergyman and developed into a saintly character, as a child he was rather naughty. Enid, who had strong maternal instincts, looked after him from babyhood and was really his devoted protector throughout his life. He had only just gone to Rugby when I married, so I never saw very much of him, to my lasting regret. I used to visit him at his preparatory school, and later at Rugby, and I wrote to him nearly every week. Once, when I was at Cambridge, at Newnham, this had amusing consequences. In our family we had the Cunliffe-Owen passion for nicknames, frequently changing them. At that moment I wrote to him, 'Dearest Doggie'. Being in a great hurry I sent this letter to my tutor, Mr Goulding Brown, and my history essay to Philip.[56] Mr Goulding Brown, a studious looking don in a remarkably high collar, returned it with a dry note starting, 'Dear Miss Hopkinson, one glance at the heading of the enclosed assured me that it was not for me'.

We were all looked after by Nanny, who came in her early twenties and stayed for over forty years. She was the daughter of Grandfather Cunliffe-Owen's coachman at Lowestoft and had delightful Suffolk expressions like 'Tie up your shoes properly or you'll fall' and 'That little old kettle is a slow boiler'. We were devoted to her. She never went out, refused to marry the grocer's assistant and so to leave the family and was always at hand for us.

We lived in an ugly house, in an ugly Manchester suburb. A brass plate, polished daily, proclaimed that the house was called Parsonage Nook and that the doctor lived here. It stood on a curious grassy eminence, at a point where six roads met. There was a garden, very much carved up, with slopes and little walls and flower beds; the one in front was filled with red geraniums, lobelias and white daisies and yellow calceolarias, in the tradition of the time. At the back there were considerable stables, a greenhouse and various outhouses, great assets to us as children. Inside it had to be very much the doctor's house. Patients really took up three rooms on the ground floor. First there was the consulting room, with a long sort of daybed–sofa, bookcases full of medical books, a roll-top desk and a bust of the Venus de Milo. Next came the surgery with a wash basin, a marble mantelpiece with test tubes of specimens to be analysed and the only telephone in the house. Enid and I often monopolised this, asking friends about homework, no matter that an agonised patient was in labour and wanted the doctor at once. While we conversed we sometimes doodled with the specimens on the mantelpiece and changed them around off their labelled stands. In the dining room at evening surgery patients waited. Dinner could not be laid till the last one had left, though mother, if it was late, rang the gong at intervals to hurry them along. There was a drawing room which was truly private and bore the stamp of our mother's taste. Finally, there was a boot room and kitchen quarters, where a splendid cook, supported by two maids, reigned for many years. One of the maids had always

[56] Bertram Goulding Brown (1881–1965) was a history tutor at Emmanuel College, Cambridge, and was notable as a formidable chess master.

to be about to deal with patients. Half way upstairs was an extraordinary lavatory with a stained glass window giving on to the staircase. This had the motto '*Deus dabit vela*', perhaps to be translated as 'God will give a covering', but that is a debatable point. My father and any classically minded friend used to discuss this outside, so that it was sometimes difficult to get out of the place. There was only one bathroom for us all. We had a huge nursery, with windows at each end and indeed at the sides. They had pale-green venetian blinds, the slats of which were taken to pieces once a week and scrubbed by a rather feeble-minded character, Edith Potts, who came especially to perform this task.

My father had a hard life, though partly from his own choice, but also from his Hopkinson upbringing. We had breakfast early, and

Figure 53 Dr Albert Hopkinson. (From <http://www.spinningtheweb.org.uk/web/objects/common/webmedia.php?irn=2625>)

then he left on his rounds. In the early days he retained a hansom cab with a driver, Mulliner, who smelt of straw and brown sugar. Mulliner was grateful to my father for befriending his unmarried daughter who expected a baby. She was allowed to sit during the hot summer in our garden every day, as long as she left before we came back from school. My father wore a top hat, a frock coat and striped trousers. His patients ranged from the wealthy to the very poor; he gave time to each, for he was doctor, marriage guidance counsellor and priest to his patients. The rich included old ladies who liked a daily visit from the handsome doctor though nothing much ailed them, and eccentrics, such as the lady whom Father urged to have her toenails cut but who said that her gardener was on holiday and her nails had to be cut with secateurs; and then there were many foreigners who were fussy about their health. The range of a GP in those days was wide. One did not readily go to hospital. When a vicar's baby swallowed lamp oil, thinking it was treacle, his wife ran with it to my father in the early hours of the morning, and he sucked it out. He particularly liked his confinement work, and every Sunday, when we had a huge joint, small boys came with basins and had a cut from it with trimmings, to take home to nursing mothers. If a poor patient needed a wheelchair or some special equipment, we held a sale of work in the nursery and raised the necessary money. My father saw patients all day, and on the rare occasions that he went out to dinner, he was on call and might have to leave the party.

At regular intervals a certain Miss Jones was installed in the dining room to do the bills. Many of the patients were not charged at all. Some never paid but sent presents instead; a few struck the shillings off the bill, made out in guineas, before paying. My father was not interested in making money, except for our education, and I never remember his buying anything for himself. My mother was like him, but she did occasionally strike a blow for the family. A man who could easily have paid had so far had his two babies born for nothing. When the third was expected, my father said he would not attend unless he was paid. The patient banked on my father's kind heart but he reckoned without my mother. When he rushed round to say the baby was imminent my mother forced him to write a cheque for the first two before she would give the message to my father.

My father and all the GPs in the neighbourhood were violently against the National Insurance Bill in 1911.[57] They met in each other's' houses and had fierce meetings and a lovely tea, when it was our turn, of which Enid and I would mop up the remains. Finally, the only local doctor who could speak Welsh was sent to see Lloyd George to present their petition and make a personal appeal in his own language.[58]

My mother must have had a dull life during those thirty years in Manchester but, like all Cunliffe-Owens, she was adaptable and never complained. In the early days, I think, she and my father went out to parties and even balls and my mother sang in a choral society; but later my father had too much to do and was too tired. She spent much of her time with us children and paid and received calls; I think two Thursdays in the month were her 'At Home Days'. Enid and I were sent for after tea to see visitors, very tidy, in clean pinafores. Sometimes our behaviour did not come up to standard. The visitor one day was Mrs Lejeune (mother of Caroline, the film critic[59]) and she asked us politely what we had been doing upstairs. 'Making ice-cream wafers', we said. 'It's quite easy, you just wet two pieces of lavatory paper' (dark frowns from my mother) 'and put mushy soap in between for the sandwich part. We'll fetch you one to see.' But Mrs Lejeune preferred to say goodbye and we were left to a lecture on drawing room conversation.

Sometimes relations came to stay, especially just grown-up cousins to whom my mother was very kind, taking them to the theatre and concerts and accompanying them when they sang 'Down in the forest something stirred' or 'Melisande in the wood', for in those days they all had singing lessons after leaving school and went abroad to learn a language. She kept up her old standards of life and always under any circumstances put on a tea gown with a lace fichu and a good brooch for dinner every night.

[57] The National Insurance Act 1911 was an Act of Parliament of the United Kingdom. The Act is often regarded as one of the foundations of modern social welfare in the United Kingdom and formed part of the wider social welfare reforms of the Liberal Government of 1906–1914.

[58] David Lloyd George, First Earl Lloyd-George of Dwyfor (1863–1945) was a British Liberal politician and statesman. He was Prime Minister from 1916 till 1922.

[59] Caroline Alice Lejeune (1897–1973) was a British writer, best known as the film critic of *The Observer* newspaper from 1928 to 1960.

As a doctor's wife, she was a great help. In fact, on one occasion, during the first war, her behaviour was spectacular. Early one Sunday morning, a distraught man arrived on the doorstep to ask for the doctor urgently; the lady of the house had gone mad and was threatening the household with a carving knife. My father was out and could not be contacted. 'Then,' said the man, seeing my mother in her Red Cross uniform, 'you must come.' My mother flagged down an early milk float for a lift and went to the house. From the garden path she saw through the window the lady, a German patient of my father, with flowing black hair, standing brandishing a carving knife. The household was grouped outside her door which they had locked. My mother opened the door, went in and shut it behind her. She told the lady in German that if she would put down the carving knife they would sing folk songs together. This they did for more than an hour until a doctor came, and she was taken away to a mental hospital.

Enid and I led a very simple life together; the boys were away at school most of the time, and Philip was too young to join in our plays. We had no expensive toys, only a few dolls, a wonderful box of hand-made bricks and a paintbox each. We had families of 'paper ladies', cut from catalogues and painted, and these had a varied life, living in colonies in cupboards and in the doll's house. We had great amusement looking out of our nursery window. The White Lion, a large and ugly public house opposite, was the tram stop for the elite of south Manchester (very few people had cars yet), and we knew many of them. There was Mrs S, who, I think, was one of the early prison visitors, very serious looking, carrying her papers. My mother, also watching, said, 'Never be a woman with an attaché case; that's why when you go to tea with her children you only have bread and butter and lettuce.' Then, there was Miss Olga Hertz, a little dumpling of a Jewess, in a fur-trimmed bowler hat, riding a bicycle to the workhouse, as she was one of the first women on the Board of Guardians as it then was. There was also always 'Pussy Rice' to watch, a poor one-armed tramp, who stood all day outside the White Lion hopefully, but we were not allowed to go near him, as Nanny said, 'Things might fly out of him'. The trams themselves had small trolley boys to turn the cable and they waved to us, as friends. If we opened the stained glass window at the side of the nursery, something that was not allowed, we could sometimes see, walking in the parsonage garden next door, the vicar's middle-aged daughter in her feather boa, on the arm of the curate.

We did not care for the vicar, chiefly because we thought he was not nice to Nanny. Her only relaxation in the week was to attend evening service at St Paul's next door, but although she hardly ever missed, and he knew who she was, he never greeted her. His white beard was stained orange with tobacco, and we thought him silly when he met us and always asked, 'And how is the good father and dear mother?'

We went to bed very early, and then we made all sorts of amusements for ourselves. Enid used to go round the room without touching the floor, swinging like a monkey from wardrobe to mantelpiece and along the picture rail. I was too fat and landed on the washing stand where some china broke. We had only candles for bedside lights, and these we used to melt down and make 'cakes' and model things from the warm wax. Sometimes we were interrupted by the young Philip, who would come to report that one of his ten dolls had fallen out of bed in his room opposite. One night in the week we sometimes had a special gala evening; we put on all our bead necklaces, pinned our hair up, tied the sleeves of our nighties under our armpits to make a low neckline, used towels for stoles, took up paper fans and settled ourselves on two chairs at the window pulling the curtains behind us. Now we were in our box at the opera. The highlight of a special evening was the arrival of the medical students from the Royal Infirmary, after their visit to the Cheadle Royal Asylum. Their brake changed horses at the White Lion, and they themselves had, I imagine, a great deal of beer there. There were lively scenes when they came out again, singing and shouting. We then bowed and smiled graciously. But nemesis would follow. On one occasion, memorable because it never happened before or after, we were spanked by our mother, I think with a hairbrush (I was intensely annoyed because Enid was so thin that she had preferential treatment and did not have to pull her nightie up and I did).

We had pets—a dog, kittens, birds, goldfish and white mice, a tortoise and even a glass case of ants—but most of these had tragic ends. The dog barked at patients and had to be given away. We dressed the kitten in our doll's clothes and put it in the pram. But it leapt out of the window in bonnet and shawl and never returned. The lovebirds died because we converted the old greenhouse into an aviary and put them in on the last day of the holidays when the paint was still wet. The white mice smelt dreadful and had babies every week and finally Philip threw the tortoise down a drain pipe for eating his sprouting flower seeds. One was of course supposed to learn the facts of life from pets but in this respect ours were unhelpful. But Enid and I were by no means ignorant about sex. She was able to glean a lot of information from the Bible, from which we put two and two together, and my father always answered any questions. My mother did too for that matter, but she was less scientific and inclined to use a rather hushed voice. During the First World War, when I was growing up, there was a great deal of talk about white slave traffic, and shocking things happened to girls. As I was driving my father in the car on his rounds (one did not have a licence in those days, mercifully, as I doubt if I was sixteen), I said, 'I'm rather afraid of being raped'. Silence—then my father said, 'If you keep your legs crossed, keep your head and there is only one man, you are pretty safe.' That was over sixty years ago.

All of us in turn went to Lady Barn House School, which was famous in Manchester.[60] I, as usual, was frightened of any new experience and refused to go; and when my mother took me for the first time, I sat down on the pavement and cried. The school was started by Mr Herford and his sister, who had studied Montessori methods.[61] It was after his death when we went and Miss Herford was headmistress.[62] She was a terrifying figure in a man's Norfolk jacket and a trilby hat trimmed with a Liberty scarf. Enid and I were convinced that she was really a man. I can see her now in the playing field holding my bat with me and saying she feared I'd never make a cricketer, a remark in which she has since been amply justified.

Nearly all the University of Manchester *Guardian* boys and girls (Scotts and Montagues) went to Lady Barn. We had a very modern curriculum; for example, clay modelling, Swedish carpentering and both reed and cane basket making were included. We had to look after, feed and generally service pigeons and rabbits and, in turn, read the rain gauge. I never could do this last but at the price of a few sweets I persuaded one of the boys to do this on my behalf. This boy, William Brockbank,[63] later became a consultant at the Manchester Royal

[60] Lady Barn House School was founded in 1873 as a co-educational school by William Henry Herford (1820–1908), the educational pioneer, Unitarian minister and one-time tutor to Lord Byron's grandson. Today, it is an independent primary school.

[61] Alice is mistaken here, as this was in fact W. H. Herford's wife.

[62] This was W. H. Herford's daughter, Caroline Blake, née Herford (1860–1945).

[63] William Brockbank (1900–1984).

Infirmary, but at school there was as yet not much sign of a distinguished career. He was nearly sent out of the room by Nanny at nursery tea with us for putting a brandy snap in each ear and 'playing with food'. We gave three plays to parents once a year: one in Latin, one in French and a Shakespearian one. In the Latin one I was one of the geese hissing on the capitol. Occasionally the staff included a master but this did not work; the last one stayed only a term and then left for the mission field, preferring to teach Hottentots to teaching Lady Barn House School children.

We girls played games and had sports with the boys, though we were quite unsuitably dressed for such games. I remember playing lacrosse in a red coat with a cape, and a black felt hat trimmed with red berries. I was told to be left attack, so joined a boy called Geoffrey Hickson (son of the Professor of Zoology[64]) and said politely, 'I think you are my partner'. He glared at me and replied, 'Certainly not, I am *on* you, that means against'. He and I were finally joint captains of the school, but I was nearly downgraded for going into the playground in my indoor shoes and thus 'not realising my responsibilities'.

After our school day, we had a relaxed and prolonged nursery tea. In the holidays we had the boys Harry and Eric and their friends round the table and revelled in their tales. There was one terrible occasion when Eric came home early from Rugby to see a specialist. He had diphtheria, his heart was temporarily affected and he had to be vetted before running in 'the Crick', the great cross-country school race in which he had come second last year.[65] I still remember the silent and gloomy tea when the doctor had forbidden him to run, and Enid and I did our best to comfort him with special cakes bought with our pocket money. After tea, our mother always read to us or sang German songs but really, to have our parents to ourselves, we had to wait till the holidays.

3.6 Boarding School

It was decided that I should have some time at a boarding school. Various reasons were given for this move; my father thought that if the

[64] Sydney John Hickson (1859–1940), was a British zoologist studying evolution, embryology and genetics.

[65] The famous Crick Run was founded in 1837 in Rugby.

boys had been to Rugby, Enid and I should also have the advantage of a public school. There was apparently a feeling that I might get spoilt at home and at my day school. Professional friends were sending their children away, and finally, I think, my mother wanted us to make new friends, as she was frightened of our marrying foreigners. Certainly the choice in Manchester would have been wide, for there were Greeks, Armenians, Syrians, Jews and Germans among my father's patients and indeed among our own friends. So my parents looked at various famous girls' public schools and finally chose St Leonards, at St Andrews.[66] Both my parents had a curious weakness for the Scots, cousins had been educated there and at least one professor at Manchester University had his daughters at St Leonards. I had to wait, however, until physically I had 'become a woman' and spiritually until I had been confirmed. Of my confirmation, I remember that the clergyman who prepared me had many personal problems and troubles of his own, and we generally discussed these at great length rather than mine. As usual, the little that my mother said made the most impression on me. 'You must never go to the communion service unless you are in love and charity with your neighbour. If you are having a quarrel with anyone you are bound to stay away till this has been put right.' I took it to heart.

I dreaded the new venture and made myself almost ill with anxiety. On the day of departure my Aunt Emma telegraphed my mother, 'Am at death's door come at once'. And though this was not an accurate assessment of her state for many years to come, my mother felt she had to go and could not see me off. I lay on the nursery sofa in my stiff new school uniform and waited for my father to take me to the school train. It was a long, long journey, and I can still remember, as darkness came down, the paper boys' desolate call of their evening news on Carlisle station. I was actually ill with a temperature for the first five weeks of term and in bed at the hospice, a sort of school cottage hospital for non-infectious diseases. I had seen nothing of the school, but the headmistress, Miss Bentinck Smith, came to see me

[66] St Leonards was established as an all-girls school known as the 'St Andrews School for Girls' in 1877 by University of St Andrews professors and their wives amid the increased demand for women's education. The pioneering spirit was in evidence right from the start—the belief of the school was that 'a girl should receive an education that is as good as her brother's, if not better' (from <http://www.stleonards-fife.org/our_school/history>).

every day, as I was technically in her house, and I came to know her very well.[67] The BS, as she was known, should have been a university don. I doubt if she really cared very much for girls, but she understood them pretty well.

No one knew what was the matter with me. I imagine it was partly psychological but at last on my fourteenth birthday my father arrived and took me home, and I went to a nursing home and had my appendix taken out by a famous Manchester surgeon. He was a nice man and also operated on my teddy bear to oblige; my father gave him his travelling rug for his pains. I have reason to think nothing much was found wrong with my appendix but, having got so far, the surgeon took it out. Then I had a delightful autumn convalescence at Barley Wood, the Wills home, feasting on pheasants and partridges and spoilt by 'Japonica'. Now I had to start school all over again, and a term late at that. I hated it and had the distinction of being the most homesick girl the school had ever known. I missed Enid and my family, I did not like sleeping in public, as it were, in a dormitory; I resented never being allowed to be alone. I was bad at games (and games were a fetish there) and, most of all, I disliked being a fag. I was such a bad fag that I had to remain one for two years instead of one; in this I was also unique.

The winter on the east coast was terribly cold, with east winds, and we had chilblains. I found, however, two redeeming features. There was a great deal of acting, for which the school was celebrated; my house did *The Rape of the Lock* and I was cast for Belinda. As ours was the headmistress's house, the production was lavish, costumes hired, a hairdresser engaged and the elite of St Andrews's society invited. As a result of this, I made some friends. The other comfort was the sheer beauty of St Andrews and the old parts of the school, the Queen's Library and the old Buchanan House. The BS's drawing room was one of the loveliest rooms I have known, and when I was summoned for some interview I feasted my eyes on the panelling, the furniture, the Persian rugs and the view from the window over the walled garden. The summer term was lovely, with primroses and bluebells all along the coast walks but, alas, it meant cricket. Hockey had been bad

[67] Mary Bentinck-Smith (1864–1921) had been a lecturer in medieval and modern languages at Girton College, Cambridge, and was appointed Headmistress at St Leonard's in 1907.

enough but cricket for me was much worse. We spent hours every day not only playing but at nets or at fielding practice. An Oxford Blue came to coach us; as he was a retired clergyman he was considered safe to be at large among us. He made me so nervous that I bowled a whole over backwards by holding on to the ball too long (I was not in the class to bowl overarm). So I was sent for a walk in disgrace. I enjoyed fielding at, I suppose, deep field, right at the edge of the field, overlooking the North Sea. Most of the time no ball ever came and I sucked a grass and dreamed dreams. Suddenly a ball would come, I leapt in the air and failed to save a boundary.

All this resulted in unpleasantness. I would be taken aside by the head of the house and told that I was a slack and not considering the good of the house and that I must pull myself together; the fact that I was quite good at lessons and in the A-stream, which would lead to the sixth form eventually, was not relevant. It seemed to me that I had now two choices: either to become a complete rebel and make life intolerable, or to endeavour to outdo everyone in zeal and enthusiasm. I decided to have a shot at the latter course. Suddenly I raced about the place, sent home for special polish for my monitor's shoes, bought a small orange tree at Wilson's nursery for her desk instead of the usual small bunch of flowers, cleaned her silver before breakfast and was incessantly at her side to do her slightest bidding. This attitude met with general approval; I found myself becoming a devotee of heartiness, and everything went better.

At the end of my first year, in August 1914, after a week of summer holiday, Enid and I looked out of our bedroom window and saw a great poster outside the White Lion with one word: 'War'. We were all set for our first holiday abroad. Our kind Uncle Edward Hopkinson was taking us all to Switzerland with his family. We went to a farm in Derbyshire instead. Eric was doing a course at Heidelberg University but he managed to get on the last boat from Flushing and went straight off to join the Cambridgeshire Regiment. Harry also joined up at once. Very soon my father was serving two military hospitals, my mother a commandant at the Red Cross hospital. When I returned to school, we saw battleships and cruisers at Rosyth as we crossed the Forth Bridge, and on one occasion someone let out carrier pigeons under suspicious circumstances and the train was stopped. Young mistresses disappeared to do war work, we made badges and knitted, we were cold and soon badly fed. My Uncle

Alfred Hopkinson was somehow in a position to land upon schools barrels of salt herrings that no one else could stomach. During the winter, we had an epidemic of measles, and some of those who had it mysteriously contracted scarlet fever in our sanatorium. Two girls died, many went home early that Lent term, but I could catch nothing and had to stay. Suddenly I was summoned home for Eric's final leave, Harry was also able to come and say goodbye to him. On the final day Eric walked with my mother to her hospital and said goodbye to her there. My father and I saw him off to France and, as we watched his train out of sight, we walked away together, silent with the same thought.

During the first year of war, week after week, it seemed the BS was sending for one of us to say that her father or her brother had been killed or drowned when a ship went down. She felt this very much and became edgy and unpredictable. There were also Zeppelin alerts and talk of evacuation. In the summer my turn came; Eric was reported 'seriously wounded and missing', mentioned in despatches and awarded the Military Cross. He had only been in France six weeks. I longed to go home, but we all stayed and were all singularly kind to each other.

At last I stopped being a fag, skipped the second social division of middle house and became bottom of the 'Desks'. This meant one had a writing table of one's own in the common room and was only one removed from the monitors. Much more important to me, Enid arrived. Being delicate, she came late. We did not see much of each other actually but sisters, and sisters only, were allowed to have baths together, and this was our club-time. We also had a splendid exeat weekend, though one I am ashamed to remember. We were invited to stay at St Mary's Theological College, actually in St Andrews, with the Galloways. Dr Galloway was the principal, and a pawky old Scot.[68] His wife was an old flame of my Uncle Alfred Hopkinson, quite young and very fond of children. We had a lovely wood fire in our bedroom but the flames kept us awake, so we poured water on the fire and all over the beautiful brass fender and made a dreadful mess. We also found a huge stuffed capercailzie in the passage and balanced it between the bath taps, so that its head and beak protruded to startle the principal

[68] George Galloway (1861–1933) was appointed Principal of St Mary's in 1915 and was active in the contemporary philosophy of religion.

at his ablutions.[69] He had upset us by forcing us to translate English sentences into Latin or French at meals. We were never asked to stay there again, only for lunch.

Soon after Enid's arrival, I had German measles; but, before being packed off to bed, I managed to see her, breathed deeply into her, and she promised to join me later in the sanatorium. Sure enough she came that evening, having rubbed glands and held a hot water bottle to her chest so that a faint rash appeared. Ten days later the doctor said that he had never seen a recurrence of German measles at the end of ten days. It was there that she told me of a rumour that I should be head of the house next term. I thought this unlikely but Enid always knew. Sure enough, my parents received a letter from the BS, saying that, although it was a risk, she had decided to take that risk and make me head of her house. Curiously enough, thirty years later, in the Council Chamber of Cambridge, Sir Montagu Butler, when I was to be elected Mayor said much the same thing about me: 'She has only been on this Council two years. We are taking a risk, but the university members of this Council have decided to take that risk.'[70]

The head of house at St Leonards held an exalted position. She went automatically into the sixth form, got to put her hair up, had a special desk with a huge silver cup with her name engraved thereon and at least one fag. Against these assets, she would be held responsible for the running of the house and for anything or anybody that went wrong. This is perhaps why Lady Reading when starting the WVS (as it then was) in 1938 once said to me, 'If I chose a young woman from the St Leonards, Roedean, Cheltenham type of public school, I know that she will probably be a good organiser, reliable and responsible. It will be a safe choice. For originality, I should look elsewhere.' We came out of a mould but it was a good mould. We, the monitors, ran the house and were called upon to make decisions. I remember a girl (call her Maisie), not a bad girl, but a silly one, inclined to dash off on the back of young men's motor bikes at weekends, for example. At one Sunday evening council meeting the BS said, 'I shall ask her father to take her away. In war time her nuisance value is too high. Unless of course you

[69] Capercailzie is a Scots word for a woodland grouse.

[70] Henry Montagu Butler (1833–1918) was an English academic. He was Master of Trinity College, Cambridge, from 1886 to 1918, and Vice-Chancellor of the University of Cambridge from 1889 to 1890.

monitors feel you can manage her. If that is the case you can try and I shall tell her father so.' We said of course that we would take her on. What the BS told her father I do not know, but he sent me a lovely opal pendant at Christmas. I remember that sadly as I lost it at a dance in Cambridge in the Master's lodge at Selwyn College. Poor vague Mrs Murray, the Master's wife, assured me that it would turn up in the spring when the lodge was cleaned, but it did not.[71]

I was by this time completely indoctrinated with the competitive house spirit. When we lost the shield for games I shared my captain's tears (she was a nice Scot and a great friend) while the rest of the house lapsed into respectful gloom. If a visitor was shown an untidy dormitory, I let it be known that the house was disgraced. Incidentally, the two visitors that I vividly remember showing around were Lady Jellicoe, wife of the admiral and wearing a trimmed hat with white grapes, and a Serbian monk, who neither washed nor combed his hair, smelt very strong and looked saintly.[72]

Enid was a great help in alerting me to the temperature of the house. She would tell me if what I had to say to them was good, only fair or plain silly. I learnt to put up notices and make lists. The worst list was that made once a month for 'hairy-man'. We all had long hair, and a carefully chosen middle-aged hairdresser waited on all the houses in turn. Actually, he was used as a kind of spy and trusted by all to find out who was going to play in what team for matches. Everyone had a set time for him, and anyone defaulting or missed out created great confusion. I wrote an ode to him in the school magazine and he was rather pleased.

I now had three fags as there were too many new girls, and it had a demoralising effect. Even Nanny was shocked in the holidays at the way I now dropped my clothes about, expecting someone to be glad to tidy for me. The Cunliffe-Owen side of me revelled in it; the other half of me realised that it was a very bad preparation for life. I learnt that people enjoyed doing things for me and that it was easier to let them than to do it myself.

The event of my year of office was a production of *Pinafore*. Why the school chose something in which the cast is an almost entirely male

[71] John Owen Farquhar Murray (1858–1944) was an Anglican clergyman and Master of Selwyn College, Cambridge, from 1909 to 1928.

[72] Lady Jellicoe was Florence Gwendoline Jellicoe née Cayzer (1877–1964).

one, I cannot think. However, the whole score was transposed to a higher key for the female voice, sailors' costumes were sent from Edinburgh, and choruses and orchestra trained for weeks. We were to perform for three nights. Enid was Ralph Rackstraw, and I was Josephine. Unfortunately, though we could act, we were bad at singing in time, and the duet 'Refrain audacious tar' had finally to be abandoned as hopeless for us.

My cousin Mary Anson, younger daughter of Uncle George and the 'little playmate', a great friend of Enid, was at school with us, so what more natural than that when her brother Ellis, a dashing cavalry officer, came on leave, he should visit her at St Andrews. He was a romantic figure and straight from the trenches, so he had no difficulty in taking us all three out to meals; the BS was very easy in such circumstances. He came to see *Pinafore* and arranged with his sister to send me a bouquet at the end of the performance. The fags got to hear of this, and I had other bouquets to my great delight. Now the giving and receiving of bouquets was not acceptable in public-school life, nor was the finale of the evening, when Ellis and I had a very fond farewell in the deserted mistresses' common room, but that was not known. Mercifully, as *Pinafore* had been a great success and the BS was pleased, she gave only a fleeting disapproval to the bouquet incident.

Now there were examinations to be taken. The staff was of high calibre; several (e.g. Mary Clarke, later Head of the Manchester High School for Girls[73]) became famous headmistresses, but the lessons were too geared to examination results to be inspiring and, if one was in the top division of the form, one never learnt anything domestic or practical. My last week at school arrived and I, the once homesick rebel, wept at leaving. One night the BS and I walked round the old walled garden and talked of what I should do. I told her that I must go home for a year and try to look after my parents, who had suffered very much over Eric's death and were tired and overworked. 'What do you advise then?' I asked.

'Go to Cambridge and read history, get a degree, then marry a man older than yourself.'

These instructions I carried out to the letter in time. For the moment I was intoxicated by the sense of freedom and the joy of living.

[73] Mary Clarke (1881–1976), a graduate of Girton College, Cambridge, was Headmistress of the Manchester High School for Girls between 1924 and 1945.

3.7 Family Holidays

We went away twice a year, in the spring to Uncle Harry Wills and Japonica at Barley Wood and to the seaside in August. At Christmas we always stayed at home. The first excitement of the Christmas holidays was the return of 'the boys' from their prep school or from Rugby. Enid and I made endless preparation, culminating in a carefully arranged table at the top of the stairs with rather queer sweets we had made and sugared oranges. We hung the traditional notice in the hall with 'Welcome Home' in gold lettering. Then the boys arrived with bursting Gladstone bags full of curious objects that they had made during the term, hyacinth bulbs just sprouting, dirty hankies, small presents for us and, carefully laid on top, their reports. The holidays had begun.

Christmas itself, as organised by my mother, was traditionally German centred as in her own home on Christmas Eve: the blinds were drawn in the drawing room all day, while 'the Tree' was decorated and tables arranged. Every year everything must be exactly the same, even the place of a special small bird on the tree. Our generation was not educated for change. Harry, Enid and I and even Philip, all had our presents done up in time, but Eric was generally out shopping till the last moment; being very generous he used to add to his presents if he had any money left. After tea we all with Nanny and the maids sang 'Hark the herald' outside the door, accompanied by mother's calling to my father, 'Albert, quick, that candle is burning' or, 'Sing louder, children'. Then Philip as the youngest led us all in. The Hopkinson aunts and uncles sent presents and of course Uncle Harry, Japonica and Tutz, who was generally with us.

My Cunliffe-Owen godmother, Gertrude Malet, used to dress beautiful dolls for me, and once, only once, Aunt Emma sent me a present, a horrible pincushion. Grateful patients sent lovely things 'to the dear Doctor's children', but this was not smiled on, as it was often in lieu of settling accounts. One Christmas a goose arrived from such a source but only addressed to Withington. By the time it had been to two other Withingtons in England, it had to be burned. The great thing I remember was that my father put on his slippers, a sign that he did not intend to go out to any more patients that evening, and he put on his velvet 'smoking jacket' to match us in our best clothes.

On Christmas Day we went to church and had our Christmas dinner at midday. My father carved a huge turkey, always telling the same story while he did so, mother calling, 'Father do go on carving while you talk'. We stuffed ourselves happily, little Philip ate orange jelly with an obvious sixpence shining through it while we had Christmas pudding, and Harry regularly ate too many crystallised fruits and was subsequently sick. In the evening we sang carols, and my father fell asleep by the fire. We were very happy.

There were parties but we were never allowed to go to large ones. It was Enid's and my ambition to go to the Lord Mayor's party and to have red-riding-hood cloaks, but this invitation was always refused. We did go to smaller and less official ones. Of these I remember two; both were fancy dress. Someone lent us a Greek costume with a key pattern stamped round the hem. I wore this with sandals and a fillet round my hair and felt very important. My mother told me to announce myself as a Greek slave girl, but I thought this rather dull and, when asked my name, said, 'Helen of Troy'. I felt puzzled when all the grown-ups laughed. The other party was at the Rutherfords. The future Lord Rutherford was then Professor of Physics in Manchester. Here my mother, having forgotten till rather late that it was fancy dress, hastily made me into a violet with a wreath off her summer hat and a bunch of Parma violets popped into a little basket. Before tea we played 'hide and seek' and were told by Mrs Rutherford that we could go where we liked but not into a certain room that she indicated. I left my hiding till too late, and the seekers had already 'coo-ed', so in a panic I went in at the forbidden door. At first I saw no one, and then an enormous man rose up from behind a desk and said, 'Who are you, little girl?'

Shivering, I replied 'A violet', and added, with my already developing sense of the dramatic, 'A modest violet and I'm hiding'.

'Come behind my desk here and you'll never be found,' said the great man in his huge voice. Alas, I never was found and nearly missed my tea. We invited Eileen Rutherford, the only child of the house and a contemporary of Enid, back to tea but this was not a success. She was very spoilt and, when we were playing trains, which had been organised by Philip, she would not cooperate, wanting to be the engine driver and not the guard as planned. So Philip, who was quite hot-tempered, bit her. This, as Nanny rightly said, 'was not a very nice thing to do'.

We were quite content to be all together at home and have the boys around to play with us, to go perhaps to the pantomime or *Peter Pan* (which I never liked) for which patients would send us tickets. Once, when I was about thirteen, I went for a week to London to stay with the Malet cousins and Uncle Hugo's adopted girl, Peggy. I saw the sights, and the highlight was a dance in the house. The cousins were much older than I was, and it was very exciting to be at a grown-up dance doing tangos and foxtrots as best as I could. The youngest of my cousins, Barnabas, was in the Royal Marines, and rather took to me.[74] He invited me to stay on for a day and to go to a ball with him at Greenwich. Parental consent was sought but my father telegraphed, 'Return for school forthwith'.

Both Enid and I were keen on acting and, I suppose, like the Cunliffe-Owens, natural mimics. During the first war, my mother was commandant of a big convalescent hospital, and Enid and I during the Christmas holidays used to go and entertain the men with comic turns. True, they were a captive audience, and competition was not great. There was no television, no radio, only an occasional concert party and perhaps an overworked gramophone to play. Still, looking back it amazes me that two teenage girls could amuse a large ward of men for a couple of hours. Our best turn was a dialogue between two Lancashire commercial travellers in a train. We had actually heard this conversation on a journey, and I had quietly taken it down. We tucked our hair under bowler hats pushed well to the back of our heads and settled down to it. Then Enid would impersonate a new born baby, lying on a table, wrapped up in a shawl and make every sort of infantile expression and noise from top to tail. 'If that doesn't remind you of home, I don't know what does,' a soldier called appreciatively. We then had various sentimental songs to sing, when the men joined in the chorus and Mother played the piano, but the favourite song was one that Enid and I made up ourselves, called 'Martha and Judy were twins'. Here we were two charwomen, with cloth caps back to front and no teeth. We managed the lack of teeth by folding our lips over our teeth, after a lot of practice before a looking glass. We should have liked to perform at my father's military hospital, but that was for officers and not considered suitable.

[74] This was Edward Barnabas Wyndham Malet (1894–1961).

Every spring we went to Barley Wood, Somerset, home of Uncle Harry and Japonica Wills, for the Easter holidays. They had all five of us children, with my mother and Nanny. My father came when he could. This annual visit was a treasured part of our lives. Barley Wood was a kind of second home and we knew every corner of the house and garden and much of the woods. The Wills were very rich, and life at Barley Wood very different from our own home. Mother loved it but father felt it rather stifling, I think. Year by year nothing changed. We travelled to Bristol and were met by the station master, who escorted us to the local train with our mountains of luggage, and we chugged off to Yatton. I had usually been sick in the train, which, as the boys said, ensured us a carriage to ourselves. At Yatton we went off in a horse-drawn wagonette, and every mile stirred happy memory. At last the tree that King Charles I hid in, the rooks flying over Presto Wood, the kitchen garden walls and the blue smoke drifting over Barley Wood itself could be seen; now the lodge, the open white gates, the pheasants strolling by the drive and, at last, the house came in sight. Japonica was always standing in the doorway with open arms, and Stokes, our friend and butler for forty years, close behind. It would be tea time, so Uncle Harry was still in Bristol at work.

The house had originally belonged to Mrs Hannah More (of 'ragged' school fame), but the Wills had added on to it.[75] On the ground floor there was a big entrance hall with a huge log fire burning, a drawing room, a dining room, a boudoir, a 'smoking room', where we children had meals with Nanny, a vast kitchen domain through the green baize door, and a collection of gunroom offices, lavatories and so on, making up a masculine world into which we were not allowed to penetrate. Upstairs, a huge billiard room and bedrooms with only, I think, two bathrooms altogether, and then attics, in which we slept when we were small and one of which was turned into a nursery for us overlooking the wood.

For a month we children led a very simple country life. After breakfast Enid and I liked to escape to Stokes' pantry and help him clean the silver, and we would draw pictures on the menu cards and exchange news of the day. He always had a day off during our stay to

[75] Hannah More (1745–1833) was an English religious writer and philanthropist. Ragged schools were charitable organisations dedicated to the free education of destitute children in nineteenth-century Britain.

go bird-nesting with Harry and Eric. Later we went to the wood with Nanny, who would settle herself under a tree with her knitting. We picked primroses to send to hospitals and looked for birds' nests. On either side of Tuckers Grove, where the primroses grew, the land sloped upwards, and Enid and I had each a separate territory at the top of these banks. They were secret places. I remember going off alone and sitting at the roots of a big tree in happy solitude. There were orchids here and an occasional white bluebell among the blue. I picked a special collection of these but I always left them carefully arranged among the tree roots as an offering. It was still and quiet up here and some child's undefined awareness of God came upon me. Beyond where the wood ended were the rolling fields and far, far away lay 'the warren', where people shot rabbits. Someone was leaving; perhaps it was I myself. I would suddenly feel sad and then rush down the slope to the companionship of Nanny, Enid and Philip.

Sometimes we went to see the gamekeeper's wife, Mrs Jackson. Here everything seemed sand-coloured, from Mrs Jackson's hair and eyebrows to the baby pheasants in the incubators and in the sacks of corn. There were ferrets too, to be seen and furtively stroked. After Manchester, it all enchanted us.

In the afternoons we sometimes went for drives with my mother and Japonica, in a victoria or high dog cart. Sometimes there was a tea party with Wills cousins coming over. On Sundays we saw something of our Uncle Harry. He went round his kitchen garden and conservatories after church, and after lunch he inspected his woods. It was then that Enid and I were sent to accompany him. He was an extremely nervous man and shy with children, and Enid and I did nothing to alleviate this. After a time, he would feel a need to retire and told us to stay on the path while he looked at a pheasant's nest; but we insisted on coming too. The walk was cut short and I suppose he complained to Japonica. My mother told us we must always make it easy for a man to 'disappear' and that we had been tactless. As we were naturally quite kind-hearted, we decided to do better, and the next Sunday made things much worse by pressing our poor uncle almost as soon as we started to go and look for a pheasant's nest and we would wait.

Actually, Enid and I were a bit in awe of him, but Philip was not. I have said before that Phillip could be very naughty, and he once ran off to the newly planted rose garden and changed all the labels round, throwing some into the pond. The Scottish head gardener came up and complained to Japonica and, after an enquiry, Philip was led before Uncle Harry in the billiard room. He agreed at once that he had done it.

'Well,' said Uncle Harry, 'that was a very bad thing to do and has given a lot of trouble; now we shall not know which rose tree is which. You must promise never to do such a naughty thing again.'

Philip, unabashed, considered and then said, 'I won't do that again, but I cannot promise not to do something else as naughty.' The result was that Uncle Harry, accustomed to having his own way, admired his spirit.

With Enid and me, there was the occasion of the basket. I had been promoted to sleep in the east room on the main floor, parted from Enid, who remained directly above with Philip and Nanny. Bereft of each other's company, we got a small covered basket which we let down on a cord and filled with notes, the odd sweet or small present. Unfortunately, once this stuck on an outside pipe, and no amount of jerking could remove it. Later Uncle Harry was taking the surveyor round the outside of the house and pointed his stick at what he thought was a bird's nest. We were chittering in the background; a gardener was sent for, but not before Enid and I had talked him into playing for time and giving our basket back to us instead of showing it to Uncle Harry. We did not see a great deal of our mother, as Japonica was so pleased to have her sister that she monopolised her. Japonica was tremendously kind and generous but had the Cunliffe-Owen bossiness, telling mother how to bring us up and emphasising our shortcomings. She had many worries, chief of which was a daily post of begging letters, especially from her sisters Emma and Vicky, and she loved to have my mother to discuss them all.

Harry and Eric had their own ploys, though Eric especially did a lot of work. He took, I think, every prize in history that Rugby had to offer. They were very good to us younger ones, teaching us to bicycle and training us in running for school sports up and down the drive. Enid was very athletic but I was too fat to be much good, in spite of Eric's encouragement.

The only sophisticated experience I remember was our visit to Bristol, when we were older—I was fifteen, I think—to stay a night with Wills cousins. True to form, a Wills cousin, P—, had fallen in love with me when I was hardly fourteen. He told my parents and they said nothing was to be said to me and, if he must send presents, he must send them to both Enid and me. Accordingly, two great boxes of chocolates or two beautiful books would be sent to us at our boarding school over a period of several years. The visit was to his three sisters and was from our point of view a great success. P—was in the army, as it was war time, but as he was delicate he was only guarding something locally. He took us to the zoo, where a monkey seized my plait and took off the bow and darted to the top of his cage. We dared the poor man to go inside and rescue it but he only sent for the keeper. At night we stayed up to dinner, where unfortunately Enid thought it proper to take only one tiny cheese aigrette, realising too late that she could have taken half a dozen. We then sang songs for the three sisters (P—leaning on the piano) and did the most polite of our comic turns. At bedtime we were asked what we liked for the night, and we said, 'Ginger beer and plum cake'. It came on a large silver tray, and Enid and I had a nice little party in bed. But the end was in sight. My last Easter school holidays, the matter of P—had to be settled, and he was told that he could 'tell me of his feelings'. Before that there was a family conference in the billiard room, to which I was called. My parents, Japonica and even Uncle Harry were there. I was told what to say and how to say it. Uncle Harry said, 'It's no good; no girl should marry a man who is not an engineer.' I could not help thinking it would be fun to be very rich. I would buy Enid a hunter. However, the whole event was badly handled as too many people involved themselves in it. I am glad to say P—married someone else and lived happily ever after.

Sometime after the war the Wills sold Barley Wood. Perhaps Uncle Harry was feeling old and tired, and certainly he felt he could do nothing more to improve it. So they moved into Clifton in Bristol. We were all sad, but we had our dreams and our memories of Barley Wood, where, for us, it was perpetual spring. No doubt this was because we never went there as a family at any other season.

The summer holiday was the only one that my father took in the year. We all went off to the country or the seaside for August. He was generally very tired and wanted somewhere peaceful and quiet. We always went to a different place: Cornwall, Pembrokeshire, the Yorkshire dales, Northern Ireland or the Lake District. When possible, we took a vicarage and, with one exception, this was very successful. The bad one was in south Wales, where two Welsh maids were left. They entertained the village nightly on the great ham and the other emergency provisions my mother had brought with her and, as they only spoke Welsh, nothing much could be done about it.

Looking back to these holidays, I realise how happily we lived on our own resources and how little organised amusement we needed as a family. We never took the car; occasionally we had bicycles and rode enormous distances but generally we walked everywhere. If it was a place near mountains, we climbed them, leaving my mother to sketch on some slope. If we were at the seaside, we swam and sailed. We forced my mother to learn to swim when she was over forty, and she succeeded at last in doing a stately breaststroke.

Harry was a strong swimmer (he swam for Cambridge later on) and he took pains to teach me. Encouraging me once from a motorboat to swim across Ullswater, he assured me that I was so plump that it was quite easy for me to endure the cold water of the lake and keep at it. What Eric and Enid did together my parents never knew. He could climb anything and was a tremendous birds' nester. At Morwenstowe in Cornwall, in order to see into sea birds' nests, he used to scale cliff faces with Enid on his shoulders. She would do anything that he told her, was light as a feather and had perfect balance. When they came to the nests, Enid would look in and report. The descent was alarming to watch; Harry and I (the stocky ones) stood in some anxiety below. Not for nothing were the Hopkinsons called reckless. Harry would sail a canoe at Howtown on Ullswater. Once a storm swept down from Patterdale, capsizing the canoe, and Harry swam ashore. 'They're an idiot family', commented the captain of the *Raven*, the lake steamer which was passing at that time. On this boat my father, with Harry, once returned to us covered in bandages, having fallen taking a short cut over a rock face. He was saved by being caught on a ledge by a rowan bush, which cut his nose and mouth badly. My poor Cunliffe-Owen mother, who liked to sit in the sun and potter looking for wildflowers to paint or press, never became accustomed to all this. She liked the evenings, when we all played paper games and there was reading aloud. What we all enjoyed was having our father in an old Norfolk suit, away from his 'little black book', which had his daily list of visits. There was one disappointing occasion when he had to return for a confinement. He had delivered seven of the lady's babies and he felt he could not let her down over the eighth. There was another occasion when, to save the life of a little boy in the Yorkshire village where we were spending the summer, father bicycled in the middle of the night miles to the nearest chemist to get the necessary equipment and save him from his diphtheria attack. Otherwise he relaxed; for a glorious month in the year, we had him to ourselves and he always took each of us children on some expedition alone.

Twice only I remember going away by myself for a weekend, once to St Annes-on-Sea to my godmother Nellie, Uncle Alfred Hopkinson's eldest daughter, who had married a man older than her father: George Harwood, a Liberal Member of Parliament.[76] I was eleven years old. Nellie was very gay and vivacious and obviously having great fun being rich and living in London with three enchanting children. They had taken lodgings in which, it seemed to me, they lived in great state. I stayed up for dinner for which they dressed, Nellie in some different confection every night. I had only a clean school blouse but I do not think I minded because George Harwood was so good with children that I felt quite at ease. It was, I think, the year of Lloyd George's Insurance Bill, and I remember a long discussion upon the subject, I putting my father and other GPs' strong objection to it. Nellie called her husband 'Mr George' or 'Sir', which amused me. The whole family were avid sightseers, and what there was to see, Blackpool Tower and the Great Wheel, for instance, we saw.

The other weekend was spent with another of Alfred Hopkinson's children, Arthur and his wife Dorothy. Actually, Arthur was nearly as old as my father, but a favourite cousin. He was a clergyman and lived near Basingstoke at the time. It was during the first war—I was about fourteen—and there was an adventure which began in the train. The only other occupant of my carriage after a time was a young RFC officer. We got on very well, except that he would play with my plait and sit rather close and finally invited himself over to see me. The cousins asked him to tea and tennis. It was very hot, and Dorothy suggested Arthur take us on the river in the punt. Whether from design or absentmindedness, Arthur put us in the punt and pushed us off. The young man soon moved beside me and started to kiss me. I had not been kissed before, and I missed Enid to share this experience. I wondered what might happen next. Not feeling sure I said firmly we must go back to tea. The whole incident, however, weighed upon

[76] George Harwood (1845–1912) was a British businessman and Liberal Party politician.

me, and I told my parents about it in detail when I went home. My father went glassy-eyed and talked about horsewhips; my mother remained calm, though she said one's husband should be the first to kiss one, but they sent a sharp letter to Dorothy to say they ought to have looked after me better. Dorothy with spirit defended herself. She said it was wartime, and anyhow not much harm was done. As Arthur was a clergyman and High Anglican like my mother, the whole matter was forgiven and forgotten, but it gave me a good deal to think about.

3.8 A Year at Home

I have only kept a diary three times in my life: once for the year 1918 after school, when I was eighteen, then not again till 1937, when we went to live at Bushy House (National Physical Laboratory), and finally in 1946, when I was the Mayor of Cambridge. I cannot account for this. I have only lately come across the 1918 diary and, seen again more than fifty years later, it strikes me as very naive. A picture of such an unsophisticated, uncritical, happy girl emerges, easily pleased and with a great enjoyment of life.

As planned, I came home to Manchester to be with my parents, stricken with my brother Eric's loss (he was of course still posted as missing) and to prepare for Newnham College Entrance Examination. The background was sad. The autumn of 1917 was the worst time of the war; we were short of food and fuel, and most people left at home were overworked. My parents were; my father worked at two military hospitals and had his private practice, and my mother was commandant at a big hospital in Didsbury. Our house was bitterly cold, with no central heating and one or at most two coal fires kept going, though curiously enough we still had three maids and our beloved nanny. Enid went back to school and I missed her horribly; the dates that she was due back for the holidays were the only ones written in red in my diary. My small brother Philip had gone to prep school. However, my eldest brother Harry had been ordered home from Mesopotamia, where he had been in the Royal Flying Corps, to continue his medical studies and was for this year at the Manchester Royal Infirmary. He came home very often for meals and was very good to me.

My mother, had it not been for the war, had planned that I should go abroad to France and then to Germany for languages, and how much I regret that that could not materialise. Though so liberal-minded, she had not yet adjusted to a new look for girls, and a programme was arranged for me, rather like that of her own day. I had singing lessons, though with no particular gifts in that way. After about twelve lessons, I was allowed to sing a song; otherwise I traversed up and down the scales, singing 'Violets, biolets' or 'Caro, baro' to enunciate clearly and breathe properly. It was boring. I also had piano lessons with an old flame of my father's, Miss Ledward; she wore a black velvet suit and a great deal of scent and was somewhat dramatic. She might rush into his consulting room, even when a patient was there, shouting, 'Albert, darling, come and hear her play this prelude.' That was all rather fun. Then there were dressmaking lessons—I was good at sewing, hopeless at cutting out and I always messed up the machine. In later life I made my firstborn son's christening robe, but my husband cut it out for me. I did not learn to cook, because with a cook in the kitchen one was not welcome, nor did I do housework, though my diary has one curious entry: 'guided the vacuum across the floor today'; perhaps they had just been invented.

I had an allowance of £60 a year from Uncle Harry Wills, out of which he had told me to save. My mother started me off as in her day, and I had to have one black evening dress and one white; they were very pretty, but there were few occasions to wear them. I also remember a jaunty black velvet cap with a veil and a 'powder blue' winter coat with a fur collar. I had to wear out my school clothes suitably adapted and, as it was so cold that winter, woolly combinations with the head of Cardinal Wolsey stamped on the thigh—some sort of trademark, I suppose—but I hated them.

My father had taught me to drive the car and, as our faithful chauffeur Moss had been called up, I sometimes drove my father on his rounds. My impression is that one did not have to have a licence in those days, and certainly I was underage when I first drove, as I was only sixteen. I loved this and was quite reliable unless dared to be reckless when, as a true Hopkinson, of course I was. Enid was rather annoyed to be too young to drive, so one day when sitting between my father and me she hissed into my ear, 'Moss never slowed down round corners; if you were a proper driver, you wouldn't.'

'All right,' I hissed back, 'you wait.' The next corner I took at speed, mounted the pavement and crashed into a stone wall. This subdued us both, but all my father said was 'Thank God it's a patient's garden'. A bill for £10 for repairs came on Christmas Eve, and I felt very contrite but, as I said to Enid, 'Probably best to have a good accident early in one's career.' While my father was in the hospital or seeing patients, I read; but, on one occasion on a dark, foggy evening, I was longing to get home for tea, and he was a very long time seeing someone. He apologised when he came out but said it had been a difficult case. 'Father,' I said, 'you've been having tea in there—I see crumbs on your moustache, you mean old gentleman.' And he had to admit it.

Part of my programme was some social work. Once a week I took a girls' club for the Girls Friendly Society, in a chilly church hall with no amenities. However, I had not been to St Leonards for nothing. I coaxed an old piano out of the authority, and we sang and danced and told stories. They were not very bright girls and my idea of producing a play at Christmas had to be abandoned; we were reduced to acting nursery rhymes to an admiring audience of parents. They were rather a giggly adolescent lot, though I had a stable aide-de-camp of fifteen, Lily, who had left school and was in a job. Anyway, we all became very fond of each other, and I kept up with several of them. One I remember, the 'soot extractor's' daughter, married very well, but the prettiest, Jane, in spite of my uplifting moral talks, became someone's mistress and was seen driving around Manchester in a Rolls-Royce.

Of course my chief activity was preparing myself for Cambridge, as I had to take 'Little-Go' in the spring.[77] This entrance examination is now abolished but in those days it included French, Paley's *Evidence of Christianity*, mathematics, Latin and Greek.[78] In some of these subjects I had qualified at school and others, like Paley's *Evidence*, I could do by myself. What I had to have help with was maths, because I was hopeless at it, and Greek, because I had never done it. The whole thing was absurd, as one could, I believe, pass on 33% and, with the Latin and Greek papers, you could have dictionaries. Somehow or other I had to acquaint myself with Greek grammar, St Mark's gospel, *Prometheus Bound*, Xenophon's *Anabasis* and all in a few months. My father in the matter of our education never left a stone unturned to get us the best possible help. He therefore went to the university for advice and was

[77] Little-Go was a special examination in Cambridge for students prior to or shortly after matriculation, to enable the university to verify the student's quality. It was abolished in 1960.

[78] William Paley (1743–1805) was an English clergyman and philosopher who wrote several works dealing with the existence of God.

recommended to the noted university coach, a Mr Grime, who was persuaded to give me individual lessons in Greek and maths.[79]

Mr Grime was a strange looking man with a rather bald, domed head and a huge, drooping red moustache; I went to him in the lunch hour twice a week. I soon found he was proof against all coaxing and feminine wiles. When I could not do my algebra equations, I looked at him appealingly and suggested that we should do them together. 'Now Miss Hopkinson, that won't do at all, you must sit there and try. I shall leave the room and give you ten minutes'; so I suppose I did them. As to Greek, I one day caused him some embarrassment. We were translating Xenophon's *Anabasis* and he asked me if there was anything I did not understand. 'Well,' I said, 'I understand all right, but why on earth should the Greek army take beautiful painted boys in front of them into battle? What could they want with them?'

'You are using an unabridged edition, there's nothing about such a thing in mine', replied Mr Grime. 'We'll leave that alone, and change your book please.'

Seeing there was some mystery, I pressed on. 'Can't you explain?'

'No.'

'All right,' I said, 'I'll ask my father tonight.'

'Undoubtedly the best thing to do' were Mr Grime's last words. Neither of my parents had done Greek, so there was no one to hear my homework. However, not without difficulty, I trained Daisy, the tram conductor (they were mostly women in the war), to hear me. She was always on duty at lunch time on the relevant days, the tram was almost empty, and the drive to Mr Grime lasted a long time. I explained to Daisy and showed her my Greek grammar book. 'Well, Smiler,' (by which name I was known to all the conductresses on that route), 'it looks funny to me, not even proper letters, just little sort of line pictures and hooks in the air; I don't think I can manage it.'

I encouraged her: 'I know what, Daisy, I'll write in what the Greek endings sound like in English, and you'll be alright.'

It worked. 'I'll tell you what, Smiler, I'll bet I'm the only conductress that knows a bit of Greek. Something to swank about,' she said as we arrived at the university stop.

[79] Presumably this was J. E. Grime, who in 1912 established Grime's Manchester Tutorial College Ltd. at 327 Oxford Road.

It appears from my diary that I always went to church on Sunday and sometimes twice with my mother. My father was of course working then. I also went to a study circle, where apparently we discussed various aspects of the character of Christ, but where and with whom I have forgotten. I was quite unquestioning; I accepted a religious tradition. My parents' unworldly point of view made such an impression on us and they had such unshaken faith that I think had they been Mohammedans or Buddhists I should have followed suit. This faith, though different now, has never left me.

I think that I was easily pleased. Day after day I seemed to go for a walk, searching for food and doing household shopping, being taken out to tea with my mother's friends and bicycling miles into Cheshire. Again and again I note: 'Had a topping time.' 'Great fun today'. Life is good when you are young. I have always had the Hopkinson gregariousness, and there were plenty of girlfriends around not yet doing war work, cousins of the same age and, above all, my immediate family. I think my mother and Nanny must have been discussing me when on one occasion I came into the room to hear Nanny say: 'Well, there's one thing, she's just as happy with old ladies and children as young men.' I was only depressed by the almost daily rain in Manchester, Manchester's ugliness and the times when Enid and Philip went back to school. I vowed to Enid that when I was independent I would never live in Manchester again; but in fact I was back, after a three-year interval at Newnham, for sixteen years.

I had my amusements. There was no television and no radio to resort to in the evenings, so when at home my parents and I huddled over a small fire and read, my father so tired that he often fell asleep. I often went to the theatre, the opera and the Hallé concerts. Manchester was still full of rich people, many of them our friends, among them German Jews. After the Royal Family changed their name to Windsor, many of the local families did the same. It was very confusing, as in one family some would change but some would not. They tended to translate their German names literally; for example, the Steinthals became the Stonedales.

These people still had their motor cars, which were propelled by gas, I believe; certainly the cars had on their roofs what appeared to be vast laundry bags, which inflated and collapsed by turns. Many of these friends were very kind to me and took me regularly to plays and concerts. My brother Harry also took me, sometimes even out to dinner first, though I note that when we went to the Midland Hotel I wrote in my diary, 'The Midland is a bad place, where men bump against you and murmur things—unpleasant'. Harry was very good company and felt responsible for me. I sometimes went to see him at the infirmary and met his fellow housemen, on the understanding that I did not 'encourage' them, as they were not suitable in the long term. He brought one or two nurses home to meals but he had a most curious technique with some of them, wishing, in the rather bossy Cunliffe-Owen way, to further their education. He set one young night sister general knowledge papers and in the middle of the night, after coffee and a little light love-making, she had to put in the answers to 'Who is Prime Minister', 'Who is Editor of Punch' and 'Where are the rivers Tigris and Euphrates?' Poor sister, no wonder she was nervous when invited to our home.

Young men began to come around. An alarming number of my brothers' Rugby and Cambridge friends had been killed, but those who were left came on leave and took me out, almost always with my parents and Harry in tow. Mother was much cheered up and amused. 'You can have flowers, chocolate or scent, but never jewellery, remember,' she said, contentedly. One or two of these young men wanted to marry me. Concerning one of these, a man much older than I, my diary reads, 'Played golf with J. At the 5th green he suddenly proposed. We sat on a gate, and it was rather bad. Why should I make people miserable on such a lovely day? He was very sporting, but I had to refuse'. It was in the holidays, so Enid was at hand to go over the matter and remind me that she had warned me that this was coming.

In due course I went to Cambridge and took Little-Go and managed to pass. But before that I'd had to have interviews at Newnham with my tutor, Miss Firth, and B. A. Clough, in whose hall I would be.[80, 81] My mother came too, and we stayed with Uncle John Hopkinson's widow, my Aunt Evelyn. My mother came to both interviews.

[80] Catherine Beatrice ('The Firkin') Firth was Director of Studies and Lecturer in History at Newnham College, Cambridge.

[81] Blanche Athena Clough (1861–1960) studied classics and was secretary to the Principal of Newnham College, Cambridge. At one time she was offered the principalship but refused. She was known to her family as Thena and to the students simply as B. A.

Miss Firth was not impressed with me. She was a queer little woman in a sloppy cardigan, and I was soon to note that the dons all wore cardigans, an ugly overworked garment. She lived entirely on nuts and fruit and did not strike us as attractive. She asked me why I wanted to come to the university, a good question after all. I replied that all the A-stream at St Leonards did.

'And why are you reading history?'

'It's really the only subject I could do.' My mother chipped in and said that I was to be prepared, only if necessary, to earn my living, and a degree was a desirable qualification. I had a feeling even then that Miss Firth could be very awkward in her relations with other people, and subsequent events justified this. Obviously, she was dubious about me. The interview with Miss Clough was much better. An urbane woman with striking pansy eyes, as my mother said, and a sense of humour; she might be a manipulator but I thought it possible that I might manipulate her. She was friendly, we questioned one another, and the meeting passed off well, mother having assured her in so many words that I was a very nice girl. The rest of the year, before I went up, was passed rather as in Jane Austen's time, in a quiet life at home. I read a great deal, mostly the English classics, and books on history and poetry. Of course I wrote poetry, the inevitable ode to spring, flowers and love; I prefer to forget them. At the time I sent a selection to Wilfrid and Alice Meynell.[82, 83] I had a gentle reply, telling me that they showed promise and that I should continue and read as much as possible. This I innocently accepted as great encouragement. Time was spent in visiting friends and relations or having them to stay at home.

The summer before I went to Newnham I went to Scotland, visiting Enid at St Leonards, where I paraded happily as a 'senior' staying with the headmistress in my old house, then on to an old school friend near Kelso. Mollie Roberton had been captain of games and was extremely horsey, and I really wonder why we were friends. I was bad at games and terrified of horses, but as usual I enjoyed being there. The River Tweed ran at the bottom of their garden, fringed with masses

[82] Wilfrid Meynell (1852–1948), who sometimes wrote under the pseudonym John Oldcastle, was a British newspaper publisher and editor.

[83] Alice Christiana Gertrude Meynell, née Thompson (1847–1922), was an English writer, editor, critic and suffragist, now remembered mainly as a poet.

of yellow mimulus, and I loved the whole countryside. My host was 'factor' to Lord Ellesmere, I think, and a rather bad-tempered man. I managed to mellow him but, after my own courteous father, he was rather a shock.

I remember I was very late for breakfast one morning, so my covered plate of eggs and bacon had been placed on some sort of heater. At any rate, it suddenly blew up and almost hit the ceiling, making a great mess. Mr Roberton's language made me blush. He also made a great scene when Mollie and I decided to see Jedburgh Abbey by moonlight with some young man on leave, who was staying in the house, with his motorbike. One of us rode pillion and the other in the sidecar. The abbey looked wonderful on the June night by full moon, and perhaps we dallied; at any rate, we ran up a hedge on the way home, had to have some repair and arrived home at 2 in the morning. Mr Roberton was waiting for us with a gun. Speaking of guns, I took my first shot there with the son of the house; I shot the tail of a rabbit with a rifle. I am not proud of the episode, as the rabbit was sitting and the gun kicked fearfully.

But Edinburgh was the highlight of that visit. I stayed with Uncle John Hopkinson's daughter Nellie, who was old enough to be my mother. She was the only daughter left, after the fatal alpine accident in her family, and had married Alfred Ewing, then Principal of Edinburgh University. They lived in one of those lovely houses in Heriot Row, with views right across to the Firth of Forth. They seemed very old to me; I was surprised that they had achieved one small boy, John. They set out to give me a lovely time. My diary records ecstatically, 'I am living on bubbles. It is glorious.' Sir Alfred was a small and rather irascible man, but he bothered to show me the sights of Edinburgh and gave me, I remember, a string of green beads, to his wife's amazement, as he was not given to that sort of thing. They had a grown-up dinner party with university people and the Scottish portrait painter of the moment, Mr Shields.[84] I was upset at the dinner, as Sir Alfred suddenly shouted, 'Nellie, there's no sherry in this soup.' However, he made up for it by kissing her shoe at the end of the evening—no mean feat as it was difficult for him to get down to the floor as he was very portly—and I was impressed. After dinner, I see, I played the piano to them all. Next day the painter took me to see the Raeburns at the

[84] Douglas Gordon Shields (1888–1943) was a Scottish painter.

Scottish Academy. What excited me most was the going to lunch in the HMS *Tower*, which was berthed at Leith, with a naval officer who turned up at the Ewings. 'It was heavenly, we drank, and I smoked a cigarette. Malcolm showed me all over his ship. The engineer lieutenant is angelic. It was all topping.' In between I played house with little John, an antidote, I should think, to his over-strict parents.

And all the time the war dragged on; but in the late summer there seemed a break. By August I write, 'Very good war news'. I read *The Manchester Guardian* avidly every day, and my diary generally records some war event as a final note for the day. Thus, from that January onwards, I have 'acted plays with Enid for a small audience. Russo-German peace negotiations. Grave meat shortage'. Or 'Managed to have a hot bath. Lloyd George puts forth his war aims' or 'Very sad, Enid back at school. Woodrow Wilson gives splendid speech and offers Peace Programme'; 'Helen to tea, and Harry to dinner. Lovely time. Munition girls strike for more food. Demonstration by them in Manchester.' 'Meeting of Guild of Helpers. Mother spoke very well. Sinking of Goeben and Breslau off Dardanelles'. 'Walked to Northenden to try and get raspberries. Ex-Tsar has been shot, poor devil'. And so it continues. By the time I went up to Newnham in October, we were within sight of the Armistice, and I had finished my year at home.

3.9 Student at Cambridge

My mother took me to Cambridge and 'settled me in'. We stayed with Aunt Evelyn (widow of Uncle John Hopkinson, who had been killed in the Alps). She lived in some splendour and entertained a great deal; during my time in Cambridge, I could go there whenever I liked. I really preferred in the end to go to her daughter-in-law, Mariana, who lived in the house next door, because she had seven small daughters and I always felt at home there. Her husband (Bertie Hopkinson) had been killed doing some aeronautical experiment in bad weather the year before. He had been Professor of Engineering in Cambridge.

Cambridge is very beautiful in the autumn. I revelled in the look of it, and the Newnham garden was a great pleasure to me. In all the fifty-odd years I have known that garden, it has always been immaculately kept. Everyone was proud of it, and I remember what a fuss there was when our wartime goat (thought to be a billy) had twins on the lawn. It was felt that the incident might spoil the grass. My room in Clough

Hall was on the ground floor and overlooked the garden. I felt rather lonely and generally apprehensive as I settled in the first day; the place seemed seething with unknown girls, and the corridors were piled with anonymous luggage. Some of the girls were much older, having returned from war work to complete their degree courses.

Gradually one identified people and made friends. There were the obvious academics, pale girls carrying piles of books and who seemed to do nothing but work, and the hearty, who joined every society and regularly bicycled out to Girton to play lacrosse and hockey. There were the rebels, running round with petitions to abolish cramping rules or sack a member of the staff, and a sprinkling of the original and unconventional. Also the pretty and sexy, only the word was not then known. As I was thoroughly ordinary, my friends have always surprised me. One, Dorothy Steven, was a serious Scot who already had a first class degree in classics at Aberdeen and was headed for the same at Cambridge. She was a pillar of the Student Christian Movement; so, under her influence, I became a rather tottery pillar too. Then there was Molly Tallerman, a square, no-nonsense girl, whose people lived in Cambridge and at whose house I was always welcome on Sundays.[85] She became an outstanding probation officer in London. The third friend, Olive Schill, came from Alderley Edge near my home; she was older and had nursed through the war and later had a career in social service.[86] These were the people with whom I drank cocoa and had those long abstract arguments at night that are the essence of college life. None of them went to dances and parties; they were all very serious minded and later, when Cambridge life became very gay and went to my head, they were a salutary and accepted influence upon me.

The first weeks were dull, and I found most of my history lectures boring. The lecturers tended to lecture (reading what they had to say) from the books they had written. Some of us soon felt it would be more profitable to stay in one's room and read the original. It has always seemed odd to me that lecturers never seem to learn how to lecture. It was and is an accepted tradition that, if a person has the academic knowledge, he must be able to pass that on in the lecture room by instinct rather than training. The lectures on economic history

[85] Margery (Molly) H. Tallerrman was a friend of Lady Bragg, whom she met in Cambridge.

[86] This is Olive B. Schill of the British Broadcasting Corporation in Manchester.

were an exception to this. Professor Clapham had not yet returned from war work, and Dr Eileen Power was giving them instead.[87, 88] She was an inspired lecturer, answered questions well and was beautiful to look at. An American, Lapsley, lectured at Trinity College in constitutional history.[89] A rather tiresome showman with a passion for the English aristocracy, he peppered his lectures with jokes and asides which, be there only two male undergraduates present, we women were not allowed to hear as we were placed 'below the salt' and ignored. It must be remembered that we were not then totally accepted in the university and could not take our degrees. I once went to hear Sir Arthur Quiller-Couch (the famous 'Q') lecture.[90] The men not yet back from the war, the audience was all female except for one coloured young man; but Q, looking around arrogantly, began 'Gentleman—'. We attended lectures in hats.

We were soon in the grip of the terrible so-called Spanish flu epidemic, which really was like a plague. I remember my father writing that patients turned black with tremendously high temperatures and could be dead within forty-eight hours. He himself worked night and day and in the end caught it, with the result that, when I went home at the end of term, he had an abscess on his sciatic nerve and had to have an operation from which he nearly died. At Newnham about forty maids were laid low with it, but very few of us students. We had to clean and do our own fires. There was no central heating in those days but each room had an open fire. I remember several times fainting with cold working in the damp autumn mornings.

The other colleges were almost empty. The only men about were the old or the wounded, housed in a great hutted hospital where the university library now stands, so we all lived very quietly. I think we had a girls' Saturday night dance and regular debates. I remember seconding the opposer at the freshers' debate on 'Criticism is of more value than enthusiasm in this world'. I was terribly nervous, but felt I ought to try.

[87] John Harold Clapham (1873–1946) was a British economic historian.

[88] Eileen Edna LePoer Power (1889–1940) was a British economic historian and medievalist.

[89] Gaillard Thomas Lapsley (1871–1949) was a Fellow of Trinity College, Cambridge.

[90] Arthur Thomas Quiller-Couch (1863–1944) was a British writer who published under the pen name of Q.

Suddenly in November it was Armistice Day. Everyone was working that morning, when a great noise of cheering, hooting and pealing bells broke our quiet. In the town everyone was running, shouting, kissing and waving flags. Old ladies on tricycles were ringing little handbells. One had to break something, so students and army cadets smashed up a shop or two run by pacifists. There were thanksgiving services in the churches and in Newnham itself, but after dark army cadets stationed next door at Ridley Hall swarmed into our grounds, lit bonfires and started to burn furniture. The authorities acted sensibly. The army were invited into Newnham and we danced, the principal and the commanding officer taking part. Everyone was violently emotional, laughing and kissing. No wonder my diary says: 'Most lovely evening'. As I went to bed, I thought of my parents and wondered whether we should now hear something of my missing brother Eric.

My second term, the gentle pattern of work, walk to Grantchester, tea with one of my friends, work, cocoa parties, and Sunday calls on Cambridge people that we knew took on a dramatic and sudden change. The army was 'demobbed', the colleges filled with people still often in uniform, and lectures were suddenly crowded. Many of the men were older than the average and arrived with 'gratuities' and an urge to spend money and enjoy life. Suddenly there was a craze for dancing—dancing in the evening, and dancing in the afternoon. The *thé dansant* was introduced. There was loud jazz music, the crashing, clashing noise of tunes like 'The wild, wild women' and 'Everything is peaches down in Georgia'. Most of our letter boxes were soon filled with envelopes stamped with college crests (for men were not allowed to wander in at any time and see us in those days), asking us to this and that. Societies of every kind sprang up. Saturday mornings, we had a pavement club on King's Parade, a sort of glorified sit-in for an hour but without specific cause, with undergraduates reading, playing cards or just chatting, filling the pavements and the road itself. Traffic was paralysed. I remember winding knitting wool round a lamp post and a policeman, but it was all good-humoured. I became very excited and restless, wandering in and out of people's rooms in the college in fear of missing something.

One night I went to a dance in the Master's lodge at Downing College, where I met a young medical, Vaughan Squires, who asked me to join a party for a *thé dansant* at the Guildhall. When I arrived there, he said he would like me to meet his cousin, Major Bragg. Beside him stood a dark man in uniform, with decorations of military OBE and MC on his tunic. I thought him good-looking. We danced together and then sat out. Of course, we knew something about each other, as he had been the great friend of my cousin Cecil Hopkinson (Uncle John's youngest; too young to have gone in the fatal alpine expedition, he had died of a head wound in the war) and knew many of the relations. He was not at all what I had expected, and I remember telling him so at once. Somehow I'd had the idea that he would be a small man with spectacles, shy and vague. The Hopkinsons had told me with bated breath that he was brilliant scientifically. He turned this aside and told me what he had heard about me, and we dallied with this pleasant theme for a while. I had not expected scientists to dance, and he said that he never did before the war but, if people dragged him to a dance, he generally enjoyed himself. Then he told me that he was a Fellow of Trinity, and how good it was of the college to keep this Fellowship open for him, after four years' absence in France when he was doing sound ranging with the army. The rest of the interval was spent in telling me about this. Finally, he said that the cousins who had brought him today had invited him to the Victory Ball next month, and he asked if he would see me there. I assured him that I would be there in fancy dress, and yes, I would save him a dance. He apologised that he was not very good at the foxtrot and could he have a waltz and, as we returned to the floor and parted, I promised this. That was our first meeting.

The Victory Ball took place at the end of term, Mollie Tallerman, my friend who never went to dances and was wedded to brogues, men's ties and fresh air, was prepared to come with me, and I joined her family party. They were a rich lot and lived in a big house in Trumpington Road; the house was afterwards taken by the Dukes of York and Gloucester when they were undergraduates. A hairdresser was engaged to do the girls' hair. Mollie agreed to be some sort of peasant woman, and I had been lent a wonderful mauve silk dress with panniers, and amethyst jewellery, and I had a powdered wig and patches. I suppose it was my first ball. Mr Bragg was there and, in spite of the Tallerman protest at going 'out of party', I danced with him.

When I went home I found that my father, having been so ill, had agreed to retire. Though only in his fifties, he had been in general practice thirty years. Uncle Harry Wills and Aunt Japonica bought us a house in Adams Road, Cambridge; it was thought very expensive at £3,500. He could not suddenly be idle, so he wrote to the Professor of Anatomy at Cambridge and was accepted as a demonstrator in anatomy. It was an experiment and a risk, as I doubt if my father had read a medical book since he qualified. Actually, it was a great success, and the medical students found it refreshing to have a man who had really practised medicine. That spring vacation I was with my family at Barley Wood as usual with the Wills.

The navy now appeared in Cambridge, heralded by an amazing poem by Rudyard Kipling starting with 'Oh, show me how a rose can shut and be a bud again', a most inappropriate simile, and ending by exhorting Cambridge to 'give them their hour of play'. There was no question that immediately they seized that avidly. There was more dancing than ever; colleges had their own dance evenings and their own jazz bands. The most glamorous-looking man was Patrick Blackett, later to become a famous physicist. I used sometimes to go to one of these clubs with his future wife, Pat Bayon, who was half-Italian and quite the most amusing girl at Newnham.[91]

Now I met Mr Bragg a third time, unexpectedly. I was going to a dance held by Professor Inglis, the engineering professor.[92] He had a lovely house at Grantchester, and I was told to join a taxi full of guests outside Newnham; and in it, to my surprise, was W. L. Bragg, holding a shoe bag with his party shoes. I think my aunt (Mrs John Hopkinson) had schemed that. We arrived early to hear Professor Inglis shouting upstairs to his wife, 'What are we going to drink tonight my dear?'

'I have told them port, Charlie'.

'Port? You can't drink port all through the evening.'

'I can, Charlie. What better? If you want something else, tell them.' Mrs Inglis was, even by Cambridge standards, eccentric. If one was to be an outstanding eccentric in Cambridge, one had to exaggerate the role, which she did; but I loved her. I danced with W. L. Bragg, and I remember our conversation in an interval between dances. He told

[91] This is actually Eva Constanza Bernadino Bayon (1899–1986), aka Dora Higgs, a student of modern languages at Newnham College. She had an Italian father but had been raised in Rome by an English couple who nicknamed her 'Pat'. Thus, to intimate friends, she and her husband were 'the two Pats'.

[92] Charles Edward Inglis (1875–1952) was a British civil engineer. He invented the Inglis Bridge, a reusable steel bridging system which was the precursor to the more famous Bailey Bridge of the Second World War.

me that he was shortly leaving Cambridge to succeed Rutherford as Chair of Physics in Manchester. I was very startled. I told him rather fiercely that he would find it wet and ugly there after Cambridge and asked him why he wanted to leave Trinity. He had, I found, strong ideas about people 'sticking all their lives at Cambridge; they get too comfortable and dug in there'. In many ways, he did not want to go himself; he said he was anxious about taking on something new and that to succeed Rutherford was an intimidating thought. I wondered what his ambitions as to his career might be; I did not know him well enough then to know that he had no specific ambitions but would take things as they came. This was so different from the Hopkinson family, who were very conscious of career structures and climbing to the top of their particular trees, that I had to digest his attitude. Presently, I pressed on and he finally admitted that he supposed 'to be Head of the Cavendish Laboratory here would be any physicist's dream' and explained that this was because it had the great traditions of Clerk Maxwell, J. J. Thomson and Rutherford directing it, thus attracting the cream of young research men.[93] He ended by saying again how very lucky he was to be going to Manchester, and I politely hoped that he would be very happy there. I did not see him again till May Week.

Meanwhile, my nemesis had caught up with me. Life had been too exciting, I had not done enough work and, in our May examinations (Prelims), I was the only one of my history year to get a third class. My tutor, Miss Firth, sent for me. It seemed probable that I might have to leave the university. Now, at that time, Newnham dons lived in a rare atmosphere. Intellect and academic achievement were all that counted. In speaking of some student's brilliant academic career, they would add sadly, 'Unfortunately, she married.' (It ought to be said that, with the passage of time, things changed, and many of the Fellows themselves married.) They had, in a curious way, withdrawn from the life of human feelings, so that personality and character were not seriously considered or understood. To me, it made them a race apart. I decided to awaken some human feeling in Miss Firth.

[93] James Clerk Maxwell (1831–1879) was a Scottish mathematical physicist. One the greatest physicists of all time, he was most famous for the so-called Maxwell's equations of electromagnetic radiation and numerous contributions to thermal theory. He became Cavendish Professor in 1871.

I took her into my confidence; she listened avidly to what I had been doing instead of working, and it seemed to wake something buried in her. At any rate, after a promise of reform, there was to be no further question of my leaving. I had promised my father that whatever happened I would stay the course at the university, as there were some Hopkinson relations who doubted I could do this.

With relief, I embarked upon my first May Week. There came a note from Mr Bragg asking me to have lunch in his rooms in Neville's Court at Trinity College, go to the boat races and meet his family. Quite a large party was gathered when his parents, who were a little late, arrived. His Australian mother seemed a very exuberant woman, bringing in with her the air of a party; she was very neat in her appearance. She seemed to me to be amazingly young to be a professor's mother, and indeed she must have been under fifty. She talked a great deal, and there seemed a disappointment in that WLB's twelve-year-old sister Gwendy had whooping cough and could not come. I was to learn later that WLB was devoted to his small sister and that he was really disappointed.

Sir William Bragg was a large man, beaming genially but rather silent. He made pleasing noises, a way of communication common to scientists, I was to discover. He grunted, made a curious sound like the half-suppressed buzzing of a happy bee and chuckled. Conversation was general and very hearty. I felt Lady Bragg looking at me several times as we ate lobster and crème brûlée. At the races WLB pointed out two of his friends, huge men, weighing down a canoe; they splashed everyone indiscriminately as they passed. They were Charles and Billy Darwin.[94] The following week I went to May balls five nights running, returning home to Adams Road after breakfasts up the river to go to sleep. Nanny said drily I might have been a night nurse. At the Masonic Ball I had a dance with WLB and, walking in the moonlight in sitting-out time, he asked me to marry him. It was agreed that we did not know each other as yet; I had only seen him about half a dozen times. We went our separate ways after that till the Long Vacation. Actually, I went to a Student Christian Movement conference at Swanwick with my friend Dorothy Steven, and for some days wallowed in mud and rain and religious controversy; but I enjoyed it.

[94] William Robert Darwin (1894–1970) was a stockbroker and younger brother of Charles Galton Darwin.

The Long Vacation brought WLB and me together again. My parents welcomed him to meals at home; we had Philip, home from Rugby for the holidays, to an enormous breakfast in his rooms, entertained Enid, and WLB generally became acquainted with my family. He and I rambled over the countryside or took a punt up the river. Sometimes it was an uneasy friendship; he was shy and felt on probation, and I felt uncertain about everything. These excursions were different from any that I'd had with other young men. WLB was very serious minded in those days, and we had very serious conversations and very seldom about people. He knew a great deal about nature and was tremendously interested in the ways of flowers, birds and animals. I remember once we watched the comings and goings of a water rat for what seemed hours. Not that I did not enjoy all this; I did, as I always loved the country. Every now and then he would come out with some whimsical remark, and I realised he had a subtle sense of humour, which struck a sympathetic chord in me. Sometimes he talked about himself and his life. He was at pains that I should know all about him and by degrees I heard the whole story. Subsequently he wrote this story in what would have been part of his autobiography had he had time to complete it after his retirement.

3.10 Engagement and Marriage

That Long Vacation had really been a rather stormy time. I did not know my mind, and WLB was in a hurry. When he really wanted something in life, he always was. It seemed all too overwhelming; at the end of the summer we parted and neither wrote nor saw each other again. My family were disappointed, as they all loved him, but his parents were assured by my Aunt Evelyn in Cambridge (who was very annoyed) that their son had had an escape; I was 'unstable and a sad flirt'.

Our family went abroad to Brittany for a holiday, on a kind cheque from Uncle Edward Hopkinson. We took Nanny too, with a packet of tea for her so that she should not feel homesick for her amenities. It was the first time Enid, Philip and I had been out of England and we all greatly enjoyed ourselves. Various Cambridge friends were around and Enid and I used to go sometimes and dance at the casino in Dinard. When I went back to Newnham, though I still went out a great deal, I worked. It was fun having Enid again; she had left school and was living at home. She shared my father's great interest in rugger; he watched the varsity play almost every Saturday, with Enid in attendance, and it was said that Enid was consulted about form in choosing the varsity fifteen. Certainly I remember being very impressed when the current captain finally gave her his international badge after he had played for England.

My parents were a bit worried about me by now but were comforted when I went off in the Christmas vacation with Mollie Tallermann to do social work at the Cambridge settlement in Camberwell. This was an experience. We students were billeted around the square where Talbot House, our headquarters, was situated. I had an attic bedroom in a deaconess' home, and very bleak and cold it was. We fed in Talbot House. The first night we had cruelly fat beef and an apple pudding into which the taps had run freely.

Figure 54 Alice Grace Jenny Hopkinson in 1918. (Courtesy: Patience Thomson)

By day and night we visited schools, evening institutes and the like, walking miles through a maze of mean streets, poor houses and endless fried-fish shops. I was fascinated by it all. I remember vividly one 'night school' as it was then called, with a Mr Bishop in charge to show us round. He was a Dickensian character, wearing full evening dress. His accent and indeed his vocabulary were very difficult to follow. The singing class I especially remember. Mr Bishop, who was not actually taking the class, joined in taking sometimes alto and sometimes bass in an amazing manner. He was then presented to take the class himself and show us how it should be done. Mr B. then seized three rulers and picked out Annie Laurie in tonic sol fa, keeping each ruler separate for each part. How the class followed him I cannot think. They were besought to sing 'God bless the Prince of Wales', shortly and crisply and not like a greasy patent bacon machine! It was very good. But there followed a minor tragedy. A young hooligan stepped among us, having just divested himself of an important part of his clothing. Mr B. dashed at him and ran downstairs three steps at a time, holding him by the ear and at the same time delivering a masterly lecture on indecency. Coat-tails flying, Mr B. boxed his ears and literally threw him out at the front door, shouting, 'Expelled my lad, expelled'. The urchin, pale-faced and leering, spread his fingers to his nose and swore richly; then he pressed his face pathetically against the glass door panel and glared at us. We then all went to the gym, where Mr B. in showing off his new patent fire alarm accidentally set it off and the bell clanged. He climbed a stepladder and clutching the bell said solemnly, 'Naughty, naughty, give over, Sir.'

The gentle deaconesses in my 'billet' asked me if I should like to be called in the morning. I said, 'Please, about 7.30.'

'But we shall be in church then, my dear,' was the reply. '6.15 or not at all.'

My room was so shabby, with the paper peeling off the wall, that I found it had been rejected by the last social worker. I think there was no electric light; certainly I had a bath with a candle. The geyser made such a terrible noise that I thought it would explode, so I dashed out into the passage, fell over a huge yellow dog which appeared from nowhere and broke my candlestick, which had been loaned by a resident district nurse. I was not a very satisfactory visitor. I wrote exhaustively to Enid and the family of my adventures of those ten days, which made a great impression on me.

Finally, tired and somewhat flea-bitten, I returned to Kensington for some relaxation. I stayed with two eighty-year-old sisters, Frances and Evelyn Redgrave. Their father had been a quite well-known painter,[95] and they were lifelong friends of the Cunliffe-Owens; I was devoted to them. Evelyn was very plain and did nothing to ameliorate her state; but she was the practical sister and had a wonderful sense of humour. Frances had been a beauty and had some romantic love story which I have forgotten. They apologised for the urban character of Hyde Park Gate, where they lived. 'When we were children there were hayfields here all round us', they said. From their hospitable house I went out dancing to various night clubs that were fashionable, to the Rectors or the Grafton Galleries and to theatres. 'Ask the young men in, my dear', said Evelyn. 'We like to see them. We shall offer them a glass of Madeira but when the right man comes, we shall ask him to dinner.'

That Easter my friend Dorothy Steven put it to me that we ought to go off together for a reading party and do some work, so Japonica Wills lent us Barley Wood, complete with a housekeeper and 'all found'. We were very spoilt, working by great log-fires, Dorothy in Uncle Harry's billiard room and I in the hall. We worked about six hours a day, breaking off in the afternoon to take long walks in the primrose woods and I showing Dorothy our favourite family haunts. In the evenings we chatted together. Dorothy said once in her clipped Scottish way: 'I have observed, Alice, that a great deal of your conversation comes round to Mr Bragg. Most subjects seem to lead to him. I wonder to myself. . . '. I said that I wondered too.

Summer term started like all summer terms in Cambridge, geared to Tripos examinations. The background is wonderful, with everything in the gardens breaking into blossom, and the smell of lilac drifting in at the windows. In Newnham around the sundial which was a memorial to my cousin, that other Alice Hopkinson, who was killed in the Alps, the little rock plants splashed purple and yellow colours. Cuckoos called so loudly that it was almost impossible to sleep after dawn on the roof, as some of us loved doing. The river lured us in the afternoons. Olive Schill—my Manchester friend—and I bought

[95] This would have been Richard Redgrave (1804–1888), a well-known landscape painter, elected to the Royal Academy in 1851. He had two daughters: Frances Margaret (1845–1932) and Evelyn Leslie (1849–1932).

a punt. We called it *Strangeways*, after the prison in Manchester; it was a suitable name as, when we let it out in May Week at vast price, it sprang a leak and flooded the feet of lovely girls and their escorts. But generally everyone was strained, working at full strength. They were strained, but I do not remember anyone at Newnham having a breakdown. It may be that the college of those days was really something of a sanctuary. The dons, however remote, were at hand in any crisis and solicitous for our health. We had not, as nowadays, men running in and out of the college at all hours of day and night and, as far as I know, there was no sleeping with them then. It was relatively a peaceful place. The Tripos day came. The principal gave us all sprays of white heather. I got a II.2; my tutor, Miss Firth, was pleased and wrote: 'It is so nice that you have retrieved last year on the examination side. Your III in May last year and a II.2 now means a lot more than the people who give congratulations know, doesn't it? In one sense it matters so supremely little, and in another so much. Your loving C.B. Firth.' My dear Uncle Charlie Hopkinson wrote, no doubt expressing the Hopkinsons' sentiments: 'It is a great pleasure to me to be enabled by you to offer my congratulations on your 2nd class honours. Frankly, I had feared that your appreciation of the social pleasures of undergraduate life might have prevented success in the more important side. We are particularly glad that it has not so resulted. ' And my old headmistress: 'Much pleased that all is well, as I may confess I have at times had anxious moments as to how the world and history might agree, and so probably have you.'

There was another hectic May Week, Enid sharing it this year; then the family took a house on Ullswater, where we each had our own boat and had fun with the Hopkinson relations massed at Patterdale. Later I joined Japonica and Uncle Harry Wills in Scotland; they always took a grouse moor, Glenprosen, near Kirriemuir. Japonica would not have any wives of the guns; she used the time for meditation. But she had to have some female company, so she had a niece to walk with her. I suppose our chief diversion was visits from the vicar of Kirriemuir (who was also chaplain to Glamis Castle, where the Bowes-Lyon family lived[96]). He was rather a gossip and carried indiscreet news from

[96] Bowes-Lyon is the name of a Scottish family that includes Queen Elizabeth, the Queen Mother.

Glamis sometimes. Japonica gave a lot of money to his church. It was a very quiet time for me.

We went out to tea with our landlady, a Mrs Ogilvie, walking over the moors in great boots ordered by Uncle Harry, in case we twisted our ankles. As we sat down to a vast Scottish tea, Mrs Ogilvie's cousins—Lady Elizabeth Bowes-Lyon (now the Queen Mother) and her sister, Lady Rose Leveson-Gower[97]—walked in with their dogs. They had the most beautiful manners, as Japonica emphasised on our way home. I thought a lot up there in bed in the vast icy shooting lodge, fishing by myself in the burn. Japonica profoundly disapproved of scientists, knowing nothing about them, and was convinced that they were all atheists.

My last year at Newnham rolled on, until an eventful day in spring, when my father came to see me and told me that 'my friend in the north', for so he always referred to WLB, had been made a Fellow of the Royal Society and that my parents had congratulated him and asked him to come and see them when next in Cambridge. I also wrote, for the first time for nearly two years, and congratulated him.

Meanwhile, during the Easter Vacation, Mollie Tallerman and I went to Devonshire for a reading party together. She said I was often very distrait: 'Everything seems to remind you of that professor. You can't see a few yellowhammers or a white violet, without mentioning him, you know, what's happening?'

'Mollie,' I said, 'he's coming to tea with me on April 23rd.'

'You are an extraordinary girl', Mollie answered. 'Will you ever do your final?'

In those days, we were not allowed to have a man alone in our rooms, so some arrangements had to be made. Louey, our faithful college housemaid, was to bring WLB up the back staircase, and then Dorothy Steven next door to me would be on guard in case of a fuss. The day came. He arrived on the minute, placing a new beech-leaf velour hat and a mackintosh on a chair. We were engaged, to be together for the next fifty years. My parents were of course delighted, and he was warmly welcomed by Enid, Philip and later Harry. WLB became devoted to my parents.

[97] Rose Leveson-Gower, Countess Granville, née Lady Rose Constance Bowes-Lyon (1890–1967).

WLB had to go back to Manchester at once but stopped in London to tell his own parents. Only his father was at home when he arrived, so he told him first. 'We'll get a bottle of champagne, Bill, and keep the news till dinner as a surprise for your mother.' His mother and the little sister Gwendy (by then fourteen years old) came into the dining room and saw the champagne. 'Is that champagne?' his mother asked. 'What can that mean?'

'Bill's engaged' said Sir William.

His mother, fearfully excited, shouted 'Who to?'

Naturally, as soon as possible I went up to London to see them, wearing my lovely diamond ring. We were all a little nervous, but tension was dispelled when we found that Lady Bragg, Gwendy and I had all washed our hair for the occasion! Gwendy was a most attractive teenager and positively squeaked with excitement. I stayed on one night with them without WLB. I recall vividly after dinner sitting between his parents on a big sofa in the L-shaped drawing room in the Ladbroke Square house and telling them all. Lady B. was reduced to tears of emotion, and Sir William grunted and chuckled and occasionally raised both legs off the ground and brought them down with a clump, a curious motion he sometimes made when much moved. The next day, Lady B. saw me off at the station, bearing WLB's first pastel, of woods near Bolton Abbey, as a present for me. His mother, I think, as she said goodbye was giving me some sort of warning. She said that life would not always be easy. 'You must make the running, my dear, and hold his hand, as I have always had to do with Dad.' I wondered then what she was really saying to me.

Back at Newnham, streams of letters arrived and people dropped into my rooms to congratulate me. Actually, most of the third years were becoming engaged, but we each thought ourselves unique. Congratulations varied. One girl stopped me in the corridor and said, 'Is it true you are engaged to W. L. Bragg? Look, I'm just reading one of his books and you don't know the first thing about science. What a waste.' I enjoyed reading WLB's letters of congratulations because I saw whom his special friends were—G. P. Thomson, Harold Hemming, Arthur Goodhart, Charles Darwin and R.W. James—people who were to be part of our life. Then, just before my Tripos, WLB saw a little house for us in Didsbury and had to buy it without my being able to see it until later. We wrote to each other every day, which must have been a strain on him. He had never had anything to do with girls before and he showered presents upon me. Always extremely generous, he seldom bought anything for himself throughout his life; but, for instance, when he thought it would be fun to buy me a hat, and I could not decide between two, he bought both. My father took him aside one day in our Cambridge garden and said, 'My dear man, you must not spoil Alice.'

To which WLB replied with spirit, 'Well, who spoilt her first, I wonder.'

My Newnham days were over, my Tripos done, with mercifully the same result as before. There was no planning a career—my future

Figure 55 WLB's sister, Gwendy, shows her artistic talent and style in this painting of a café interior in South America in 1932. (Courtesy: Patience Thomson)

was settled. Could I have foreseen that one of my daughters and one of my daughters-in-law would one day come to Newnham, I should have rejoiced. The Braggs took a summer house in Brittany, and I spent a month with them all. 'Too long' my mother said and, as usual, she was right. I think it was a strain for them all, for I was very possessive and wanted WLB all to myself. However, we all got to know each other. Sir William had quiet enjoyments of his own: drawing, reading, playing with Patience and, as it were, looking on at family life. His wife often spoke on his behalf. I thought he was a wonderful man but even then I sensed that he found the love between me and WLB rather overwhelming. He did not care for emotion. Once, years later, he said to me, 'Most people cannot expect a love like yours and Bill's'. Lady B. puzzled me; coming from a family who were all direct to bluntness, I found her ambivalent—utterly warm and kind but I was never quite sure if she was expressing what she really felt. Gwendy I loved, and she was to become a lifelong friend.

WLB and I were to be married at Christmas, though I felt I ought to go to a domestic science college and learn to cook; but WLB thought we had waited long enough. I had a last visit to the Wills in Scotland to the grouse moor. Japonica Wills had come round to scientists since her great friend, the Metropolitan Bishop of India, had told her not only that they were much to be admired in their search for truth but that he would particularly like to meet the Bragg father and son. I was very pleased about that and she planned to give us some lovely furniture and Cunliffe-Owen diamonds, and her complete blessing.

We had to furnish and decorate our house. WLB had much better taste than I had, and he already had very fine old furniture bought for a song for his rooms in Trinity from a rich undergraduate who became bored with eighteenth-century mahogany. I had furniture from the Wills, but I had to choose carpets and curtains. Lady B. had set her heart on shopping with me for these. My problem was either to go firmly by myself or agree to her coming with me and, if I did that, it was impossible not to let her choose everything. I decided to give her the enormous pleasure of letting her choose our sitting-room colours. The result was charming. WLB was much amused. My trousseau was another matter. My mother had this in hand: a dozen of everything, and undies embroidered by nuns; and how startled they were when I told them I would like golden plovers, WLB's favourite birds, worked on some garments. They met this challenge nobly after a detailed sketch of the bird from him.

In order to be married in Great St Mary's Church (our parish church was ugly and too 'high'), the vicar insisted I sleep three nights within his parish. Accordingly, I had bed and breakfast in the Master's lodge at Caius College, with Sir Hugh Anderson, a dear little man and a great power in the university.[98] Enid and I were close friends of his children. Maisie, the daughter, married Trenchard Cox, who became the director of the Victoria and Albert Museum.[99] Sir Hugh was very amusing. One morning at breakfast he said, 'I'll show you what I do with this pile of letters.' He placed an enormous paper basket at one end of his long dining room, put the letters in, went to the other end

Figure 56 The wedding on 10 December 1921 of WLB to Alice Hopkinson. Left to right, back row: Gwendy Bragg, Vaughan Squires, Enid Hopkinson and Alice Hopkinson. Left to right, front row: Molly Thynne, WLB, Felicity Hurst and Alice Bragg. (Courtesy: Patience Thomson)

[98] Hugh Kerr Anderson (1865–1928) was a British physiologist and educator. He was Master of Gonville and Caius College from 1912 to 1928.

[99] George Trenchard Cox (1905–1995) was active in museum studies and was the director of the Victoria and Albert Museum in London from 1955 to 1966.

and took a run, and a flying leap landed the little man inside the paper basket on top of his correspondence.

The wedding was duly celebrated on 10 December. It was a Saturday, so that my brother Philip, then fourteen, could have the full benefit of a weekend home from Rugby. The weather was so fine that roses were still out in the garden. As a bride, I had everything traditional: a white chiffon–velvet dress (short, according to the ugly fashion of the day) trimmed with swansdown, a veil over my face caught by clusters of orange blossom, a train of old lace lent by Aunt Minnie, a bouquet of roses, and satin garters for 'something blue'. In those days, we were not bothered by make-up or 'perms'. Lipsticks had not been invented. I had long hair, which was 'water-waved', and simply powder on my nose. The bridesmaids were Enid, Gwendy Bragg, my cousin Alice Hopkinson and two small girls to carry the train. The older girls had golden velvet dresses, girdled with pheasant feathers (in defiance of my mother-in-law, who did not like this flight of fancy) and picture hats to match. The small girls had champagne-coloured silk dresses and bonnets. The best man was WLB's cousin, Vaughan Squires.[100] No less than three clergymen officiated. My cousin Arthur Hopkinson married us (and I said 'obey'), E. S. Woods (later Bishop of Lichfield) gave the address,[101] and the vicar said prayers. WLB always says, in the excitement of his family dressing themselves, he was almost forgotten. I sent him a buttonhole of lilies of the valley.

Among all the relations, there were a number of WLB's scientific friends, and I invited them all to come and stay. We left for a honeymoon in the south of France but first we had a week in Somerset, in a dower house that Japonica lent us, near Barley Wood. We crossed the channel, mercifully on a calm day, as I am a very bad sailor. I remember everyone was jostling to put chairs for themselves on the sunny side of the boat. I urged WLB to hurry to do this. 'Someone has got to be left without the sun', he said. 'We can't all have it.' This was typical of him, but he added laughing, 'But of course, now I've got you, I shall have to fight.'

We spent a few days in Paris, which WLB knew well, and he enjoyed showing me the sights and introducing me to one or two of his friends. There was a very grand lunch at the Duc de Broglie's, with

[100] Vaughan Squires was a medical student at the time in Cambridge.

[101] Edward Sydney Woods (1877–1953) was an Anglican bishop.

footmen in white gloves. The Duke was a scientist and, while WLB and he had a scientific talk, the Duchess showed me about the house. I was surprised that she had two dressing rooms: one full of lovebirds in cages and one full of beautiful hats on stands, like a milliner's shop. We also saw Lucien Bull, a great friend, half French, half Irish, who had worked with WLB during the war. He and his sister took us to the theatre, to a box. I had an idea that one should dress up in Paris, so not only did I wear a trousseau evening dress but carried an immense ostrich feather fan (given by Japonica). I don't think anyone else was in evening dress in the theatre but I was much too happy to care. I also carried a large tapestry bag. We went out to supper afterwards but, when Lucien said he would get a taxi for us, I told him that I knew WLB would love to walk back to our hotel. 'But,' Lucien said, 'my dear young lady, you have beautiful brocade shoes.'

'You look', I answered; and then and there, to WLB's delight, I opened the tapestry bag and produced a pair of most serviceable brogues, which I put on at once at the table and walked out fluttering my immense fan.

I have noticed people often have reservations about honeymoons, but we had none. From the moment we left Paris in the deluxe night express, its sides shining like the flanks of a cow, I was to be enchanted with everything. I had never had a meal on a train, and the dinner was so good that, when the waiter snatched my plate away with three mushrooms still to eat, he had to bring it back. Both WLB and I liked good food, and wine, but WLB really liked simple cooking with perfect raw materials—the 'very best farmhouse', he used to say. Early in the morning, he woke me to see olive trees and vineyards covering the dry brown ground.

He was proud of all the arrangements he had made, as well he might be. We stayed at a lovely hotel at Valescure, near Cannes, with a sitting room to ourselves. The warm sunny days we spent walking in wooded foothills till we reached a spot where we could see the Alps, and at night he read aloud to me. At Christmas, when he thought I might be homesick, he made a surprise Christmas tree, digging and potting a tree from the woods. We had all the children in the hotel to see the decorations we had bought in Nice and we ate sugared melon. He loved having children around. We were discovering each other. I found out two somewhat disconcerting things about him that I had to understand and which had to be explained from time to time to others. We would be deeply talking, I would say something and look at him for response, and suddenly he was far away. 'Willie, you're not listening,' I would cry, rather upset. He tried to explain. Sometimes he would have been diverted by seeing or hearing a special bird or noting some effect of light and shade that he would like to paint; sometimes something that one had said earlier had triggered off a train of thought of his own so that he did not listen any longer—again, some quite irrelevant but exciting idea had leapt into his mind, scientific or philosophical. 'I am sorry, do begin all over again.' During our lives, people felt they were boring him, perhaps that he was 'too clever' for them or that he wanted to get away; but that was never (or very seldom) the case.

Figure 57 Predator and prey. (Sketch by WLB; Courtesy: Patience Thomson)

The other strange thing I was to discover was that he often fell in with a plan or an arrangement that he did not really want to do or liked. His family were the same, but with my direct, blunt relations, I was puzzled at first. 'You say what you want to do and we'll do that, why don't you?'

He would think very carefully and then explain. 'It is like this; I believe I like to please everybody, so I get let in for things I don't really want. Everyone seems to think so much more quickly and know what they want to do. I very often don't know at the time, but often after I've done what was planned, I realise too late it was just what I did not want to do.'

I said that was very complicated, but I would try and find what he did like, in time to help. I had a great deal to learn. We laughed at the same things; he had a subtle and dry sense of humour peculiarly his own. I loved to make him laugh and say, 'You'll be the death of me', but he made me laugh with his original way of looking at everything.

Our lovely honeymoon had to end, and early in January we had to go to Manchester for the beginning of term.

3.11 Life in Manchester

Our first home was a small house in Lapwing Lane, Didsbury, a southern suburb of Manchester where a great many of the university people lived. Lapwings had long left the neighbourhood but there were beautiful trees lining the road, and big old houses and gardens, reminders that not so long ago Didsbury had been in the country. Our house had a small front garden with a few shrubs, and a strip at the back with one very large lime tree. We arrived there on a dark wet winter afternoon, to be greeted (though greeted is hardly the word) by a curious old woman in black, muttering in German and clearly none too pleased to see us. This was Charlotte, the Bragg nanny, who had been housekeeper to WLB in his bachelor days and had been persuaded by my mother-in-law to settle us in. She disliked me on sight and, during the four or five days she remained in the house, she hardly spoke to me but went about muttering, 'Ah, poor Mr *Villy*, God help him.' She was also angry with him as, in the excitement of getting married, he had not sent her a Christmas card. There was some story in the family that she had once, long ago, dropped a baby in her charge, and this fatal accident had somewhat deranged her. My mother-in-law, who had managed her somehow for many years, had an idea that a honeymoon couple would cheer her up, but this was not the case, and I felt very discouraged.

The house was in perfect order and very charming. WLB had very good taste in colour and arrangements and the rooms looked very pretty, but he had to start work at once, so what more we had to do I had to do for myself.

The pattern of our days was set at once. WLB set off immediately after breakfast, walked the short distance to the 'Terminus' and made his way to the university by tram. I did not see him again till about six o'clock. My days seemed very long and strange after Cambridge. Charlotte did everything in the house, and she was followed by a huge gaunt woman called Mrs Mate, of whom I was rather frightened. It was, in those days, a disgrace not to have a maid and showed that a young woman 'could not keep staff'. Twice a week in the afternoon I had to be 'At home' to callers from 3 o'clock onwards; a stream of ladies came, leaving cards in a bowl in the hall and staying about a quarter of an hour. Mrs Mate served an elegant tea, with our wedding cake cut in pieces on a silver dish. They were the wives of professors, doctors and Manchester business men, all, or so it appeared to me, very old; indeed, many were friends of my parents. One rather nice lady, I remember, asked me if we'd had all that we wanted for wedding presents. I said, laughing, 'Yes, everything but a crumb scoop.' The next day, a beautiful silver one arrived with a card: 'To complete your presents'.

The result of these calls was a series of dinner party invitations invariably enclosing a note asking me to wear my wedding dress; after Cambridge I found these rather dreary and formal. Manchester was still very rich, and people seemed to vie with each other in the splendour of their dinners. There would often be a dozen people, all in evening dress, and a procession to the dining room and which I always had to lead on the arm of my host. There was one occasion when he happened to be a very old man called Donner, who was the chairman of the university council.[102] We processed to the dining room down a long passage and he opened a door. 'Tut, tut' he said. 'No lights, that's queer.'

[102] Edward Donner, First Baronet (1840–1934) was a British banker, philanthropist and supporter of Liberal causes.

'Sir Edward,' I said, 'I think we are in your china cupboard,' and so indeed we were!

I can see now in the dining room the long table with a beautiful damask cloth, flowers in silver vases, menu holders, a crystal and silver epergne supporting a pineapple and hothouse grapes and, on a side table, finger bowls, each with a flower floating in the water. The dinner itself would consist of at least six courses, with probably six different wines.

At that time there was a new principal of the College of Technology, as it then was: Bernard Mouat-Jones; he was very amusing, but after our fourth encounter, conversation wore thin, and I remember he said that we should have to fall back on religion or politics, subjects which were not allowed at dinner parties then![103] WLB of course took in his hostess, who invariably told him that she did not know any science and that he must be very clever. We seemed hardly to have finished dinner before we had trays of china tea handed round and could go home. Then of course, WLB and I had all the fun of comparing notes and discussing the party.

Such dinners were not given by our fellow professors, with the exception of one or two chemists, who appeared well off. There were comparatively simple Sunday lunches and suppers. The ones we enjoyed were with Ernest and Shena Simon (later Lord Simon of Wythenshawe), who served huge dishes of kippers on Sunday nights to an interesting mixed bag of guests, and the talk was always stimulating.[104] There would be people from *The Guardian*, bright young men from Metropolitan Vickers, teachers from the two famous high schools, and such like. Lady Simon was very much the public woman, very doctrinaire and a strong feminist. In those days she thought me a complete nitwit and attacked me fiercely for saying allowances should be made for girls taking their examinations during their monthly periods, 'a most defeatist attitude'. I also upset Sir Ernest Simon (then treasurer of the university), who was explaining

[103] Bernard Mouat-Jones (1882–1953) was a British chemist famous for identifying the chemical in mustard gas.

[104] Ernest Emil Darwin Simon, First Baron Simon of Wythenshawe (1879–1960) was a British industrialist, politician and public servant. He was Lord Mayor of Manchester from 1921 to 1922. His wife, Shena Dorothy Simon, née Potter (1883–1972), was a politician and educational reformer in Manchester. She founded the 'Women Citizens' Association' in Manchester.

to me a clever business deal he had made. I said I thought that it was rather dishonest but then added, 'But you see scientists, except chemists, are very pure.' This I was never allowed to forget. We always enjoyed those Sunday evenings.

WLB has always preferred small gatherings, which was just as well as, when it came to returning all this hospitality, our dining room would only hold six at most. Cocktail parties had not yet come to Manchester, so we had to give simple dinners. An intimidating guest who asked to come alone was C. P. Scott, the famous editor of *The Guardian*, who'd had us to dinner. I rang up his doctor, an old friend of my father, and asked for help. 'Easy' said the doctor. 'Get on to Lloyd George at once and you'll be alright'. Mr Scott arrived in a hansom cab wearing a long cloak, and I kept rather quiet while WLB talked about the university. At dinner, when soup came and I had already realised that small talk bored him, I started, 'Mr Scott, you know Mr Lloyd George, I believe.' We never looked back and continued happily till 9 o'clock, when his hansom cab came to bear him to the *Guardian* offices. But I had been at school with his grandchildren, and I really was finding everyone rather old and so earnest.

But there were lighter moments. Some younger people appeared at our table, and then I relaxed, put a goldfish in each finger bowl to make a change and was only just stopped by WLB from blackening the face of our cleaner's boy and dressing him up as a native bearer of hand coffee. We also went to a ball or two in the famous Assembly Rooms with wonderfully sprung floors and splendid Victorian furnishings.

On Saturdays in that first winter term, we used to take a train and walk in Derbyshire; it was very cold and bleak, with long black walls and a sugaring of snow on the hills, and sometimes we beagled there.

I should, of course, have found myself something to do—voluntary work, a part-time job, or taken classes—but it was not done sixty years ago. One 'started a family', but I had not as yet. So I moped and felt homesick, missing my family, returning endless calls in the afternoon, generally in that soft Manchester drizzle, and then waiting to hear WLB's key turn in the door. Actually, I did take part in a small project for the Women Citizens; the project entailed calling at every house on a new housing estate, asking how the tenant used their bathroom and then recording the answer. Sometimes they used it for its intended purpose but sometimes they kept coal in it or used it as a large wastepaper basket. It seemed to me a rather pointless exercise.

All this was hard on WLB. True, he had not a reputation to make, he already had his FRS and the Nobel Prize, but he had a big laboratory to run, and he was at an exciting point in his research. It was a story that I have now heard many times among young scientists' wives and have often been able to explain the situation with understanding from personal experience. The husband comes home in the evening, tired, his head buzzing with his ideas, and itching to get on with his research. He feels guilty, unable to respond to his wife, who had been waiting for the moment of return ready to amuse and be amused. This is what WLB had to say later about our situation:

> Then again there is a difficulty particular to scientists. When such a one is absorbed by some problem of research, he gets sunk into a deep well of concentration and outside events produce no impression—he really is 'lost to the world'. He is then a dull companion. This is hard on his nearest and dearest, who cannot resist pulling him out of his well, from time to time, to see if he is still there, and it takes a long time for him to get back to his problem again. I know I was often torn between the research and Alice, feeling that I was failing her wretchedly as a companion when I was hot on the chase.

Mercifully my family, my parents, my two brothers, WLB's mother and a school friend or two came to stay that first winter and relieved the pressure. Best of all, Enid arrived in the locality. My father was determined both of us should be trained to earn our living (should need arise), and Enid undertook a children's nurses training at the Princess Christian College in Fallowfield, a suburb rather nearer to Manchester, so I had the great pleasure of having her for her free times. Our Aunt Mabel was on the committee.

Gradually, we discovered people who became great friends, for example, the economist Henry Clay and his wife,[105] and the Unwins. Professor Unwin was an economics historian, shy, who made a noise in his throat when he finished a sentence, like a hen clucking, and Mrs Unwin was a pleasure to Willie as she painted very well.[106] Then there was Sir Kenneth Lee, the chairman of Tootal Broadhurst. He was interested in science and admired scientists and believed in research. It was his dream to find out how his products

[105] Henry Clay (1883–1954) was a British economist originally born in Germany.

[106] George Unwin (1870–1925) was a pioneer in economics and business history at the University of Manchester.

could be made uncrushable, and he told WLB that he would be prepared to retain a physicist for that purpose. WLB explained to him that it would only be a sound proposition for a good scientist under certain conditions. Such a man must not be hurried; in fact he might need anything up to ten years in time, and he must be free to publish his results; and, he warned K. Lee, the success of these results could not be certain. Kenneth Lee agreed and, on that basis, a Dr Willows was engaged.[107] The scheme was justified. Willows used to report progress to WLB when they played golf together, finally partially undressing in a bunker to demonstrate that his tie, shirt and underwear were now uncrushable. It was an enormous scoop for Tootal Broadhurst, and I am glad to say Willows was made a director.

William Temple appeared on the Manchester scene as our bishop, and his fascinating wife Frances took us all by storm.[108] WLB and Temple got on very well, as the latter had an understanding of the scientific point of view. Japonica Wills sent Albert Schweitzer to lunch with

Figure 58 Lukyns; Kenneth Lee's house in Surrey. (Sketch by WLB; Courtesy: the Royal Institution London)

[107] Dr. R. S. Willows was the first to develop crease resistance for cotton in 1932.

[108] William Temple (1881–1944) was a bishop in the Church of England. He served as Bishop of Manchester (1921–1929), Archbishop of York (1929–1942) and Archbishop of Canterbury (1942–1944).

us; he was giving a Bach organ recital in the cathedral to collect funds for his leper hospital in Africa, to which Japonica gave support.[109] He impressed us as a great man.

I must also mention Sir 'Tommy' Barlow and his wife Esther, who were our neighbours in Didsbury.[110] We used to go to Sunday lunch with them and share his superb wines, of which he was a connoisseur. He had his own works, cotton I think, somewhere in Lancashire, but he had wide interests besides business ones. One was certainly art and, after lunch, we used to look through his collection of Dürer drawings. His wife Esther was a classical scholar and was said to write Greek poetry; but she combined this with a passion for icing cakes, she once told me.

WLB, after two years as Professor of Physics, had played himself in, and the worst of his immediate post-war troubles were over. The tough and highly critical war veterans had finished their courses, and a younger and easier lot of students appeared. His lectures to them still caused him anxiety and he took tremendous trouble in preparing them but at least he had no worries over discipline. Gradually, he had collected a loyal and devoted staff. But here in Manchester and later at the Cavendish in Cambridge, he only had one secretary, who remained throughout his tenure of office. At Manchester he had Mair Jones, a little robin-like woman, who soon became his watch dog and, for day-to-day events, his memory. Anyone wishing to see the professor had to pass, or be challenged by, her. Although rather small, she became formidable if she thought WLB was being put upon and would rise to her toes and speak her mind. We all became very fond of her and she was soon a family friend. Once, I remember, he set off on a dark foggy morning, in the brown trousers of one suit and the blue jacket of another; Mair Jones quietly rang me up and asked if I could bring into the lab one or other of the missing partners.

Then there was Kay, the lab steward. Rutherford had tried to persuade Kay to go with him to Cambridge but Kay wanted to stay in

[109] Albert Schweitzer (1875–1965) was a German—and later French—theologian, organist, philosopher, physician and medical missionary. He was awarded the Nobel Peace Prize in 1952.

[110] Thomas Dalmahoy Barlow (1883–1964) was an art historian and businessman. He was chairman of the firm Barlow and Jones. He married Esther Sophia Gaselee (1886–1956) in 1911.

Manchester and remained with WLB till, in his turn, WLB left for Cambridge. It was Kay's responsibility to prepare all the experiments for lectures and, as like WLB he had great enthusiasm, the two got on splendidly. Kay always had the right equipment at the right moment. He was blunt and quite capable of saying, 'You didn't get that to them, Sir,' and WLB always took his advice. Whenever I visited the lab, I would make a beeline for Kay's room and, when among all his impediments he had made some place for me to sit, we would have a long chat and I would hear how things were going.

WLB had two senior staff members upon whom he could lean heavily on the administrative side. He was the artist and creator and was much too vague to enjoy that side of running a laboratory. R.W. James was just the right person for a 'second', and quite unflappable; the two were old friends from the war and now did various research projects together. James had been with Shackleton on the last polar expedition. In later years he became a professor in Cape Town and was the only man in South Africa to be an FRS. He was often in our house and, being a bachelor then, was able to come at all hours and discuss matters with WLB. The other man was Nuttall, who had suffered nervously from his war experiences but was a great support to WLB, taking over all the organisation of lectures, students' courses and the like. These two, with one or two others, were what WLB called his 'guide dogs'. He always acknowledged he had blind spots, but throughout his life there appeared people to guide him over them.

Ernest Scott Dickson was another of them. A man of great charm, he had a splendid sense of humour and a lovely singing voice; actually, the lab was a very musical place. When the students left at the end of the day, the building would ring with Scott Dickson singing 'King Charles, King Charles'. At the beginning of the year, when Scott Dickson's class had to sign their names, he started by saying, 'Mary Pickford, Lloyd George, Harry Lauder and other celebrities need not to sign,' thus anticipating the students' rather crude annual joke. He had a nice hearty wife, also very musical, who lived near us. There were clever research students, several of whom later became members of the staff; people like that dynamic little Welshman Williams,[111]

[111] Presumably Evan James Williams (see WLB's account).

W. H. Taylor,[112] and H. Lipson. I remember how Lipson with his colleague C. A. Beevers, as young students working in Liverpool, used to come over to Manchester with their current crystal model to get help over the structure, and how Beevers used to bicycle all the way to save the fare.[113, 114]

There was only one member of his staff who was a real thorn in the flesh to WLB, and that was a wild elderly man called Bell.[115] He was an insecure and frustrated character and against everything. Not a good lecturer himself and not much good at research, he was critical of everyone else, and especially of WLB. Throughout his life one person of this kind would get on WLB's nerves to such an extent that the person in question could in the end do nothing right, and the sight of him became anathema. But mercifully there was never more than one at a time. I know now I was not much help in the matter. Instead of soothing my husband and pointing out that the man was not really so bad, as I learnt to do later on, I added fuel to the fire by being violently partisan.

These are some of the people I remember in the lab when I first knew it. I have never done any science so, although WLB sometimes 'cleared his own head' by telling me about the research there, I could only get a feeling of the general excitement. Throughout our time

[112] William (WHT) Hodges Taylor (1905–1979) had been a graduate student of WLB in 1926 and remained in Manchester until 1934. He then went to Cambridge to work with J. D. Bernal and then to the Davy-Faraday Laboratory to work under WHB, before returning in 1936 to Manchester as Head of the Physics Department of the College of Technology. He was the first person to make a full description of the crystal structures of feldspars in 1933, one of the most common minerals on earth (and later found on the moon). He later moved to the Cavendish Laboratory in Cambridge with WLB and was instrumental in running the newly created Crystallography Laboratory, where he remained until his retirement in 1971.

[113] Cecil Arnold Beevers (1908–2001) was a British crystallographer who graduated from the University of Liverpool in 1929. He later invented with Henry Lipson the famous Beevers–Lipson strips, a means before the age of computing to enable crystallographers to make complex calculations to solve crystal structures. Many boxes of these strips were exported all over the world. A convinced quaker, Beevers later went to Edinburgh and established a company using disabled workers to manufacture crystal structure models.

[114] WLB's own account with a list of researchers at Manchester was published at *Acta Cryst.* (1970). **A26**, 173–177.

[115] There was at the time a member of staff called H. Bell. Could this be the person referred to?

in Manchester, scientists from all over the world came to work with WLB and play some part in what became known as the 'Manchester School', and I very soon was infected with the research atmosphere. I sometimes asked WLB whether he would like to do only research but he always said that he felt it ought to be combined with teaching; what he would have gladly avoided was the public side of administration. There were people ready to do it for him domestically as it were, but he had to represent his case for plans or finance at the university senate. This he really hated and he used to come home after some of the meetings completely exhausted. I think he probably exaggerated, but he would often tell me that he could have put his good points much better, that he had become indignant in the face of opposition, that he had not been able to think quickly enough when questioned. Yet I would be told afterwards he was generally right and eventually would gain his point. All his life he dreaded large committees. I remember Rutherford saying to him about the senate, 'Don't let those chaps bully you, Bragg. You give them hell.' But that was something he could not do.

When summer came, we had the first of our wonderful holidays abroad. But first, in June, WLB took me to the Norfolk Broads, which was perhaps the sort of holiday he liked best. We lived on the boat and sailed along the beautiful Norfolk rivers and across the various broads. I had given WLB a paintbox and, for the first time, he started watercolour painting instead of using the pastels and charcoal or pencil in which he had always worked, and he loved this. Also, there were birds for him. The Broads in June were so quiet then, with no motorboats with blaring radio sets, that we could steal round a bend in the river and surprise a bittern poised like a church lectern on the top of a post or, entering a broad, there would be a grebe on her nest. WLB was a competent sailor and before we were married used to sail with Patrick Blackett or G. P. Thomson, who had their own boats. Friends such as Arthur Goodhart and David Ritchie made up the crew.

But the thrill of that summer was a trip to Scandinavia. WLB had never been to Stockholm to receive his Nobel Prize because, when he and his father won it in 1915, the war had made a visit impossible. Indeed, his father never went. Now WLB was invited to receive it at a special ceremony. When this was known, he also received invitations to go to Denmark and Norway and give lectures. I was very excited at the prospect. Always fond of clothes, I veered in those days towards the spectacular and had no idea of selecting a small and suitable wardrobe. I took everything I had, including a tulle ball dress and a navy-blue satin picture hat with white feathers nearly two feet across. The only time I could wear this confection, which had to travel in a huge hatbox loaned by Japonica Wills, was at a small lunch at our Embassy in Oslo, where its sheer size created a sensation.

We stayed first in Copenhagen, with the famous Danish physicist Niels Bohr, the father of nuclear physics, and his wife and their six small sons. Margarethe Bohr was an enchanting woman, an international model for all physicists' wives. In looks, in brains and in charm, she had a peculiar distinction. We became friends at once. They gave a big dinner (sixteen people) for us in their flat and Niels, by whom I was sitting, told me that he would make a short speech about WLB and welcome me too. I was very proud when he alluded to WLB as 'this brilliant young man who has also so much modesty and charm'. I had not understood that a little later in the dinner WLB would have to reply, and I became very nervous; in fact, Niels said, 'You are suddenly quite pale.' I need not have worried. It was always very hard to understand what Niels said, even in Danish I believe, but he explained to me that scientists now tend to have their great research idea in their twenties when they are reasonably free of responsibility, and they then spend the rest of their lives explaining it. But he added, 'I think your husband may have a second one.'

We went on to Stockholm to get the Nobel Prize at a ceremony arranged just for WLB. The king was in the country, so we did not have the traditional royal banquet and tremendous social trappings, but I greatly enjoyed it all. People were half amused and half shocked that WLB had not thought to wear a tail coat and striped trousers, as it was an afternoon ceremony. The papers noted he had a 'grey walking costume with blue tie, and velours hat and his young wife'—but here in my letter to my mother I add: 'You will think I am getting conceited if I repeat what they said.' I had my first taste of reporters, who all reported on our youth, which seemed to make a great impression. WLB had of course to give a lecture. This he did with great fluency, making his impromptu drawings on a board, in explanation of points. As I did not understand, I sat nursing a bouquet of red roses and amused myself by studying the audience, mostly white-headed. Some of the reporters made splendid caricatures of the 'young lecturer'. Afterwards, one of the younger and gayer professors took us shopping, bought me a pink boudoir cap that I admired and let us taste smoked reindeer.

We were entertained every night with all the Swedish formalities and customs. I remember the first one at the house of the head of the Nobel Institute, Dr Arrhenius, next to whom I sat in my white velvet wedding dress. He started off by saying that he was sixty-three, had been married twice, had had so many children by his first wife and so many by the second. (Time has obliterated the actual numbers from my mind.) 'Now, tell me about you', he said.

I replied, 'I am twenty-two, I have only been married once, and for seven months, and I have no children.' After that we managed very well. It was said of him in Stockholm that he did something rather brilliant when young, for which he got the Nobel Prize, and had gone round all the universities of Europe ever since accepting honorary degrees and making witty speeches. When we went home at midnight, he insisted on lending me and wrapping me in a huge white shawl over my thin one. He then gave a jovial push as the car went off, and his son in mock admiration spread his mackintosh for the car to pass over.

We completed this lovely trip by going to Oslo, where WLB gave a big popular lecture; but he said it was a bad one, because it was not popular but much too specialised. Anyway, we had enjoyed ourselves immensely. It was so much fun being shown the sights of these capitals, especially the museums and galleries, by experts, being driven round the countryside and, most of all, sharing all our experiences. We ended with a short holiday by ourselves, staying on the shores of a great lake among silver birch trees that were reddening with autumn, and bushes of wild raspberries. Then we came home from Bergen, crossing Norway by the famous mountain railway, which was so curving that you could see the guard's van parallel with the engine in some place. WLB loved it but it made me very sick.

3.12 The 'Manchester School'

Those years that we spent in Manchester were ones of tremendous scientific activity for WLB. There was a ceaseless flow of scientists of all nationalities in and out of the laboratory and of our home, contributing to what came to be known as the 'Manchester School'. The Hopkinson family had been engineers and I had not really met scientists. I suppose the first one to come to stay was the famous Dutch physicist

Lorentz with his wife; they were a very dignified elderly couple, she dressed in sombre black, and I felt rather young and foolish. He, I think, spoke seven languages and made a great impression on me. In the middle of the night, I heard a bell. WLB did not wake, so I crept out in my nightie, only to confront Dr Lorentz in a very long nightshirt, with a high-peaked purple nightcap on his head. The most courteous of men, he was appalled to have disturbed me, having mistaken the bell cord for the light.

We also had Professor Sommerfeld from Munich. I remember how he came straight in to supper and at once began a discussion on crystal structures with WLB, using all the knives and forks and silver, generally, to illustrate his points, and alas, scratching mysterious lattices on the tablecloth with a fork. Long after the meal ended, I had to ask them to break up so that we could clear away. Then there was Charles Darwin, later Sir Charles Darwin, a great friend of WLB. They had written a scientific paper together and climbed in Skye, and he came and stayed to discuss something. In the evening as he settled to talk he threw the cushions from his chair on the floor and said he hoped that I was the sort of person with whom he could put his feet on the mantelpiece (metaphorically at any rate). I realised this was a compliment. Professor E. Milne came to Manchester to the Chair of Mathematics when he was recovering from sleeping sickness. Poor man, he had an operation for double mastoids and, as he was living alone in digs, he came to us from hospital for a spell and (quite inexpertly) I did his daily dressings. Linus Pauling, who later became a famous man, came from the States to work with WLB for a time. Pauling already had that dedicated air as he used to wander about Didsbury, hand in hand with his young wife, with a small boy in tow. I gave a tea party for the 'physics wives' to meet Mrs Pauling. She came in a long gown and a picture hat on top of long curls. The rest of us were wearing our ordinary summer frocks, and over tea she gave us a full account of the birth of her little Linus; this topic seemed to us all highly unconventional, in those days.

There were so many foreign research students who came and went that I no longer remember them all or in what order they arrived to work, and they have long become professors in their respective countries; some have retired and some are dead. There was Ito,[116] from

[116] Tei-ichi Ito (1898–1980) was a renowned Japanese mineralogist and crystallographer from the University of Tokyo.

Japan, who brought me a wonderful royal blue silk shawl with great gold and yellow dragons on it, Namier,[117] from France, who got me into trouble by pressing his trousers on his landlady's mahogany table and burning it, and a Romanian, bearing a green scarf with storks all over it, with obvious implications. Zachariasen from Norway was with us one winter, trying to make his fellow researchers skate as fast as he did on the frozen lake at Wilmslow; Waller from Sweden, and dear Bert Warren from the States with his splendid sense of humour; Peierls and Bethe, all writing papers jointly with WLB or with each other, making scientific history. Professor Bernal of course was in and out from London, and I must say he took a little getting used to, but what a splendid talker upon anything, until the small hours! Bernal would talk about anything—love, sex and politics, as well as science. He would suddenly come into the lab, WLB would bring him to supper, but with no luggage (it was always said that he had a toothbrush permanently in his pocket, but I doubt it), and he would stay the night, because it was too late to go anywhere else. His hair was so long and unkempt that it reminded me of the last bit of a field of corn being cut—you never knew what might have taken refuge in there. WLB loved having him, and I have seen Bernal running along besides a moving train, still talking physics to WLB, who would be hanging out of the window to catch his last words.

I have kept a visitor's book since the day WLB and I were married, and I see repeatedly the names of scientists who have since become famous: G. P. Thomson, Blackett, Ralph Fowler, Eddington, Hartree and, of course, Rutherford; they all came to examine or to lecture and stayed with us. Actually, shortly after we came to Manchester, Hartree came to the Chair of Theoretical Physics; his arrival was a great pleasure to WLB, as he and Hartree did work together. The Hartrees lived close by and made a mild sensation by installing two pianos in their living room—both husband and wife were very musical.

Albert Bradley came to work in the lab and contributed some brilliant research on metals. He was always very tense and seemed to live for his work but, alas, burned himself out, though not before he

[117] Alice is probably referring here to Lewis Bernstein Namier (1888–1960), who was a historian. He was born Ludwik Niemirowski in Wola Okrzejska, which is in what was then part of the Russian Empire and is now in Poland. He served as Professor in Manchester from 1931 to 1953.

had made his reputation and became a Fellow of the Royal Society. I was very fond of his wife, who was a great support to him. She has written me a letter with some memories of the time, from which I quote:

> Albert joined WLB in 1922 after taking his degree in chemistry. He had attended some of the Prof's lectures on crystal structure, and was then offered a grant to work on the subject, and thus became the Prof's first research student. We were married in 1929 and so my own knowledge of the Physics lab does not go back earlier than that. I remember Albert being pleased when I chose a ruby for my engagement ring, because the Prof had determined the crystal structure of rubies. When we were first married, I was dismayed to find that Albert automatically returned to the lab every evening after our meal, coming home about midnight. So I took to going with him, and read or sewed until he was ready to come home. There was a camp bed in the lab so that students could stay the night to watch experiments. I remember how eerie that part of the Physics lab was. After a while I persuaded Albert to stay at home and invited the students to join him in the evenings. One of the pleasures of being connected with the Physics Department was meeting so many people of different nationalities. At one time Albert had two Chinese students working with him, Lee and Cheng, who had to speak to one another in English because they lived so far from each other in China that they could not understand each other's dialect. Then there were the German refugees, Helmut Goldschmidt,[118] and later Bethe and Peierls. I remember you saying how good it was to see them all talking to us without fear of political or racial recrimination, there was such a feeling of friendliness and belonging in the Physics Department.
>
> In those days we used to make models of crystal structure for use in lectures. The models were made of different coloured glitter-wax, and connected with glass rods. We used to have model making sessions at home in the evenings. I remember the Prof's advice to Albert when he was going to read his first paper before the Royal Society. 'Hold up your model, and say, "This is the body-centred cube." After that 90 % of the old gentlemen will go to sleep.'
>
> I realise now how hospitable you both were to us all. I loved coming to your dinner parties, and learning how to do things—you gave us some lovely parties. Looking back I think those Manchester days were the happiest in my life, and Albert's.

[118] Helmut Julius Goldschmidt (d. 1970) was a German-born metallurgist who obtained his degrees at the University of Manchester.

Altogether there was a galaxy of talent and, apart from the administrative work of the lab, it was a very happy time for WLB and so for me. Looking back, I feel that we did not do much for the students, as distinct from the research men; there simply did not seem to be time. Sometimes the honours students came to tea on Sunday, as they would have done in my Cambridge home, but certainly at first we were not successful over this; neither of us realised at what effort and expense these poor chaps came. They lived in all parts of Manchester, sometimes outside, and when they came they expected the 'high' or even 'meat' tea of the North around a table, not sandwiches and little cakes in the drawing room. They were very shy. I remember one of the first lot coming in and being appalled that so far he was alone. He told us he had examined the path to the front door and, seeing footprints, had assumed that he was not the first. I made light talk and was astonished to hear one say, 'Mrs Bragg, I cannot allow that remark to pass unchallenged', for they were very serious. Then, of course, they did not know how to go, and nor did we. The first time, on a Sunday, I said at about 6 o'clock (to WLB's great surprise) that we must go to evening service, and we both arose with an air of finality. 'That's all right', said one of the students. 'We'll be quite happy till you get back. I see you've got a piano!' In the future, we managed better.

One honours student I especially remember soon after we were married was T. S. Littler, because I connect him with a scientific interest of WLB's that few remember now, and indeed I doubt if many knew, and that was the teaching of the deaf.[119] Mrs Ewing was a lecturer in the education of the deaf, and herself very deaf, but a splendid lip-reader.[120] Alas! Not always. She especially wanted to meet Sir William Bragg, so when he stayed with us we asked her to come and see him. Poor Mrs Ewing, she could not lip-read his words as his moustache drooped over his mouth, hiding the movement of his lips. She and WLB, I think, met at lunch in the canteen and became friends,

[119] Thomas Simm Littler was known for his work on the design of hearing aids for the National Health Service. He graduated in Manchester in 1925 with 1st Class Honours.

[120] Alexander William Gordon Ewing (1896-1980) was Professor of the world-renowned Department of Audiology and Education of the Deaf at Manchester University. He and his wife, Irene R. Ewing (d. 1959), were pioneers in enabling deaf children to communicate in the hearing world using spoken language.

and he was anxious to help her. Her husband, Sir Alexander Ewing, has written to me:

> The help that he gave to our contribution towards the early development of what came to be called audiology was most valuable. I believe, he first became interested in our field when he learned of Irene Ewing's work as lecturer in education of the deaf and her personal problem of progressive deafness. I remember an experiment that he organised in his Department with her as subject, to find if vibration of the body in such a case as hers would improve auditory acuity. The result was negative and contra-indicated a hypothesis that because of vibration due to loud noise, patients with conductive deafness seem to hear better when it is present.
>
> It was when T.S. Littler became the third member of the team that Sir Lawrence's help became extremely important. Our previous clinical work with hearing-impaired children had reached a point at which collaboration by an expert on electro-acoustics was obviously necessary to further progress. Sir Lawrence was good enough to lend us a laboratory in which T.S. Littler and I carried out, as a first step, experiments with voluntary staff members, to study effects on human hearing of stimulation at the very high sound levels that we found to be needed if severely and profoundly deaf children were to be helped by hearing aids. Sir Lawrence brought Lord Rutherford to see us at work and to watch some of our experiments. It was after this that the Royal Society made us a grant. T.S. Littler became a pioneer in designing electronic hearing aids.

Altogether, this was a very exciting time scientifically, and it was a mercy for me, and so for WLB too, that I had settled down in Manchester and had many friends in and outside the university. I led a domestic life, enjoying my babies, entertaining in a small way and having my family to stay. My Aunt Mabel, (my dear Uncle Charlie had died) came in and out, and she registered some disapproval. She thought I led a rather frivolous life. Actually I even played afternoon bridge in those days and did no public work whatsoever, with one or two exceptions. Once Aunt Mabel told me to propose a vote of thanks in the town hall at a large meeting of the Gentlewomen's Association. This was dreadful; I blushed, stammered and forgot words. At the end, the redoubtable Miss Phillips, the great lady of Prestwick Park, Manchester, and aunt of the famous G. M. Trevelyan, bustled up in her hard felt hat and hairy tweeds and said, 'My dear young woman, let this be a warning to you, never speak in public again. I can see it's as painful to you, as for everyone listening.' I went home and cried; but my father-in-law, who was staying with us, told me I must speak again as soon as possible, but the opportunity did not arise at once. My only other effort was to let myself be persuaded by Mrs Stocks (later Baroness Mary Stocks) to help at what must have been the first birth-control clinic in England.[121] It was held in a room at Salford, a stronghold of Roman Catholics and so at risk from local public opinion. The small Catholic boys used to urinate against my car and write rude messages on it as a protest. I had to take notes and records of the patients as they came in. I learnt a lot. 'Have you had any abortions, miscarriages?' I asked.

'Why dear, I got rid of the last with a crochet hook, if you want to write that down.'

I did not do it many times, as I wanted to be with WLB in the evenings. My public work did not start until we left Manchester, when I was thirty-seven.

During those years in Manchester, our first three children were born. In 1923 Stephen arrived, not without considerable difficulty, and I was very ill. My mother came and was a great comfort to WLB. Nowadays I should have been taken to hospital or had a specialist and certainly been given a blood transfusion. However, I was young and strong, and all was well; and the births of David in 1926 and Margaret in 1931 were uncomplicated. I think there can be few experiences so blissful as being left alone with one's newborn baby beside one, after all the pain and trouble, gingerly exploring its feather-soft feet and hands and dropping off to sleep together.

These babies gave great pleasure to the grandparents, since they were the first grandchildren on either side of the family, and they could always be left in Cambridge or London, assured of a warm welcome. We were very lucky in that my parents had a suitable house and garden for them, and the Braggs had room in their 'cottage' in Surrey or in the London flat.

Soon after WLB and I married, Sir William Bragg, in 1922, had moved from UCL to the Royal Institution, where a large and beautiful flat was provided for the director. On Friday evenings, WLB and I often went there to attend the famous 'Discourses', which were preceded

[121] Mary Danvers Stocks, née Brinton (1891–1975), was a British writer and radio broadcaster.

by a dinner party for the lecturer. Situated in the heart of Mayfair, the flat was wonderfully central for sightseeing, shopping or theatres. Our small son, Stephen, used to refer to it as 'The Royal Destitution'.

Perhaps I should say here that these visits sometimes had their difficulties. Much has been said and even written in later years about my father-in-law, and my husband's relations with him over their research work. I am myself inclined to think that difficulties which were inevitable and natural have been exaggerated. The truth is that father and son did not find it easy to communicate with each other, certainly not about their feelings. I learnt to know that expressions of feelings and emotion embarrassed the father. They were both in the same line of research, and their minds worked in a similar way, so that, in the case of Sir William, he really came to think that some idea of WLB's was his own; or WLB would tell him about a special line of research, and his father, later, would think about it himself and might see a further development. At any rate, some visits of ours were rather spoilt by this situation.

My mother-in-law, as it were, buried her head in the sand in the matter. She liked to feel that everything was perfect, so she would only say to me, 'Dad is so proud of Willie, he is always telling people about his work.' I myself played rather an inflammatory role (until I learnt better) and would go into the study and try and have it out with Dad. I repeat that the real point was that the two minds worked in the same way, and that had to be accepted. I remember Professor Robinson, Professor of Chemistry in Manchester, and a good friend of ours in those days, once said to me, 'The relationship between your husband and his father is the admiration of the scientific world.' I believe it had the effect upon WLB of his falling over backwards always to see that his own research students received maximum credit for their ideas, whatever share he himself had had in them. It had the effect on me of warning my eldest son Stephen, as soon as he showed his scientific leanings, not to become a physicist and court further complications. In fact, he became an engineer.

3.13 Travels Abroad

WLB did not care for conferences; he much preferred to discuss in small groups and preferably in his own time. Because of this, in all our married life we hardly ever went to the British Association meetings unless there was a special reason, as for instance once when he was president of Section A. We also went once to Glasgow, when his father was president, to Oxford, when Charles and Katharine Darwin had a house party of physicists of all sorts, and to Aberdeen, because his great friend George Thomson had the Chair of Physics there. Another and basic reason for this omission was that the British Association is held in September and, as soon as our children reached the stage of school holidays we never went away at these times. But small conferences and lecturing trips, especially if they took us abroad, he loved. In those happy days, the appropriate foreign university always paid my expenses too. We have been in this way to many countries, though we always regretted that we never went to South America. WLB especially wanted to complete his lists of birds of every continent! Our trips abroad would fill a book with little trouble. When on our travels, I noted our doings, with descriptions of people and places and in the greatest detail every day, and sent them to my mother, father, Harry, Enid or Philip, who kept them all. When I say that I wrote them sixty pages on our first visit to Scandinavia for the Nobel Prize and that I kept it up on this scale until my mother died in 1942, one can guess at the volume of these efforts. Obviously I must be selective now and pick out some highlights. It surprises me that the university let us go, generally in term time, so that in vacations we could take our children on family holidays, of the bucket-and-spade and seashore variety, generally in Wales nearby, and later further afield.

Our first trip to Canada and the States took place when Stephen, our son, was only seven months old, and according to all the pundits this was very shocking as we were away several months. He had a faithful nanny, and he was received with open arms first by my family and then by the Braggs, and he has always been as stable as they come. The loss was only ours. The families sent letters and photographs, and general news of him most faithfully.

Nowadays scientists from this country think nothing of flying over to the States 'just to ask a question and get the answer', as one said to me. But just after the first war, this flow had hardly begun, and to make the sea voyage worthwhile the scientist would stay probably some weeks lecturing in various universities. We started off by sea, bound for a month's lecturing in Ann Arbor (University of Michigan) via Montreal, laden with trunks, the inevitable hatbox, golf clubs and tennis rackets—we were enthusiastic players of both games in those days. I spared my family no detail of that voyage on S.S. *Regina*: menus, the appointments of the cabin, deck games ('Willie has won the cock-fighting competition and got a silver cigarette case.') and minute descriptions of the passengers, their clothes and behaviour. Of American men I wrote, 'American men seem rather odd. You find them shouting with laughter about absolutely nothing, sitting in little corners of the bar, always in "plus fours" of an English cut, slapping their legs and shouting "Atta Boy". American women have lovely feet and ankles, are mostly shingled, wear beautiful clothes, but they are very pale, and rather intense and earnest. Everyone is very friendly to us.'

Ann Arbor in those days was a small and rather dull place; certainly I was rather bored at the beginning and inclined to go back to my baby. We were paying guests with the Professor of Chemistry, we had our meals out and I felt at a loose end. However, soon that was all changed. American hospitality is well known, and soon we were never eating alone and were shown everything there was to see, though there was nothing very much in fact. It was terribly hot. I learnt that scientists had a rather higher social status over there than at home: we were made a great fuss of, we were news, and this of course is tonic. One day our Scottish landlady, Mrs Smeaton, said, 'The papers have called up, asking for your photographs and to have a list of those who are entertaining you, but of course I put them off and said that you both hated publicity.'[122] I gasped and WLB had a good laugh. I immediately counteracted this, and the reporters appeared. WLB did his best, but all over the States it was difficult to find a reporter who could take in even what atoms and molecules were (though this ignorance is not confined to American journalists!).

I had a great time with them, giving my reactions to baseball games, women's clubs, the students, the food. . . anything. I positively wore them out but I was careful not to be critical. For I discovered almost at once that Americans cannot bear criticism. When I said that I thought their sleeping car arrangements were nothing short of barbaric, the reporter was so upset that I retracted. The matter of the photograph led to a scene with the doyenne of the science faculty, a kindly fat lady, Mrs Randall. She bustled round and told me that it was not very nice, and certainly not in keeping with the image of a professor's wife, to see a photograph of one with bare neck and shoulders in print. There was

[122] Mrs Smeaton is presumably the wife of Professor William Gabb Smeaton.

a great streak of puritanism in the Midwest, and of course there was also Prohibition then.

We had a fabulous weekend while there, in Chicago, staying with a millionaire family who became lifelong friends and typically supported us with regular wonderful food parcels all through the Second World War. That was Mrs Schaffner and her son Joseph, a young man about my age, who was patron to scientists and proud to help in any way. I was interested in the fact that the rich American woman was far from ostentatious and managed her household very cleverly. The Schaffners for instance gave a dinner for us of twenty people, and one Danish girl cooked and served it. True, she was paid £500 a year (a lot in those days).

The Chicago letters home are full of parties, country clubs, the open air opera, the look of the high buildings and the beauty of the great lake. 'We feel very stimulated, which is the typical reaction of any visitor to America. WLB was a great success. Women would say to me, "Why, I'm just crazy about your husband. Isn't he the loveliest person?"' I passed this on with pride to his parents, with whom we joined up in Toronto for the British Association meeting. A whole boat load of British scientists came for this, of whom the stars were, besides my father-in-law, Sir John Russell, Sir William Beveridge ('Bev' of the London School of Economics) and the Rutherfords, with their daughter Eileen.[123, 124] There was also Sir Richard Paget, who had made dramatic experiments on speech production.[125] The Braggs brought Gwendy, my teenage sister-in-law, magnolia-skinned and wearing grown-up clothes. Toronto was completely en fête, and we had an orgy of dances, dinner parties and, one supposes, lectures, but I cannot say that I remember any of them. While there we had a letter from G. P. Thomson to say that he wanted us to be the first to know of his engagement to Kathleen Adam-Smith, a most enchanting person as we were to find later.[126]

[123] John Wriothesley Russell (1914–1984) was a British diplomat and ambassador.

[124] William Henry Beveridge, First Baron Beveridge (1879–1963), was a British economist, noted progressive and social reformer, largely responsible for the setting up of the welfare state.

[125] Richard Arthur Surtees Paget, Second Baronet (1869–1955), was a British barrister and amateur scientific investigator who specialised in speech science and the origin of speech.

[126] Kathleen Buchanan Adam-Smith (1900–1941), married George Paget Thomson in 1924.

Now we had a choice of putting in week or two crossing the continent to Vancouver in a special British Association train, with all 'laid on', or going camping in Canada by ourselves; and we chose the latter. A guide was found for us, a competent young man, with red hair and a water lily in his slouch hat, and he organised food for ten days, two tents and two canoes. Algonquin Park, among lakes and often virgin forest, is probably like a Butlin's holiday camp nowadays but, at that time, throughout our trip we only saw one other camping party. Again, enthusiastic letters about this flowed home; but perhaps the best description comes from a Toronto newspaper, dated August 1924, headed 'Intrepid Scientist and Wife Near To Nature and Wolves':

> We decided that when we came to Canada fifty years from now, we might travel across the West in a train as elderly people should, this time we wanted an adventure', said Mrs Lawrence Bragg to the Star today. This was why Mr and Mrs Bragg diverted from the rest of the British Association party, and set out on an expedition of their own into Algonquin Park. With 'a pleasant scarlet-haired guide', they travelled nine days by canoe and portage through Algonquin and if they did not precisely have adventures of the Fennimore Cooper kind, they had at any rate any number of experiences denied to dwellers in the cultivated

Figure 59 A famous photograph of father and son taken during the British Association meeting in Toronto in 1924. (Courtesy: Smithsonian Institution Archives SIA2007-0340)

area of the British Isles. They are a very charming pair, as affable as British royalty, and they are entirely free from the venerable tradition that clings to the British Associationists. Mr Lawrence Bragg, winner with his father, Sir William Bragg, of the Nobel Prize for Physics in 1915, is young, lively spirited and dark. Mrs Lawrence Bragg is young, lively spirited and fair, and their impressions of their Canadian trip are of the happiest possible sort. They went forth into the wilds, armed with a copy of 'The Drama of the Forests', presented to them by the author, Mr Arthur Hemming, and a piece of mosquito netting, which came from the same experienced source. Mrs Bragg wore khaki breeches, according to the Algonquin tradition, and confessed to be 'suffused with blushes' until she had adjusted to the Algonquin Park point of view, which is that no woman has any reason to be suffused with blushes, unless she is to be found wearing a skirt. They paddled all day, and camped on balsam boughs at night, and defying tradition, used the mosquito netting for a pillow. And for the first time in their lives they heard the howl of the wolves in the open. They were real wolves, and not scenic accessories provided by a thoughtful management howling operatically in the wings. They were less than a mile away, and after listening to them for one night, the party arranged thereafter to camp on islands 'though the guide told me they could swim', said Mrs Bragg, 'he liked to assure us that if they did track us there was nothing between them and us, but his little hatchet.'

She was convinced however, that the scarlet-haired guide was not as truthful in the matter as the wielder of a little hatchet should be. For nine days they travelled through a wilderness tempered to their comfort by the commercial providence that looks after the tenderfoot. There were three well-cooked meals a day, plenty of shelter against the weather, and hot water produced for shaving every morning. 'They certainly do you awfully well here on trips of this kind', said Mrs Bragg.

They saw almost every species of wild animals known to Canada, except bears, which took fright when they heard the party approaching. And they were much interested in the dams built by the Canadian beaver. 'But we never saw them building', said Mrs Bragg, 'they threw up work as soon as they saw us'.

They are leaving Canada entirely delighted with their experiences here, they have encountered what the old world undoubtedly regards as the twin terrors of the new, wolves and prohibition, and have suffered as Mrs Braggs energetically pointed out, as little from the one as from the other.

Back in Toronto we stayed with the Masseys, the millionaire family who made tractors and agricultural implements. Like my own, they were a particularly united family, and in the Massey Park, beside the old parents, each married child had a large house and garden and the latest type of swimming pool. There was the usual round that I was beginning to expect by now everywhere, of tennis parties, golf, dinners and dances. But I remember best the droves of small Massey children, who were drawn to WLB as to a magnet, and I saw one four-year-old girl, in a golden-coloured smock, coming up to him confidingly and saying, 'Mr Bragg, will you take me to the bathroom please?' When WLB said, cravenly, 'Shall I call your nanny?' she replied in astonishment, 'Oh! Mr Bragg, can't you manage me?' And, of course, he did.

Then there was Schenectady, home of the General Electric Company, where we had to call, for WLB to lecture. I mention this because I suppose it was the first time I myself had spent a day 'touring a great laboratory'. Later I was to do this many times all over the world. I knew no science, but I liked to get the smell and feel of a place where WLB was so completely at home and hear him talking or listening to the research men. It always touched me that the people took such pains to explain to me what I could never understand. All my life, scientists have explained their ideas to me. I think they forget their audience and clear their own heads while I look appreciative, which is all I can do.

Philadelphia was holding some great celebration at the Franklin Institute, and most of the famous British scientists were there before they left for home after the British Association. We were all put up by American hosts; WLB and I stayed with a distinguished Quaker couple and had the usual round of lecturing and festivities, with a wonderful organ recital as one of them.

We were especially happy in Princeton (my favourite American university), as we stayed with Gus and Esther Trowbridge, and Gus had been the American opposite number to WLB, doing sound ranging in the war. He was to be our eldest daughter's godfather later. We all four became great friends and, when he took an important post in Paris, we used to go over and stay there. Gus had a hawk-like, Red-Indian type of face, and it would have been easy to see him with feathers cascading down the back of his head. He knew WLB very well and used to say, 'The trouble with you, Willie, is that you're not tough enough for this world.' He was! Then to Wellesley, Baltimore, Boston, Harrow, Yale and New York, and we were nearly home.

It was October, and 'the fall' delighted us with all the trees suddenly turning scarlet and gold, and great stretches of wild Michaelmas daisies and golden rod. I thought New York rather noisy and vulgar but spectacular! The policemen were not polite and, compared with London, everyone seemed in a rush. Many pages home describe the harbour, the skyscrapers, the swarming streets, marvellous picture galleries and a glorious 'first-night' or two. It was great fun as we stayed with our friends Arthur and Cecily Goodhart, from Cambridge, who were having a long-drawn-out honeymoon in the States so that Arthur could show Cecily to his American relations. Before we were all married Arthur had been sailing several times with WLB, G. P. Thomson and others, while Cecily had been at Newnham with me.

We went off from New York to stay with Arthur's relations, the Lewisohns, who had a lovely white colonial house with pillars and green shutters and which stood in grounds that sloped down to the New York Yacht Club on Long Island Sound.[127] We shared with the Goodharts a sort of suite that had rooms leading into each other, and a French maid. Cecily and I used to have breakfast in bed together, in lace boudoir caps, and have a good gossip—and sometimes a good cry from sheer fatigue, the result of the splendid American hospitality. Our host and hostess were a cosmopolitan and cultivated couple, and we had some glittering parties, where politics, art and books would be discussed.

Finally, before we sailed home from New York, we went to the Rockefeller Foundation Committee for lunch, and they said to WLB: 'Look, you are the only man in the world who has made crystallography his department, and all science is watching you; so, if you want money for your lab, tell us and you can have it.'[128] So WLB left walking on air. I noted to my parents that 'sometimes Americans tend to speak in superlatives—but it's very encouraging.'

We had of course many other trips to the States and Canada, but one's first impression is peculiarly clear-cut. In future I would go out and join WLB for part of the time, as when we had the children I did not want to leave them for long. In 1928 he accepted an invitation to

[127] The Lewisohn family of New York was established by Adolph Lewisohn (1849–1938), a German-Jewish immigrant born in Hamburg and who became a New York City investment banker, mining magnate and philanthropist.

[128] How times have changed!

go for some weeks to Boston to MIT. He was asked to give a series of lectures and to 'start at least half a dozen men on some fundamental research', an alarmingly tall order. He also wanted the chance to get away and write his own book. Among the research people, he discovered Bert Warren (later a professor), with whom he did some original work and who came to us in Manchester later.

I had a terrible voyage out that time, with great storms in the Atlantic. I had been ill before I left with some sort of paratyphoid and unable to eat on the ship. WLB said when we met that I looked like a wrinkled white prune. At Boston I was first down the gangway and was dashing in ecstasy into WLB's arms when a bossy American official darted between us shouting, 'No, you don't, not now and not here,' and I was hustled away to be interrogated by a customs officer. WLB had been lent for our stay a huge Buick car, and he enjoyed my surprise as he drove me off to the hotel, having passed his driving test, which is the sort of venture that is always something of an effort to him. In this car, with chains on wheels, we used to go off and explore the snowbound countryside by ourselves; but were soon never alone in the evenings, as we were always out to dinner.

The institute seemed to me as big as the whole of Manchester University and far richer. We used to go to huge dinners with the president, whose official house reminded me of a cinema or a stage set, and we dined off a service that had belonged to Napoleon. Strong cocktails were served in spite of prohibition, although we had iced water at dinner. The guests were all very rich and of impeccable lineage; they seemed all to have come from that overcrowded boat, the Mayflower. Enormous reports flew home: WLB's success as a lecturer, the look of Boston under snow, with the old purple glass in the windows, the Sargant watercolours in a picture gallery, the gorgeous lobsters and clams we ate, the High Church we used to go to, where 'I was turning East the whole time, almost a Roman Catholic', and the enormous red roses people used to send. As usual, I was delighted with it all.

We had a weekend skiing with Canadian friends in the Laurentian Mountains. WLB loved that, and the skiing holidays in our life, though not many, were always a great pleasure to him. I had never done it before but a determined and very competent young man took me in hand, and I loved it. Another weekend we went to Mount Holyoke, a women's college, for WLB to lecture. The women professors 'seemed rather frightened of WLB, so I am sorry to say that he became rather wild and flippant'. By this time he was getting confident about popular lecturing, and developing his own techniques like doing lightning sketches on the blackboard; the American audiences were very appreciative. The only thing that he could not stand was a reception immediately beforehand, when he would be introduced to literally dozens of people. We tried to avoid this and yet not hurt anyone's feelings, but it was not really possible.

We went once more to the States in our Manchester days in 1934. Again, WLB wanted to write a book, I think the one on minerals, and again I went for part of the time. I came out on the ship with Sir Arthur Eddington, from the Observatory in Cambridge, the shyest man I have ever met. Actually, I was of great service to him. Every day I had tea with him by arrangement, and he read his detective story, while I headed off the passengers intent on being able to say that they had spoken to the great man; we formed a pleasant if silent friendship. He stayed with us in Manchester, but away from his devoted sister (he was a bachelor), he left all his dress clothes behind, crammed into a chest of drawers, and they were rather a bother to send.

This time we were based at Cornell University at Ithaca, where, until I came, WLB had the unique experience of staying in a most select graduate fraternity house and being very feted and spoilt by those hand-picked graduate boys, who regularly entertained a visiting professor. I was not allowed to stay there for more than three days, for, alas, one wife had committed the unforgivable sin of trying to organise and mother the thirty young men, so never again, they said. WLB and I took a little flat and equipped it from Woolworths at minimal cost. I insisted on bringing our water jug home, an extremely ugly thing, of thick green glass, costing about 6 d. I carried it in my hand as a souvenir, and fifty years later it is still intact.

Every three years these 'Telluride' students had a sort of gala weekend, for which invitations were eagerly sought by girls all over the States. The girls chosen stayed in the top floor of the fraternity, but they had to have a married woman as chaperone. They did me the honour of inviting me to come in and stay and fill this role of 'housemother', for which apparently there was great competition. I rang up the dean of women and, as I thought, very reasonably, asked how far American girls were supposed to go. She seemed rather taken aback and asked me to come over and 'visit with her', so that I could hear the form. They proved to be rather spoilt beauties but somehow very endearing. Their weekend was crowded with a wonderful concert, a baseball game and, of course, a ball. This latter was held in almost total darkness with alternating bands from New York, and the dancers were moving around cheek to cheek but with no other parts of their anatomy touching, producing a curious effect.

WLB attended most of these functions but preferred to go home to bed, and even I baulked at going off to see the sun rise at dawn with them. About noon I would be invited to take a shower with some of the girls and hear their views on life and on their partners. One had danced with WLB (she was 'majoring at some university in sanitary engineering, which he took to be plumbing'). She pronounced him to me as 'just too darling' but said she would never marry an Englishman (a view with which they all agreed), as he would expect his slippers fetched, to have first look at the newspaper and not to be kept waiting. I thought they treated their escorts in a very cavalier fashion.

WLB finished his book. We took a lightning and most exhausting trip to Virginia to lecture in various universities there and have a glimpse of the south, and then home again. Could I have known that twenty years later, after the war, I would be returning with WLB, with no hatbox, no trunks, no tennis rackets, nor golf clubs, but with a briefcase full of notes to give lectures myself, I should have been greatly surprised.

A scientist who is breaking new ground is in great demand for lectures and conferences and, as WLB loved travelling and found it very hard to refuse attractive invitations, we went to many places besides America. It would be no doubt tedious to describe them all, although the family letters still record most of these visits. But there are one or two special trips to Europe which always stood out in our memories. One certainly was the Solvay Conference, held every three years in Brussels. It was small, international and intimate, since members of the Solvay family took an active interest in it. Later on, after the war, WLB was President for about twelve years, when we generally stayed in the Solvay home, Chateau de La Hulpe. The Solvay business, on the proceeds of which the conference was run, was the equivalent, I think, of Imperial Chemical Industries in this country. The subject of the first one that I attended was chemistry, and all the great chemists were there. I remember some, Sir William Pope, from Cambridge, Sir Eric Rideal and his lovely American wife, Sir William and Lady Hardy among them.[129] The hospitality was fabulous. We were all put up in some big hotel, where the old Madame Solvay herself received us on arrival. The men read and discussed the latest scientific papers, while wives and daughters had a stream of activities planned for them. At night we always had boxes at the opera for us all or splendid receptions. I remember we, the ladies, did not know how to thank Madame Solvay. After the first opera we sent her a bouquet of Parma violets, after another gorgeous entertainment a sheaf of roses, but on the last day when Lady Hardy ran up to the rest of us and said, 'Do you know all our hotel bills have been paid, even extras like baths?', our gratitude could only express itself in an enormous orange tree in a tub, with suitable note.

Another great event for scientists, in the twenties, was the Volta Celebration in Italy. It must have been his centenary; he was the man who invented the Volta pile, and the word voltage is coined from his name. As he was born in Como, the conference and celebrations were to be held first there and then in Rome, and those invited were to be guests of Mussolini and the Italian government. A galaxy of scientists and wives appeared from all over the world. The Rutherfords were there, Eddington, Aston (of isotope fame), Charles Darwin, Arthur Compton (an American Nobel Prizeman) and the Kramers, from Holland, are among those I remember.[130] The Americans in several cases brought their mothers as well as their wives, as this was certainly an occasion not to be missed. I think the conference part was rather languidly conducted. It was wonderfully warm, and there was swimming and boating on the lake of Como, steamer trips for us all and wonderful firework displays at night. Great houses around were open, and Italian hostesses entertained us. Unfortunately, the British manners were not always impeccable; some rather unscrupulous delegates tended to slip away and swim when the Italians gave papers, and even the ladies were not always at their best. I remember one morning party for us in some countess's beautiful garden. Among the refreshments were dishes of marrons glacés, WLB's favourite, so

[129] Sir William is presumably William Bate Hardy (1864–1934), who was a British biologist and food scientist. He was married to Alice Mary Finch.

[130] Hendrik Anthony 'Hans' Kramers (1894–1952) was a Dutch physicist who worked with Niels Bohr to understand how electromagnetic waves interact with matter.

I could not resist slipping some into my handbag for him. 'You are taking some home, perhaps?' I heard the countess at my elbow. Of course, when I explained, she gave me a boxful for him. The other sinner that morning was Lady Rutherford, our doyenne, as it were. She was never anything but forthright, to use the kindest word. We were all presented with sweet little bouquets of garden flowers. 'What a pity to have cut them', said Lady Rutherford, when the rest of us thought she was thanking on our behalf. 'English people so much prefer to see flowers growing.' No, our manners, even to each other, left something to be desired. We went swimming one day at the famous Villa d'Este, where there was a splendid diving board and chute. Aston, who was well known in Cambridge to be very careful of himself, felt sure that there were splinters on the chute, so someone other than he must go first. I rashly was persuaded to do this, and sure enough sliding down into the water I encountered one and felt my bathing dress (as it would then have been called) rip. I called to the others, lolling in the boat alongside, that I could not come out without a safety pin. Of course no one had one. After a pause the shy Eddington said haltingly, 'I wonder if you could use my tie pin?' and I did. Alas! The last night at a banquet both Rutherford and I drank orangeade instead of champagne; he was not allowed by his wife to drink, and I was just very thirsty. We both became very ill, Rutherford had to go home, but I made an effort to go on to Rome with the others and keep the trouble at bay with medicines.

I would not have missed Rome for anything, and as usual everything was made easy for us to see a great deal with special guides. So many people have described Rome that perhaps I should concentrate on the people. Here, about four young Italian secretaries were responsible for us all; they seemed to us as if they lived in considerable fear of anything going wrong. One was a very attractive young poet, de Bosis, about whom some mystery had gathered. He was subsequently, and rather secretly, engaged to Ruth Draper,[131] the famous American diseuse,[132] it was always said. When we were there he had rooms, perhaps indeed lived, right inside a part of the Roman wall, where one night some of us went to drinks with him. There was a strange feeling

[131] Ruth Draper (1884–1956) was an American actress, dramatist and noted diseuse who specialised in character-driven monologues and monodrama.

[132] Diseuse: storyteller, dramatic-singer.

that it was some sort of hideout. After we left he was shot down, presumably on Mussolini's orders, while flying his own aircraft dropping anti-fascist leaflets over Rome.

The scientists were going to see Mussolini at a special party, but the wives were not included. I, and indeed other wives, thought this a shame and were not to be put off by the secretary promising us an audience with the pope instead. I argued with the secretary, saying rather tactlessly in my enthusiasm that we could always see the pope, but Mussolini, well one never knew what might happen. Anyway, somehow or other we were included with our husbands in the Mussolini party. He gave the party in the beautiful grounds of some great house, where we understood his mistress was lodged, and it was a very impressive affair. We all lined up to be received by the dictator, and one of the secretaries, pale with anxiety, gave our names as we moved up to him. To each scientist Mussolini said a few words in his own language, certainly French, German and English, and those words he seemed to have learnt by heart. When it was WLB's turn he said that he knew he was doing work on X-rays and crystal structure and that he had shared the Nobel Prize with his father. The next name was then called so quickly that there was no time for WLB to say anything. Our host was, as it were, modelled on Napoleon, short and stocky and seemed to force his jaw into a fierce line. His grasp of my hand crushed my rings into my fingers. Later a message was sent around that Mussolini would talk to scientists if they presented themselves, but somehow or other, only the American ones did. Unfortunately, the food poisoning that I'd had in Como became much worse, and we had to go home after that. I just managed the journey, but once back I was ill for several weeks, attacked by something very like typhoid.

We visited Russia in 1931.[133] Scientists and technologists have had preferential treatment and been welcomed to Russia ever since the Revolution of 1917, but when we went, over fifty years ago, it was still quite an adventure. A few British scientists were invited to an international conference in Leningrad and Moscow but actually it was cancelled before it began. The British group started blithely unaware of this until they arrived, which made for difficulties. Paul Dirac from Cambridge, already a famous mathematician though looking almost boyish, (Professor) Ralph Fowler, Bernal, WLB and I went off together

[133] WLB in his autobiography gives this as 1932.

on a Russian boat under the wing of Kapitza and his wife Anna. Kapitza, a Russian of course, was the blue-eyed boy of Rutherford with whom he was then working at the Cavendish in Cambridge. He spoke an engaging broken English. Anna was charming, good-tempered and alert, yet placid and quiet, with brown eyes like two beads and very short black hair. I was to see a lot of her. Before we left I had a postcard from my mother marked urgent and saying simply, 'Do be very careful, and on no account take your Bible.'

The voyage, until we reached the Baltic, was very rough, and we were not helped by the fact that the captain was anxious to make a speed record and impress his inspector who was on board. He hoped to win a prize, though this I gathered would only be a bust of Stalin, but it could have been money, which would be scrupulously divided between the members of his crew. The food was wonderful, though at first we were all too sick to eat it, but in ordinary amenities, like bath plugs and lavatory locks, the ship was lacking. We took a lot of medicine with us, and biscuits and chocolate, which we had been advised to do, but Dirac brought quite a large gramophone in a bright blue case, which seemed odd. I asked WLB what was Dirac's field, but he told me that he was a mathematician so pure that he could not begin to explain.

We anchored in the harbour of Leningrad early one morning in a damp autumn mist. I suppose Leningrad has been greatly spruced up by now, but then in spite of the intrinsic beauty of its position on the edge of the Baltic, its lovely buildings and golden cupolas, it was very shabby. There were broken windows patched if at all with rags, peeling paint, scarcely a car to be seen, and trams so packed with people that they sometimes dropped off. It was soon pouring with rain, but no one had an umbrella, and hardly anyone a mackintosh. Our hotel had once been very grand, but now it presented only a fading glory: the brocade chairs were torn and dirty, there was no paper lining in the drawers and there was generally no hot water and very bad food. Anyway, we were lucky, I suppose, to be anywhere, because when a physics professor was somehow summoned, he explained that we were not expected and there was no conference. Worse still, we had no money. Mercifully Dirac had just had one of his books published in Russia, and he was not allowed to have the proceeds at home; he had a nice nest of roubles in a Russian bank, and we went to him for our day-to-day expenses. I remember getting a handful of roubles from him for mineral water for brushing my teeth; it was not safe to drink the water. However, the prospect brightened and, as the scientists were already there, some sort of conference was arranged, with the help of Kapitza. While the men were at their hurriedly organised discussions, I, when not with Anna Kapitza, went off on my own and saw things that I was not supposed to see, just by accident and through lack of the language. I wanted to see a maternity and child welfare clinic, but I landed up in a hospital after a tremendous walk (no taxis of course), and how I found it I cannot think. I went in by the main entrance, causing a general flutter, and after repeating 'English, English', a German woman doctor appeared and I got across to her that I wanted to see round. She asked me if I were a doctor and if not what was my profession. I saw that I was going to lose out on this, so I conveyed to her that I had an honours degree and that I could work, but having (as I then had) three children I stayed at home. 'A kept woman?' she said, scornfully. However, she reluctantly led me off to a babies' ward where the nurses were not in uniform, there was an unattractive smell, and the poor little patients were terribly white and wan with summer diarrhoea; many also had rickets. 'Are they all going to die?' I said, partly because I could manage to say this in German. Naturally, she was annoyed and said that certainly some would get better. She had her revenge then because we came to an operating theatre, and she said that I could watch a big operation. I managed to escape, and someone took me to a wing of the hospital which was indeed the children's' crèche or clinic and in every respect a show place which could with confidence be shown to an outsider.

We all used to go sightseeing, and some of these sights will never be seen again, as during the war they were destroyed or taken away perhaps by the Germans. For instance, in a wonderful palace at Tsarskoye Selo, we saw a room where the walls were a mosaic of amber and tortoiseshell of the greatest beauty, but it has all disappeared, I am told.[134] Peter the Great had given Frederick of Prussia horsemen and soldiery in exchange for this Baltic amber. We saw room after room of treasures in exquisite taste. No wonder I told my family that 'Buckingham Palace seems a lodging house in comparison'. We also went

[134] This is a reference to the famous Amber Room, which was taken apart by the Germans during the Second World War and probably destroyed. It was subsequently rebuilt in 2003 by Russian artisans, financed by donations from Germany.

to Peterhof, rather like a Russian Versailles, where all the fountains played for us. They had lovely shapes, one like a mushroom, another a holly tree, all made in metal. But all the original ones have gone, after German bombing, I believe.

A unique occasion was the centenary of the Alexandrinski Theatre, for which occasion Anna Kapitza managed to get presentation tickets for herself and me. I think no other foreigner was there. The enormous theatre was packed with workers, who were wearing collars for such an occasion, reading papers and chewing dried sunflower seeds. We were first harangued for some two hours by party members in khaki-coloured uniform. A number of them were on stage flanked by the oldest actresses and actors who had played in the theatre's early days. Most of them were wrapped in shawls and rugs as it was cold. Later several of the plays performed a hundred years ago were given in their original settings, followed by modern plays with clever scenery made out of really nothing but suggesting its nature by some imaginative trick. There was an interval and, as we were ravenously hungry, Anna and I joined a long queue for a tatty refreshment bar, where we only succeeded in getting two small chocolate creams. At the end of the interval it was announced that 'our comrades the tram drivers and conductors would remain on duty to take everyone home' and that it would end at about 2 am.

This Russian visit was so exciting that my diary to my family was fifty pages long, and when I came home I wrote a series of articles for *The Yorkshire Post*, as Russia was high news. One could dilate on the wonders of the Hermitage or the ballet, but they are still, fortunately, to be seen. It was the making of a new country, or at any rate, a new society that was so exciting. There was dirt (no soap), poverty, queues for food, fleas and discomfort—everything was sacrificed to the great five-year plan of growth. We met an English foreman from a BSA factory at home, commissioned by the Russians to teach his skills, and he told us, 'I don't know how they'll manage—no good with their hands. They can't put a bicycle together yet. Stalin or someone high up called for a parade of bicycles from our factory on a certain day, and we just had to prop them on lorries, most of the parts tied up with red ribbon to keep them together.' Well, that was fifty years ago.

We went to Moscow, because the scientists had to lecture there to get some money for the journey home. WLB and I stayed with Monkhouse, the head of Metropolitan Vickers, who were doing a project there. He had always a loaded revolver in his car, and no wonder as, shortly after we left, he and his colleagues were arrested on trumped up espionage charges; although finally he got away with expulsion from Russia for five years, at the time, he told us afterwards, he thought that he and the others would be shot. That was the frightening side, of which we saw some evidence. But there was no doubt it was a most stimulating visit and, in the light of the future development of the USSR, it was in a sense like seeing the making of history, at a time when foreigners were not encouraged to visit there and watch it.

3.14 'Farewell to Manchester'

We had been more than ten years in Didsbury when we decided to move further out of Manchester. Both of us longed for something more like real country. We were not alone among our colleagues; university people were beginning to move out now that cars were more general. We rented a house in Alderley Edge in Cheshire; we could manage this, we felt, if we did not send our two boys away to boarding school. Anyway, WLB, having been brought up in Australia, had no tradition of boarding preparatory schools, or for that matter public schools, and thought the whole business unnatural. There was a boys' school on the fringe of Alderley, which, although it was not all that we could have desired, answered the purpose, and it had at least the great advantage that mathematics was taught by one of WLB's own students. With his help, Stephen, when the time came, won a scholarship to Rugby.

By the time we left Manchester in 1937, we had four children: two boys and then two girls. Naturally, one is tempted to say a great deal about them, what they were like as small children, their amusing sayings and their funny ways. But they are long since grown men and women with families of their own, and it would be distasteful to them to have their childhood thus recorded. I think parents must cherish these memories in private. I confess that I could not say that I have not still their mementoes, the books of snapshots, their first letters, their home-made cards, their kindergarten reports, the odd raffia mat and, yes, even their first shoes; but I must leave it at that. We had our private jokes and private vocabulary—all those bonds important to family life.

Stephen and David went to my old school Lady Barn House. It was much less rigorous than in my time but otherwise very much the same, still full of university and *Guardian* children. Apart from school, the children lived a very quiet, well-ordered life compared with nowadays. A visit to the zoo, shopping in Manchester, a cricket match at Old Trafford, a day out somewhere or a children's theatre were treats. The children invented their own amusements. It was clear very early that Stephen, with his absorbing passion for trains and how things worked, would be scientific and, equally clear, that David's bent would be art. He very soon showed interest in the shape and colour of things—the way things looked, not how they worked—and he loved to paint. When he was about two years old, a young nanny, Hilda, came to us and stayed till she was married some seventeen years later. She had a very special place in the family.

For a long time, they were the only grandchildren on either side and so were in great demand at the Royal Institution and in Cambridge, as well as with the young aunts and uncles. While we were in Didsbury, Enid and Harry both married: Enid married a barrister, Malcolm McGougan, and Harry married an old school friend of Enid's, Greta Strohmeyer. Gwendy Bragg also married; her husband was an architect, Alban Caroe. Philip, who after Rugby broke new ground in the family by going to King's College at Cambridge, became a clergyman and remained the bachelor uncle all his life.

The house at Alderley Edge was definitely rather pretentious. It stood on a hill at the top of quite a steep drive and really looked like some huge dovecote without its stalk, standing in a lovely sloping garden. An astonishing feature of this house (Windy Howe) was the billiard room, complete with a full-size table, and equipment. This room became WLB's study, and the billiard table his delight. When he was writing a book, he could spread all the chapters around the table, and he himself sat on a high stool and moved round. Of course, the trouble was, people staying and friends coming in, and especially my father, often wanted to play billiards, which was an unusual game to be offered in a professor's house; so when this happened, WLB put each chapter in position under, instead of on top of, the table, safely outside the range of any kick from a player. This worked well. When the staff or students came out, we all used to play racing demon and the like on this splendid table.

There was an excellent room below of the same size, where the boys could lay out their trains permanently. There was a great deal of space for the family: a balcony, a verandah and, leading out of the dining room, a conservatory full of pot plants—even the common kinds of orchid. It was thoroughly Victorian, right to the lamentable attic bedrooms for maids, and the labour-giving kitchen. What really charmed us was the view; we looked right across the Cheshire plain to Beeston castle and, in clear weather, to Wales. Our faithful nanny Hilda, and Florence, our cook, agreed to come with us. We had in fact three maids and a full-time gardener, but in those days in this respect we were not unique among our colleagues. John, our gardener, was rising eighty years old, but it would be hard to imagine a more active or more skilled man. In his spare time he gave talks around the place on bee-keeping. Unfortunately, he suffered from diabetes and kept his insulin in the potting shed under a flower pot. Occasionally it was, not unnaturally, lost, and I would have to rush him into hospital for supplies. In those days I do not remember WLB, in later years such an enthusiastic gardener, doing any works in the garden, but he loved discussing and planning with John. Every sort of vegetable and fruit, including white raspberries, was grown. There was a tennis court, a small pond into which the children sometimes fell and wonderful shrubberies for hide and seek. The bus, very important to everybody, stopped at our very gate, at the bottom of the hill.

We were not dependent on the people who lived in Alderley, which was just as well, as they gave the impression, and I am sure they felt, that a professor, and especially a scientific one, was a non-starter socially. They were a conventional lot, nearly all merchants, in shipping, cotton and the like, and the standard of life was geared to money. We were also thought very odd in that we kept our children at home instead of sending them away to school. However, I was at least the niece of Mrs Edward Hopkinson, our dear little Aunt Minnie who though widowed still lived at 'Ferns', about twelve miles from Manchester, and I scored high on this relationship. In winter the boys used to go to lunch with her alone, on Sundays sometimes, just as I and my family had done when I was a child, and in the afternoon I would have a message that they were staying till after tea. After lunch the little aunt put her feet on the fender rail and went to sleep, and the boys had an orgy of reading *Punch* and illustrated magazines.

Figure 60 Gwendy, 1963. (Sketch by WLB; Courtesy: Patience Thomson)

But we did have two families locally who were great friends. Opposite our gate lived Hubert and Joan Worthington, both architects, and their two children, who were the same age as our own, and they were in and out of our house.[135] Hubert was a member of the well-known firm of architects in Manchester and also London and, like WLB, a tremendous enthusiast for anything he undertook, so they soon became great friends. Joan was many years younger; she had been his student. We loved her—she was so definite about everything and much in demand as an adviser on colours and interior decorating, generally. The two were very popular in the university, where Hubert designed

[135] John Hubert Worthington (1886–1963) was an English architect. One of his achievements was the design of the war cemetery at El Alamein, which contains over 7,000 graves.

Figure 61 There are very few known published photographs of WLB and WHB together. This photograph has been extracted from a Bragg family home video and shows father and son enjoying a stroll at Windy Howe some time in the mid-1930s. (Courtesy: Patience Thomson)

various buildings. Few days passed without our all seeing each other. Once, we all four gave an evening party together in our garden. Hubert, in his optimistic way, managed to borrow some of the Manchester Corporation workmen's braziers to dot about and look cheerful and we hung Chinese lanterns in the trees. But the climate is against such festivities, and it was too wet to go near the garden when the night came.

The other family who gave us great amusement and pleasure were the de Ferrantis of electrical fame. They lived with their five children in Alderley, and the first time we went to dinner with them, we had never in our lives seen so many electric fires pouring warmth: in the fireplace, in the ceiling, on the walls—everywhere. Vincent de Ferranti was of Italian origin;[136] when I asked Dorothy de Ferranti,[137] on arrival, if I could see the children, the young Sebastian and Basil were in their bath,[138] and I am reported to have taken one look at their dark

[136] This was Vincent Ziani de Ferranti (1893–1980); using his father's inventions, he made a successful business, Ferranti Ltd.

[137] Dorothy Hettie de Ferranti, née Wilson (1895–1993).

[138] Basil Reginald Vincent Ziani de Ferranti (1930–1988) was a British businessman and Conservative Party politician and was the grandson of the electrical engineer and inventor Sebastian de Ferranti.

hair and eyes and asked 'if all their children were black'. As they are all still alive, I had better not enlarge on the various escapades, many of which were colourful in the extreme, of this whole family. They supplied us with a great deal of fun.

Soon after we moved, WLB developed a curious mannerism while lecturing, a sort of stretching out of his arms that became very marked, so much so that one of his staff told me about it anxiously. Apparently WLB's whole arm had the sensation of 'going to sleep', and one of his hands was slowly losing its grip. I told Sir John Stopford, the then Vice-Chancellor, who was an anatomist, and when we went into the matter it was traced to the accident WLB had had thirty-five years before in Australia. He had fallen off his tricycle and broken his elbow.[139] After all these years the muscles, or nerves perhaps, were giving trouble again; it took Sir Harry Platt three hours to operate and rearrange the nerves in WLB's elbow, thus restoring the use of WLB's hand. From then on, when I was asked where WLB was, I enjoyed starting the answer by saying that he had fallen off his tricycle, and then making a dramatic pause.

Figure 62 Two examples of sketches made in 1930 in Fitzwilliam, New Hampshire, USA, by William Henry Bragg, showing his keen eye for detail and artistic ability. (Courtesy: the Lady Adrian)

[139] John Sebastian Bach Stopford, Baron Stopford of Fallowfield (1888–1961), was a British peer and physician.

We went off to convalescence with his arm in a sling, to Cecily and Arthur Goodhart, who had a house for the summer at Folkestone. We took David with us, as in age he just matched Philip, their eldest boy. David was also convalescing but from scarlet fever, caught from one of the Ferranti boys. As a token of contrition as she said, his mother Dorothy came to disinfect our boys' bedroom. She was what WLB called 'quite approximate' about this, though confident. I think we just closed the windows, blocked the fireplace with brown paper, lit a lot of fat and strong smelling candles about the room and then went off to a gossipy tea.

WLB used to go off by car now every day to the university, and I would spend most mornings pushing Margaret out in her pram, like most of the other mums, to shop or just walk in those lovely Cheshire lanes. We all had nannies then, and one saw the children in the morning and after tea. Rather ceremonious compared with nowadays, certainly, but there seemed plenty of time to read aloud, play games and sing songs, and WLB was very good at making things with them. There were generally children about the place, sometimes staying. Enid's two little boys or Gwendy's child would be parked on us, but of course nannies came with them. We used to give lovely children's parties, of which WLB was the life and soul. Once he made a shadow play from the children's book *Winnie-the-Pooh*. Together, we made a little stage, which we placed on a table, and figures of the various animals cut out from stout cardboard. These were constructed so that their limbs and, in the case of Eeyore, his tail and ears were all jointed, and we could move them about. There was a light behind the stage, and the scenes were shown as shadows. I remember in the storm scene I had to waggle the fringe of a big shawl, thus making the most effective heavy rain. I read aloud the appropriate parts of the book while WLB moved the characters. The children loved it, and we always had to do it a second time. When he was in the States or somewhere and I had to run a party alone, the boys felt it doomed to failure from the start.

I decided to read for the Bar. I had in mind one day being a poor man's lawyer, as it was then called, operating from the Citizens Advice Bureau in Manchester. Now when WLB was working in the evenings, I would be steeped in criminal, Roman or constitutional law. He used to hear me. There was a very kind Professor of Law called

Eastwood, at the university, who used to help me.[140] He was rather an isolated man, as he was deaf from a boxing accident in his youth.

I ate my dinners at Lincoln's Inn (where my Uncle Alfred Hopkinson had once been treasurer). Once I had an amusing experience. Wishing to chat with my neighbour, I asked him if he did not agree that the old bencher who had said grace was like a caricature of all barristers. He agreed but added, 'It's my father'; my companion was Quintin Hogg, Lord Hailsham.[141] I passed Part I of my examination, but alas! I remain on record at my inn as a student still, I believe, as we left Manchester, the war came and I never finished. I also took a course in journalism with some institute which gave tuition by correspondence on the understanding that if you had any talent whatsoever, you would repay yourself for the course within six months. I entered thus with great enthusiasm and was in the middle of typing my last homework article for the course (it was on bull fighting, I think) when I had to be taken hastily upstairs to have my fourth baby, Patience Mary. She was the only one of the children to be born in Cheshire and not Manchester. The Worthingtons had their last child, Olivia, at about the same time, very companionably. While I was doing my journalist course, WLB as usual took a great interest. He himself wrote very good English and he used to go through my homework, deleting what he called 'purple patches' and cutting out superfluous adjectives.

Scientific friends, as well as family, came to stay, as our house was such an attractive place in which to spend a night or two. WLB's relationships with his scientific friends varied. He was always most at ease with G. P. Thomson, as they had so many links: sailing together, the difficulty of famous fathers, and an interest and understanding in each other's work. Also, his wife Kathleen and I got on very well, and we enjoyed staying with them in Aberdeen. When she saw our house for the first time, its dimensions convulsed her and she sat on the steps and rocked with mirth. 'What a place, Alice, you'll be the death of me.' Especially as she glimpsed though the window a great tray of potted chrysanthemums, in the manner of the time, as big as babies' heads.

[140] Reginald Allen Eastwood (1893–1964) was Professor of English Law at Manchester from 1924 to 1960.

[141] Quintin McGarel Hogg, Baron Hailsham of St Marylebone (1907–2001), Second Viscount Hailsham, was a British Conservative politician, Leader of the House of Lords from 1960 to 1963 and Lord High Chancellor of Great Britain 1979 to 1987.

We enjoyed having Ralph Fowler to stay, but he was, like Blackett, a very forceful character. He was the kindest of men, as I had reason to know: years later, despite an extremely busy life, he used to come in and see if I was lonely when WLB was away for a year on a mission to Canada, when we were in Cambridge. He used to try and involve WLB in affairs of state but, though WLB listened and admired, he had to go his own individual way, whatever his dynamic friends expected of him. In the end, they realised this.

Looking back, I know that WLB was very happy in the Alderley days. The look of the countryside was of great importance as a background to his life: the shapes of the trees, the stretch of water meadows below us in the plain, and the rooks streaming home at night counted for much. He loved having his children all at home. It was a period of great research development with a continuing group of international students. He was fast developing his skills of lecturing, and especially popular lecturing. All through our life he used to go through the latter with me, and indeed any article, as opposed to a scientific paper, that I could understand. I tried to go to all his 'popular' lectures. He was made to feel, I can only say, cherished in Manchester; after all, we had been there sixteen years, and the unforgivable sin with Mancunians is not to stay. They are very sensitive, and I remember on several occasions how upset they were when some man could not be coaxed from the amenities of Oxford or Cambridge to the university or research in industry because he could not face Manchester. I regret to say it was sometimes the wife who felt it was wet, dirty or too far away. However, our time in Manchester was now drawing to a close. Early in 1936 a surprising offer came to WLB. He was invited to become the director of the National Physical Laboratory at Teddington, which meant leaving university life and becoming a civil servant. I myself had only a vague idea as to what this meant: the only fact I did know was that all clinical thermometers at that time were tested for accuracy by the National Physical Laboratories and marked to that effect. Obviously there must be more to it than that! It was, and indeed still is, a complex of national laboratories for scientific testing of many kinds, like the Bureau of Standards in the United States, as was explained to me. Had this offer come earlier, or later, I doubt if WLB would have considered it; but he had been a professor here in Manchester for eighteen years and, as I had heard him say to other people, 'After about sixteen years, men often feel an urge to change jobs.' We

Figure 63 WLB and family. Back, left to right: David, Patience and Alice. Front, left to right: Margaret, WLB and Stephen. (Photograph from the album of Albert Hopkinson; Courtesy: Patience Thomson)

had great heart-searching, and he decided to go up to London and have a talk with his father. Rather to my surprise, Dad was keen on his accepting and advised him to look into the matter very carefully.

WLB went off to Teddington and reported back to me. He had gained the impression that the laboratories, under the long reign of their late director, Sir Joseph Petavel, had become somewhat uninspiring places, doing standardised jobs and very little original work. It was suggested to him there that if he came, he would boost research and generally inject new blood into the place. He was promised a minimum of administration, as there was a most competent deputy who would take care of that side. There was a large salary and a lavish

Figure 64 WHB with HRH Princess Marina (Duchess of Kent) in 1940 with the Anglo-French Ambulance Corps. (Courtesy: the Lady Adrian)

allowance for upkeep of the director's flat and for entertainment, through some Royal Society trust. There were two floors of a lovely Georgian house for the director, set on the edge of the laboratory site in the middle of Bushy Park. We decided that while he was considering it all I had better go and see the place for myself.

I shall never forget my initial impression of it. First a very bad one, as I walked along the road from Teddington station to the laboratories, an unprepossessing dormitory town all around me. I turned in by a porter's lodge, following a long drive with laboratories on either side, and finally I suddenly came upon the most beautiful great Georgian house, overlooking the park. It was July, and one wall was covered with an old magnolia out in its full glory of huge white flowers. I fell in love with this house on the spot. The deputy (Higgins) was there to show me round. I remember standing on the broad sweep of steps to the front door and looking about me before we went inside. There was a great sweep of lawn dotted with old cedar trees, and in the distance there were deer grazing in the park. Actually a ha-ha or invisible ditch divided the garden from the park, but to me that day the deer seemed to be wandering on the lawn itself. The house had been built, or rather begun, in 1715, and Lord North had lived there. WLB used to say it was no wonder that he lost us our American colonies when he had Bushy House to complete and enjoy at home. In the last century it became a royal residence, when the Duke of Clarence (later William IV) lived there with his Mrs Jordan (the 'Little Romp', the actress) and their numerous children. When the Duke became King and had so shamefully shed Mrs Jordan, he and his Queen Adelaide retained Bushy House. We went inside, and I remember Mr Higgins saying, 'I'm afraid, the ground floor belongs to the laboratories; the director has the two top floors only.' The first floor had no less than seven sitting rooms, of which the drawing room had been Queen Adelaide's bedroom, and her dressing room leading out of it made a lovely little boudoir. They were most beautiful rooms, with moulded ceilings, elegant cut-steel fireplaces and, folding over the windows, the painted shutters of the period. The domestic quarters did not match up to this magnificence. The kitchen was at the end of a long sloping passage, forty yards from the dining room, along with bedrooms for staff. That this set-up would pose problems, it was clear.

I brooded over the whole question in the train going home. I said to myself (as I so often have had to do!) that I must keep my head; that the royal residence had bewitched me, but I had a feeling that we ought not to go. Instinct told me that WLB would never do research there; scientifically it might break his heart. There was probably nowhere, outside Cambridge, better for physics than Manchester, and in Cambridge Lord Rutherford was in command of the Cavendish Laboratory and was hale and hearty. Other questions, such as the children's schools rose to my mind; everyone would be uprooted.

However, WLB decided to accept the offer and leave Manchester that autumn. Now strange events overtook us. We were within a month of moving south that October and in the midst of farewell parties, when suddenly and unexpectedly Lord Rutherford died. Who would succeed him in Cambridge? Echoes of that talk at the Inglis dance when I was a girl in 1921 returned to me, and I heard again WLB admit that to be Head of the Cavendish in Cambridge would be the physicist's dream. It seemed now any chance for WLB had gone. He himself was philosophical, assuring me that Cambridge would want a nuclear physicist to carry on Rutherford's work, and he was not that. I wrote to his father, expressing my fears. Dad wrote me a letter starting, 'It was obvious that when the news of Rutherford's death came to me this week, Willie should wonder if it would have been better for him at Cambridge than at Teddington ... We must be realistic and take things as they are for there is no chance of altering your position now, i.e. of going to Cambridge instead of Teddington, at any rate there is no step that he can take.' He goes on to say, 'What a worthwhile job the NPL will be, and what a great deal WLB can contribute there.' The letter is signed, touchingly, 'Love in quantities, Dad.'

We therefore applied ourselves to the job at hand, which was altering Bushy House from a bachelor establishment to a family home. Looking back, this was really a very amusing, though frustrating experience. Up till now we had always had 'little men' to do our painting, carpentering and plumbing. Now I had to go to the Office of Works for everything. First I had to deal with the formidable Sir Frank Smith, secretary of the DSIR, to whom I suppose WLB was finally answerable. I gave him my plans for making the place possible to run, and these included making two of the seven sitting rooms into a kitchen and a pantry-serving room, respectively. 'You cannot move the kitchen, Mrs Bragg, it would cost too much.' Our eyes met. 'Sir Frank,' I said, 'I shall move the kitchen.' I did. But many minor battles with the Office of Works underlings raged on. There was the question of wallpapers. These were graded in price and we chose the expensive ones for the main rooms and cheaper ones for the bedrooms; the cheaper ones were sometimes the prettiest. I had to learn that the Office of Works liked everything of the most expensive. I remember their man saying, 'but you cannot have this cheap wallpaper in the spare room, who knows you might have a "Sir" staying there.' In the same spirit, we had to have a marble bath in the children's bathroom, so I gave up trying to save the taxpayer. It was a new experience, having to be serviced in every detail by a government department. I think I am right in remembering that it issued even our lavatory paper, which was stamped OHMS (on His Majesty's Service). Anyway, everything was very beautifully done, and our dear friends and architect neighbours, the Worthingtons, came down, planned the alterations and helped with the furnishing.

We now embarked upon a very different life. We had a most efficient cook-housekeeper, who not only engaged the three maids, but cooked, shopped and in fact did everything. The garden was done for us; I had only to say what flowers I wanted planted. Stephen went off to Rugby to school, and David to a prep school near Malvern. There were only the two little girls, with the nanny, in the house now. WLB was at home much more. He had only to cross the drive to his office every day and, for the first time in our lives, he was always back to lunch. As a top civil servant, he had invitations to private views, dinners, the royal garden party and so on, and we seemed to drive up to London a great deal. In a way we all found ourselves isolated. We had no neighbours there and no one dropped in; the children especially missed this. On the other hand, we were now close to both of our families, and they and friends were always coming for the day or to stay with us. Weekend child guests had a lot to see when the labs were officially shut on Sunday, such as the big wind tunnel, and the great tank where experiments on ships were carried out. Somehow there was one eel in this tank, and we all used to try and fish for it. I spent a lot of time with the children, especially ferrying them about for riding lessons in the park, visiting other children and sightseeing in London in the holidays when the boys came home.

Then there was the entertaining. Hardly a day passed without some official visitor for lunch. I tried to have all the wives from each lab (physics, engineering, metallurgy and so on) to 'At homes' and a look round our new home; there would be anything up to about eighty guests at a time. They were very appreciative but I realised sadly that I could never remember so many people or weld them into a group. We had a huge party at Christmas for their children at which WLB was the life and soul as he invented wonderful competitions. When an important committee, such as the Executive or the Admiralty Board, met on NPL affairs, I gave them lunch, and our housekeeper excelled at these occasions. These lunches would now of course all be done in the official lab dining rooms; in our house they had a character of their own. This was fun for me, as the guests were interesting people from science, industry or government circles.

All the time, the question of Rutherford's successor nagged our minds. More letters came from Sir William Bragg saying that 'the whole matter would be taken out of our hands, and if the Cambridge electors wanted WLB they would put their heads together with Sir Frank Smith (our boss) and settle the question. We must forget it.' But it was very unsettling, and we could not but be aware of the general talk about it among the scientists around us. Added to our feelings of insecurity, there was now the threat of war.

When we returned in February, there was the letter from Cambridge inviting WLB to succeed Rutherford. Sir Frank Smith came to see us both and said 'though it's cutting off my nose to spite my face, I advise you to go to the Cavendish.' I was immensely relieved that WLB was going to get back to university work and, above all, research. But I was rent to leave our beautiful home, which we had now decorated and furnished to our own taste, and especially my own little boudoir leading out of the drawing room with its view right across the park. I used to spend a lot of time in this little room, writing articles for papers, a history of Bushy House for *Harper's Bazaar*, some travel ones for *The Yorkshire Post*, and some rather silly ones for evening papers, with titles like 'Let's be born at home' and 'Pass along Praise'. Furthermore, we should lose our large salary, our lavish allowances and all the supporting maintenance to which I had now begun to be accustomed. If the iron went wrong, a lamp bulb burned out or a tap leaked, I had only to ring the maintenance office on the site, and someone came. It had been a very spoiling experience, to be repeated once more in my life, but not for another nearly twenty years.

The rest of the year our lives were very restless. For myself, I seemed always to be driving off somewhere in the car, going to Cambridge to my parents to look for a house for the family, going to spend weekends at the boys' schools (once summoned to Rugby late at night to bring Stephen home for an emergency appendix operation) or dashing to London. It was not just we who were unsettled; the whole country was in a state of uncertainty about war. WLB began planning the construction of trenches in case of air raids, and even the children were fitted for gas masks. Our good friend Sylvia Fletcher-Moulton came over one evening and asked me to go and see the Dowager Marchioness of Reading, with a view to roping me into some vast voluntary scheme (the WVS) she was launching for the Home Office.

I shall never forget this visit to Lady Reading. I think no woman has ever made such a strong impression on me. Her appearance alone was remarkable, as I first saw her in office. She looked oriental, Persian perhaps, with her black hair and aquiline nose.[142] I saw her in imagination with a helmet and spear and draperies, perhaps woven on a tapestry. Within half an hour I found myself compelled to be one of her half-million women recruited for ARP (Air Raid Precautions) work. My immediate assignment was to go the north of England, to Liverpool and Manchester, making contacts and preparing the way for her to address large women's meetings in those cities. Arrangements made, she and I with one or two other henchmen then went up together for a night.

[142] Actually, Lady Reading had been born in Greece.

Thus, that summer I got to know Stella Reading, and I learnt a great deal from her. Her methods were, like her looks, somewhat oriental and I did not find them entirely acceptable. My north-country spirit jibed at getting what one wanted by first 'buttering the paws' and softening people up, men especially, till you had what you wanted. 'You must always go to the top people', she would say. Still, I am grateful to her for setting me on an unexpected career of voluntary, public work, and for many words of wisdom about public speaking, committee work and organising. But, as WLB said, 'We never really went all the way with her, much as we admired her tremendous achievements.'

Meanwhile, WLB was having a trying time. People in the labs felt unsettled with a director who, within a year, was leaving them, and they did not all go along with his plans for possible war. But he had his trenches dug, and soon we had a stream of visitors, ranging from officials from the London Parks to government people coming out to see them. He interested himself in the disposal of scientists to secret placements, should war come. It was all somewhat unreal. On the other hand, people from Cambridge were coming to discuss the building programme of the extension planned for the Cavendish Laboratory. He became very tired and slept badly.

We had to press on with our personal plans. Thanks to my father, we secured a lovely house in Cambridge which was along the famous 'Backs' and which we rented from Caius College; we were to move before the autumn term. For the month of August, we took a house at Studland in Dorset, and all enjoyed ourselves. In September, war looked imminent, and we sent our little girls off to Grandfather Bragg and Gwendy to their country house in Surrey to be safe in an emergency. Enid's little boys joined them; then suddenly the tension lifted with Chamberlain's visit to Hitler at Munich, and life returned to normal, for the time being.

Going back to Cambridge was a homecoming for me. My parents were there and, though my mother was now very ill, they welcomed us with delight. The university also welcomed us, but in that typical Cambridge way, in fact, as if we had never been away. Wherever we had been in the interim was ignored, in a curious way. As WLB dropped back into his Fellowship at Trinity, one Fellow greeted him by saying, 'You've been away for a bit, haven't you?' I picked up my connections with Newnham, and we were part of it all again.

I soon came to know another Cambridge, the town which is not the university. Lady Reading, as soon as we arrived, put me in charge of starting the WVS in Cambridge, and I began to know the community of tradespeople, college servants, railway men and the workers in light industry; a world flowing out and around the university but definitely not of it.

Meanwhile, the children were happy to be where friends were easily accessible and could drop in. I well remember the small bicycles and tricycles parked or, more likely, carelessly thrown round the great beech tree on our lawn, as the children came to play with Margaret and Patience. We were to have open house for generations of undergraduates who came to meals, to talk, to borrow anything from eggs to suspenders and to beg beds for girlfriends for a dance.

No wonder, during that first autumn that, as WLB and I wandered among our old haunts by the river, through college gardens and along the Backs, often hand in hand, he repeated, 'Aren't we lucky?'

3.15 Wartime in Cambridge

Our first year in Cambridge was of a pattern that would never be repeated. As a new professor and his wife, we were entertained by our colleagues to lunch and dinner parties, and the university wives called on me wearing hats and gloves and left visiting cards. I punctiliously returned these calls; maids opened front doors and served tea. We ourselves had a cook, Nellie, a temperamental and awkward little body and who remained with us for ten years, two maids, of course, our nanny Hilda and a full-time gardener who produced magnificent fruit and vegetables; Margaret and Patience settled down at day schools, and our lovely old house and garden were filled with their friends at weekends.

I tried to get to know the wives of the Cavendish Laboratory staff but there were about sixty of them and it was not easy. I do not remember that in those days any of them had jobs, and certainly some of them were very lonely. The husbands were attached to colleges and dined 'in Hall' several times a week while the wives stayed at home 'alone with a poached egg on a tray', as they told me. We had a Christmas tea party for them in our house and even gave a dance, which was great fun. Scientists are notoriously bad dancers, so WLB invented a complicated intellectual paper game as an alternative entertainment.

Many of the men sat on the stairs absorbed in this, so much so that when I offered refreshments to one professor he said, without looking up, 'Oh! Damn it, leave me alone.'

The threat of war had not receded, and emergency plans were being made. During the year, scientists from the laboratory were slipping away to start radar work and other projects. Shelters against air raids were being constructed. Air force bases sprang up from Cambridge to the coast, and consideration was being given to receiving evacuated children from London should the need arise. I was instructed by Lady Reading to start the WVS, working under the local authority, and to enrol as many women as possible for Civil Defence. I cannot say that anyone was very enthusiastic, but I took over a shabby office in the town and, with my first batch of recruits, dutifully made lists and filed cards with names of possible helpers should they be wanted.

That summer of 1939, in June, WLB and I had a wonderful holiday in the French Alps at Lauteret, looking for wild flowers. The valleys were full of narcissus, globeflowers, lilies and gentians, and high up in the mountains we found rare species of forget-me-not, primula, anemone and ranunculi. While we edged our way along the precipitous paths, we would see the marmots running about below us, whistling shrilly. That holiday became a treasured memory, for it would be five years before I, at least, would go abroad again.

Later in August, we all went off to Wales, where we had been lent a house by Hugh Lyon, the headmaster of Rugby (where Stephen was currently at school), on condition that, if war seemed imminent and orders came through to evacuate our cities, we would leave the house the same day, as the Lyons would want their house for their own family. A few days before war was declared evacuation started. And even as we were packing up to go, we saw trains filled with children, steaming inland. War was declared three days later.

Returned home, we were immediately faced with an agonising decision. Should we send our children to America for the duration of the war? Offers poured in to members of the university to give homes to their children, and many were accepted, most of us thinking that the parting would last perhaps six months. To us also came cables to send our children with or without me or Nanny. WLB thought we should accept. His imagination ran riot; he saw Hitler in East Anglia, and nameless horrors awaiting us. Nanny was persuaded to get permission from her mother to go with the children, as I flatly refused to leave. I could not bear the thought of indefinite separation from WLB and the two boys and thought we should all stay together. We decided to go round and see Sir Will Spens about it. Soon to be nicknamed 'the Will of God', he had been appointed Regional Commissioner, to take over should our region be cut off from central government. How he enjoyed that role! He was emphatic that we should not send our girls abroad; it would spread alarm and despondency, and already the townspeople were beginning to feel that the university had some knowledge, withheld from them, about immediate danger of invasion. He advised us to send the children to the west country for a time. Lady Cawley, sister of our dear friend Sir Kenneth Lee, offered a refuge, and they went off with Nanny for a few weeks to a life of luxury at Berrington Hall in Herefordshire.

Life in Cambridge changed rapidly. In our own home, the two maids and the gardener were called up. The town was packed with evacuated mothers and children, students and staff from London colleges, and personnel from various ministries. We ourselves gave accommodation to a mixed bag of people, including at one time or another, a girl student from Bedford College, a pregnant bus conductress, a friend's child with pneumonia, and a professor from London, to mention but a few. My dear sister Enid had had to give up her home in Highgate when Malcolm, her barrister husband, went to a post in the air force and, while he remained in London, she kept house for him and some of his friends in a series of borrowed houses; the youngest boy, James, was sent with his nanny to us for the winter.

Both WLB and I found it a strain having so many people around, and there were practical difficulties. The house was big and rambling and thus hard to heat, and food and clothes were rationed. Actually, we were very lucky in having regular and splendid food parcels from our old friends in Chicago. The colleges emptied, many undergraduates went off to the services, and the Fellows to various forms of war work. Art treasures were despatched to places of safety, and we ourselves housed in our cellar some of the glass from the windows of King's College chapel. This prompted one of those curious comments for which the provost of King's was famous:[143] 'Take care of our glass, dear friend, and it will surely take care of you!'

[143] John Tresidder Sheppard (1881–1968) was an eminent classicist and the first non-Etonian to become the provost of King's College, Cambridge.

I sometimes think, looking back, we rendered some services by remaining, all through the war years, in the same place. We could always be found, so that we had a great many visitors, some of whom to stay, squeezed in somehow, and others dropping in. There were those too who came on business to the laboratory, where certain secret work was going on, so WLB was always there in charge, except for one difficult year when we were without him, as he was sent on a scientific mission to Canada and the States. He also sometimes went to Scotland briefly on advisory work for the Admiralty and always returned with a most welcome present of kippers.

I will give just one example of these visitors; he came from the Russian Embassy in a great black chauffeur-driven car. He had lunch with us after a visit to the laboratory and saw our youngest girl, Patience (now back from the country). She reminded him, with her two fair pigtails, of his own child in Russia. His eyes filled with tears and he said he had to give her a present. Fishing in his wallet, he handed her a slip of paper, saying, 'Take that to the grocer, darling, and you will get something nice to make things with.' Patience politely took it, trying not to look disappointed. In fact, it was a food card heavily stamped with the word 'Diplomatic' and entitling the holder to literally pounds of sugar. I took it to the grocer, who assured me that it was perfectly honest and legal, and the children and their friends had an orgy of toffee-making.

Every day, I bicycled off to the WVS office in the green uniform that Lady Reading had had some haute couture dressmaker design for us, and this was a great saving of clothing coupons. I often stopped on the way to call in on the shoemaker Mr Carter for a moment's chat. Like so many, he had sons in our local regiment, whose members had been taken prisoner at Singapore by the Japanese and were working on the notorious Burma railway, from whence came no news.

When I think back, I am astonished at what our WVS achieved. We were instructed 'never to say no' to any request made to us. Our activities included running a canteen, open night and day, for service personnel at the railway station, running a leave centre for troops, supplying car drivers and collecting salvage. Most taxing of all, we billeted all the evacuated children, which was really the responsibility of the local authority. We did it, and they took the credit. In this sphere,

Hester Adrian (wife of the Professor of Physiology[144]) took on this unenviable work on our behalf. Planting out children on often unwilling hostesses was a difficult job. To return to Hester for a moment, we were to work together for many years, and she became one of my dearest friends. She had sent her two children to America and missed them very much. Early on in the war, she lost her leg in a climbing accident in the Lake District, and I became close to her as I was able to help her in various ways. It would be tedious to give a detailed account of the WVS work but I will give just two examples of it that I shall always remember.

One Sunday night, in the summer of 1940, I was told to get at least twelve volunteers to be ready the next day for a job, the details of which could not be told except that about half a dozen must be prepared to stay away for twenty-four hours. In fact, there had been a leak in information about the movements of our ships, and several hundred people from the region were to be rounded up and taken away somewhere. Next day, at a few hours' notice, the women and children involved, all I think European refugees, were brought to Cambridge by the police, herded into a church hall and looked after during the night by my volunteers. It was a heart-rending experience; these poor women had no time to make any arrangements and had no idea where they were going, what would happen to them or where their husbands had been taken. Enid was living with my father at the time, and she was among those who stayed with them and came to the railway station next day. She gave me a harrowing account: they had forgotten vital things like baby food, extra nappies and so on. Above all, they needed messages. She did what she could, and then early in the morning these unhappy people were put on a special train, and I and a fresh team of helpers and two uniformed policewomen accompanied them. No one knew our destination, which turned out to be Liverpool. From time to time, we stopped to have our coaches filled with such people coupled on to our train. Our charges were nearly all educated women, including one or two wives of scientists whom we knew, and they plied us with anxious questions which we could not

[144] Edgar Douglas Adrian, First Baron Adrian (1889–1977) was an English electrophysiologist and recipient of the 1932 Nobel Prize for Physiology, won jointly with Sir Charles Sherrington for work on the function of neurons. His wife was Hester Agnes Pinsent. He was Chancellor of the University of Cambridge from 1967 to 1975.

answer. All that long gloomy day, we could only try and take their minds off their anxieties by singing, playing cards and comforting them. Darkness was falling as we drew into Liverpool Station at a deserted platform, lined with policemen. We had heart-rending goodbye scenes and then watched them being herded into buses bound, as we heard later, for the Isle of Man.

The other example of our work is more cheerful. I was suddenly asked if we would prepare a house for Belgian refugees arriving next day. The military had just vacated it and left it so filthy that no paid helpers would touch it. I had by then found myself a splendid deputy in Mrs Frances Clode (finally to be Dame Frances and succeed Lady Reading as Head of the WVS).[145] She had all the qualities that I lacked, so we were rather a good combination. I now summoned Frances, and we decided we would make a joke of this assignment and of course go ourselves. I was a bit worried as I had never scrubbed anything in my life, but as usual Frances said that she would see me through. I put on a huge hessian apron and a man's cap back to front as the women did in Lancashire and, armed with a large scrubbing brush, set out to supply light relief to our team. The house turned out to be incredibly dirty; it was full of rubbish and smelt. I was shut into a larder where I could not do much harm and the others could not see my feeble efforts. Finally, an immaculate lady from nearby came in and said that she so admired our efforts that she had prepared supper for the eight of us. She had the sort of house where everything that can stand on a white lace mat does so but, dirty as we were, we thankfully accepted.

At night, since our office was in the middle of the town, we had to take turns in fire-watching. This meant sleeping in the office and, with gas mask and helmet, turning out and patrolling the street when the first air-raid siren wailed. I used to make the others laugh by undressing completely and folding my clothes neatly on a chair before seeking my palliasse on the floor; a public-school training dies hard. I was always very frightened inside, walking about the streets in the middle of the night, with whizzes and bangs and crackles all round me. I had many interesting conversations with other fire-watchers, one of whom, Mr Woolaston, ran an antique shop much patronised by Queen Mary. Meanwhile, at home, the children would go down to the cellar to sleep beside a crate of glass from the King's College chapel windows.

[145] This was Emma Frances Heather Clode (1903–1994).

I would have enjoyed sharing my experiences with my mother, now that we lived close by my parents, but she was very ill, with two nurses, and could no longer take any interest in things. She died in 1940. If ever the phrase 'her children rose up and called her blessed' applies to anyone, it would be to my mother. My father lived on alone in the same house for some years, and we tried to cherish him. Both Enid and Philip, who was now a priest and an air-raid warden in the heart of London, came to see him as often as possible. My brother Harry was with the RAMC and was in medical charge of one of the boats evacuating the army from Dunkirk. Greta,[146] his wife, was based in Surrey, helping a doctor and his family. We all met as often as we could. To complete the family picture, Stephen left Rugby with a scholarship to Trinity College in 1942, to do a degree course in engineering in two years instead of three, as engineers were in short supply and badly needed. David, after Rugby, was directed to the coal mines, where he had a wretched time and alas! had a bad breakdown.

WLB was sent on a mission in 1941 to Canada and the United States and was away for the best part of that year. The children and I missed him terribly, but I tried to remember that many of my friends had to contend with much worse anxieties and partings and that he was not in any danger. Shortly before he left, he received a knighthood. How well I remember getting a message one morning from his secretary (dear Brenda Smith) asking me to call at the laboratory. I went and, when he told me, we danced the polka all round his room to the astonishment of Brenda Smith, who could not yet be told. It was decided that as his father was 'Sir William', WLB, to avoid confusion, would be called Lawrence (his second name). Alternatives like Sir William junior or minor were not really acceptable. I was sad not to go to the palace with him, but I was ill with flu. He celebrated his return from America by buying a wild pony for Margaret; the gift gave her enormous pleasure, as she had a passion for riding.

While WLB was away, I was co-opted on to the town's Civil Defence Committee; my being on the committee was a great help to the WVS, as I could hear what was going on and give our point of view. The members, all men, were, with I think two exceptions, all town councillors. The exceptions were members of the university. I should explain that,

[146] Greta Hopkinson, née Strohmeyer.

in Cambridge in those days, the university elected six members and two aldermen to the town council. One of those on the committee I now joined was Sir 'Monty' Butler, Master of Pembroke College and father of R. A. Butler, the politician.[147] He was a formidable figure on any committee, in spite of being extremely deaf. I generally sat beside him and, with all my experience and sympathy for my deaf Hopkinson aunts and uncles, I took to jotting on a pad anything important that had been said and which I thought he had missed, passing it to him. Then one day the town clerk, Mr Kemp, who was secretary to the committee, said drily, 'You need not bother; Sir Monty knows what all the members here are going to say before they say it.'

I noted with interest the feelings between 'town and gown'; it was a mixture of respect coupled with irritation on the part of the townspeople at what they felt to be the arrogance of the university. I myself scored by now knowing a great many townspeople, as well as people from stretches of the town where the academics never penetrated, such as down by the railway, and yet I was of the university.

I had not been on this committee long before I myself was invited to become a town councillor representing Newnham, the ward in which we lived. There were no elections during the war, and the two political parties agreed to having me, as officially I belonged to neither. I accepted and joined a small band of like-minded members, the Independents. This move had a startling consequence. The end of the war was in sight, heralded by our landings in Normandy in the summer of 1944, when I was invited to be the next Mayor in the autumn. Without hesitation, I refused. My children were too young, I had a home to run and, finally, I had only been on the council for two years and was totally inexperienced. 'Five years hence, perhaps,' I said when I had made suitable thanks for the honour done me. 'Five years', echoed the elderly councillor who approached me. 'Why, by then your looks and your mind won't be all the same.' This seemed hard since I was only forty-five, young by mayoral standards then. I discussed it with WLB. To my surprise, he had no hesitation in telling me to say yes. 'One must not miss these opportunities. If you refuse you will regret it.' Thus, with a quaking heart, I accepted.

[147] Montagu Sherard Dawes Butler (1873–1955) was Governor of the Central Provinces of British India between 1925 and 1933. He was the father of the noted British politician Richard Austin (Rab) Butler.

I resigned from the WVS to prepare myself. It was sad to leave the members from whom I had learnt so much but whose friendship was to stand me in good stead during the coming year. They indulged my passion for surprise presents by having a farewell party and presenting me with two beautiful Queen Anne walnut chests of drawers and a period tray. Mr Woolaston, the antique dealer and my co-fire-watcher, had apparently supplied these and was heard to say that he'd had to hide them from Queen Mary on her last visit.

My final act was to take part in a ceremonial march of the members of all the branches of Civil Defence. The town clerk telephoned me one day: 'Would my ladies', he wondered, 'care to take part in the March Past?' I told him candidly that many of us—that is, the members of my organisation—had neither the shape nor the feet for marching. However, I did not wish to refuse outright and so asked him what was involved. It was to be a ceremonial march, on Sunday afternoon in three weeks' time, of about two miles in from the city boundary, around the square to the town hall, where General 'Blank' would take the salute from the steps. I considered for a moment. 'You can count on us', I said with dignity. I then sent cards to all members, inviting those able and willing to march to let me know at once and, as time was short, the first twenty to reply would be chosen. This was perhaps rash, as they might not all have been the best judges of their suitability for the purpose, but in fact only twenty volunteered.

Now, mine is not a military family, though I must say that individual members have acquitted themselves well when necessity demanded, and probably for that reason I decided that we had to be properly trained. I spoke to the Lord Lieutenant and told him that we needed a guardsman, at least, to help us. It was arranged that Sergeant Brown, retired, should take us in hand and he was warned by the Lord Lieutenant to deal softly with us. I gathered my ladies in to a tea party and unfolded this plan. We were to meet Sergeant Brown on a sequestered school playground, at dusk, when it was hoped that the most dawdling child would be off the premises. I said that we had to compete with professional bodies accustomed to this sort of thing and that we must be very careful, not least of our appearances. Knowing my ladies, I explained that we should not be able to talk during the march, that it would be best not to wear high heels or jewellery, however well it matched our uniform suits, and that we should not greet friends in the crowd. In the end, mellowed by tea, everyone was cooperative, except that objection was taken to the length of the route. It was finally agreed that we would join the procession on the corner by the hospital.

We all turned up at the school playground to find a tall, slim and immaculately equipped guardsman awaiting us. My problems crowded in on me at once. Should I say 'Sergeant', 'Sergeant Brown' or perhaps 'Sir', as he was presumably our superior officer for the time being? Or should it be simply 'Brown'? I compromised: 'You are Sergeant Brown, no doubt?' And that was the last time I spoke in my normal voice for half an hour. Attention was concentrated on me as leader, and I came in for the full force of the sergeant's violence. First he told us the words of command, roughly how to stop and go, and then he stepped out beside us round the playground, barking at me to throw my voice right from my stomach.

But, apart from having no notion how to do this, I was so afraid of attracting attention from the road that I could only squeak and trust to the loyalty and good sense of the members ranged behind me. We ended, exhausted, by the sergeant taking the salute standing magnificently between the net-ball posts. He was gloomy about us; he felt that on the day we should never get off on the proper foot. The ladies crowded round him and said that he must be there; in fact, it was suggested that he should actually lead us. At that he said something dreadful but apologised at once.

The day came all too soon when, supported by our husbands, children and dogs, we assembled at the hospital corner. Our escorts stripped us of handbags, furs and a bracelet or two—in fact, anything unmilitary, like umpire receiving sweaters—and here now was the band coming into view at the head of the procession. It would soon be our turn to fall in with it, I realised with sick apprehension. Suddenly, like a lifeline, Sergeant Brown swung out from the crowd. We restrained an impulse to hug him. Somehow he got us all into position, spacing me out in fearful isolation in front and, with a sharp bark, started us. Never have I felt so abandoned, so forlorn. It was with infinite relief that I heard the pair behind me, presently, discussing what they would buy in the sales. It showed me that they were still there. Now I must begin thinking about 'eyes left'—I was nearly at the town hall—I yelled the command and found myself gazing, it seemed for ages, into the eyes of a general whom I realised I had once sat next at dinner—I marched on trying to remember if it was five hundred

I was to count before we could turn our heads again. It seemed long afterwards that a familiar and fierce voice detached itself from the crowds. 'Eyes front.' The ladies obeyed Sergeant Brown. They never even heard me.

We sent the sergeant a presentation box of cigarettes with a gracious note and hoped for a line of congratulation on our performance. He thanked us most politely and did allow himself to add that he would never forget our march.

3.16 His Worship Mr Mayor

I was to become Mayor on 9 November 1945, so that summer I had to prepare myself and put my house in order at home. I felt my family might suffer. Nanny Hilda would have been my great support but, though still living with us, she had married a petty officer soon to be de-mobbed and was expecting a baby the week before Mayor's Day! Not without misgiving, I engaged a Miss P. to be a sort of backstop, ready when Hilda had to leave us. This arrangement was not a success. Miss P. had the highest reference as a supervisor in an aircraft factory but perhaps this had been too much for her, as she suffered from nerves and sick headaches, and neither Nellie the cook nor the children took to her. Finally, we imported a wild and lively Irish girl as housemaid, who added to the gaiety of the household. When we all played mixed hockey on the lawn in the holidays, and Hannah was put as goalkeeper, she used to cover her eyes and yell, 'Mother of Jesus, here comes Sir Lawrence!' She allowed herself to be put by the girls on Joey the pony, who promptly threw her, and she had to be bandaged for cuts on legs and arms.

The family were scattered. Stephen was now with Rolls-Royce, working at Barnoldswick in Yorkshire. David, still in uncertain health after working down the mines, was at an art school in London but very often recuperating at home. Margaret was about to start, in September, at a boarding school (Downe House). I think now that this was probably a mistake, and that she and Patience should have remained at day school. I was influenced by tradition. My father had been anxious that Enid and I should have equal opportunities educationally with my brothers and only the best would do. The best was a famous boarding school. Thus, it was natural that I should feel the same when we had to decide about our daughters. So it was that

Patience was the only one at home in term time and I am sure felt the strains of being bereft of Margaret and her beloved Hilda and putting up with an overbusy and often tired mother. In my mayoral diary, almost every day I noted, 'back to put Patience to bed and read to her before going out again'. It is the 'before going out again' that matters. There is a price to be paid for everything, and the price a woman in public affairs could pay is a child's security. At least my job was only going to last a year.

My father was another responsibility. He of course was very proud of me. His own father had been the Mayor of Manchester, and he loved seeing me 'robed and chained'. We found him a housekeeper, and I dashed in and saw him whenever I could.

Most important of all, of course, was my husband. It was he who had persuaded me to accept office and he loyally took part in my civic affairs, when possible; but he was having a very tiring time restarting the Cavendish Laboratory after the war years. My first weeks in office, he went to Portugal for two honorary degrees and various university festivities, and I could not go with him, a great sadness for us both.

I began seriously to ask myself why I was doing this. The only person who actually did ask me that was the late King George VI. I was sitting with him at a party in Trinity College, and he was told that I had just been the Mayor of Cambridge. After a pause he said slowly, 'Why did you do a thing like that when you did not have to?' Well, I suppose I am given to adventure, I have a strong Hopkinson sense of civic responsibility and there is a great deal of the actress in me. I frankly enjoy making public appearances, all dressed up, taking salutes, thanking people and always going first. I may say this habit dies hard, and I am still inclined to thank the choir after church, like signing cheques with one's maiden name long after one is married. I had really no idea of what I was undertaking. The fact that I was a woman and thus had no Mayoress meant that, besides being chairman of the council and of the chief committees of the council, I would have a tremendous social round, opening and shutting things, and was expected to say 'a few words' wherever I went, sometimes three times a day.

My first immediate problem was clothes. We were still rationed with clothes coupons and, like all mothers at the time, I had spent these on the children; so I wrote to our old friend of Manchester days, Sir Tommy Barlow, then the head of the board of trade, and sought his help.[148] He wrote back in his usual explosive way and told me that I could have no extra coupons and that it would be bloody good for Cambridge to have a shabby Mayor. Friends came to the rescue. Sir Kenneth Lee's wife, Giulia, sent me a lovely fur coat of her own, and I had a parcel of beautiful clothes from America from Betty Compton (wife of Arthur Compton, the celebrated scientist), who wrote that she had grown too stout for them! Finally, Tommy Barlow released some coupons for all Mayors so it was not too bad in the end. The official robes presented some problems for a woman. The hat, shaped like an admiral's, was very heavy but I managed that, only insisting on a clean lining. I inspected the scarlet robe, in which I was somewhat drowned, and promptly caught a moth in the fur trimming. Most difficult was the beautiful chain, which has prongs to grip into a man's coat. In evening dress, these prongs were either going to grip bare flesh or tear delicate material; but somehow we even managed to overcome this.

Most Mayors have served on their council for some years, but I was so new to it all that I was at the start particularly dependent on the Guildhall staff. Closest to me, in that he was always around, was Ingle the macebearer, a handsome ex-grenadier guardsman, who had served a succession of Mayors and upon whom I could rely. He never let me down. As I became starved of exercise, I was obliged sometimes to do physical jerks in my huge Mayor's parlour. Ingle used to tell callers that I was 'in conference' and not available. At all civic functions, he walked before me in scarlet and lace carrying the mace.

The most important official was the town clerk, a very small man—fussy and sometimes pompous but, as Sir Monty Butler said, 'a wise old bird '. I discovered that he, obviously anxious about his inexperienced Mayor who was a woman at that, read both my incoming and outgoing letters at first; in addition, he suggested drafting my speeches, but I would not have that. We really got on very well. He liked to call on me every day at noon to discuss matters and have a glass of sherry. One day he arrived carrying something wrapped in a bath towel. 'Are you going for a swim, Town Clerk?' I asked joculantly. For answer, he unwrapped the towel, disclosing a bottle of sherry. 'A return of hospitality,' he replied, 'but I did not care to be seen in the corridors carrying

[148] Thomas [Tommy] Dalmahoy Barlow (1883–1964) was an industrialist and banker.

a bottle.' The city treasurer and the surveyor were very good to me in explaining what was necessary for me to know about their work.

Undoubtedly the Chief Constable, Mr Bebbington,[149] was the most distinguished of the officials. I saw a good deal of him, as I was the chairman of the Watch Committee and he had to brief me in all police matters. The very first week of my mayoralty, I found myself heading a deputation to the Home Secretary, Mr. Chuter Ede, to ask him to allow us to retain our own police and not have them merged into the county force.[150] I greatly took to him and I think he liked my honesty.

'You are leading this deputation?' he barked, raising his great bushy eyebrows.

'I am,' I said, 'but I really know nothing about the matter yet. Sir Monty Butler will speak for us.'

We kept our police force; but later, when Chuter Ede met my husband, he said, 'Ah! Yes, I let your wife seduce me over the Cambridge police force.' During the year there was a conference for both chief constables and chairmen of watch committees, and I decided to go with Mr Bebbington. Unfortunately, the guest speaker at the big dinner was the Lord Chief Justice, Lord Goddard, who said that, if any woman came to the dinner, he would not make the speech, as it was not suitable for female ears.[151] So I did not go; and my chief constable loyally refused to go as well and took me to the pictures instead.

Mayors at that time were ex-officio justices of the peace, and however unsuited for the job, took the chair if they chose to attend court, which sometimes embarrassed the bench. I was the first Mayor in the country, I believe, to waive this right and said that I intended never to take the chair. The clerk, Mr Harding, and the magistrates were delighted, but those who were likely to be future Mayors felt I had let the side down. In the event, I was appointed to the bench in my own right during my year of office. Finally, I had to have a Mayor's chaplain, and I asked my father's friend Cannon Weekes to act.[152] We only

[149] Bernard Bebbington.

[150] James Chuter Ede, Baron Chuter-Ede (1882–1965), was a British teacher, trade unionist and Labour politician. He served as Home Secretary under Clement Attlee from 1945 to 1951.

[151] Rayner Goddard, Baron Goddard (1877–1971), was Lord Chief Justice of England from 1946 to 1958 and known for his strict sentencing and conservative views. He was a controversial and much criticised figure.

[152] George Arthur Weekes (1869–1953) was Master of Sidney Sussex College from 1918 to 1945.

Figure 65 Mr Mayor in 1946. (Courtesy: Patience Thomson)

fell out once, when he remonstrated with me for taking the chair at some arts festival on a Sunday morning in church time.

My main business was taking the chair, in full regalia, at the regular council meetings, and I found it always very alarming, however well prepared I might be. There were fifty-four councillors, four of whom were women. The meetings lasted three hours without a break; the men left the chamber if they wanted but I had to remain rooted to my great carved chair. The town clerk sat beside me and steered me through the agenda and sometimes, what seemed to me, a sea of awkward amendments. On one occasion, a councillor went on and on talking, and everyone became restive. I wrote on a piece of paper, 'What will stop this man' and passed it to Mr Kemp—the answer came, 'only death'. Actually, I stood up, red in the face, and surprised even myself by saying, 'Councillor. It is clear to me that you are boring everyone. Please sit down.' He did. It used to astonish me that the council would argue for an hour or so about whether a tree should be cut down and yet pass a half-million sewage scheme almost without discussion. I was always addressed as 'Sir' and announced as 'His Worship the Mayor'. I was, I was told, an office, not a person. Thus it came about that, over the maternity and child welfare minutes, a councillor, at the end of an impassioned speech, declared, 'After all Mr Mayor, you are yourself a mother, Sir.'

The council decided on two courses of action which led to my making my most important speeches of the year, and over which speeches I took immense trouble. One was the ceremony of making Dr G. M. Trevelyan (Master of Trinity) High Steward of the Borough; the other was conferring the Freedom of Cambridge to the local regiment. We have had a high steward here in Cambridge for 400 years. The office is nowadays regarded as an honour without responsibilities, but originally the borough appointed a high steward to intercede with the Crown in its sometimes violent differences with the university. It is still resented that James I called upon the high steward to see to it that the Vice-Chancellor of the university should always take precedence. On this occasion, G. M. Trevelyan was delighted to accept, the more so that his great uncle Lord Macaulay had, in the last century held the office.[153] I had read that, in the past, the borough had offered some present to the new high steward, such as a piece of plate or some delicacy in the way of food. I considered 'fat sheep, or a dish of lark's tongues' difficult gifts to arrange, but a local pike or two seemed possible and had been an acceptable offering in the past. With the help of the local angling society, we therefore presented him with a large pike tied with blue ribbon to a specially made oak platter. Mrs Trevelyan subsequently made it into a fish pie but it was rather tasteless and full of bones.

The other big occasion was the giving of the Freedom of the Borough to our local regiment, which was done in the marketplace, with speeches from the balcony of the Guildhall.[154] The whole town turned

[153] Thomas Babington Macaulay, First Baron Macaulay (1800–1859), was a British historian and Whig politician.

[154] This occasion was filmed and can be viewed here: <http://www.britishpathe.com/video/a-cambridge-freedom-to-county-regiment/query/01564000>. Notice how Lady Bragg is referred to as Mr Mayor!

Figure 66 Mr Mayor giving the Freedom of the Borough to the troops outside Cambridge Town Hall in 1946 (photograph by permission of British Pathé Ltd).

out to watch. The regiment had been drafted to Singapore, where within two weeks they were surrounded and taken prisoner by the Japanese. The rest of the war had been spent working under appalling conditions on the notorious Burma Road project. Somehow or other, they had hidden and managed to bring home the regimental flag. I found my speech from the balcony difficult, as the exploits we celebrated were not the usual acts of physical valour but of moral courage and endurance. The worse part was taking the salute, standing alone on a small platform. I feel faint standing too long at the dressmaker, and here I was in heavy robes on a hot September afternoon, motionless while a long wave of soldiers marched past me. The Lord Lieutenant urged me never for a moment to stop moving my toes; it worked. It was a most moving ceremony, especially as my brother Eric had been killed serving with this regiment in the first war and, at tea with the men afterwards, some of them remembered him.

The government during my year of office was Labour, with Attlee as Prime Minister. Various ministers came to visit the town and be entertained and, of these, Chuter Ede, then Home Secretary, was easily my favourite. He was very well read, and we had such an interesting talk about books at lunch that he missed his train to Darlington, I remember. Aneurin Bevan came, but I was a bit uncomfortable with him, as he squeezed me at lunch and called me 'my dear'.[155] He told me how he had cured himself of stammering, by walking over the Welsh hills, saying poetry with his mouth full of pebbles. The one I had most fun with was George Tomlinson, then Minister of Education.[156] He did not drink and was a lay preacher on Sundays. He came with two civil servants in striped trousers and large horn-rimmed spectacles, who stood in the background, plied with sherry by Ingle. 'What can I offer you, Minister?' I said as I received him in my parlour. 'As we are both Lancashire, would you like a nice cup of tea?' Yes, he would. When it came, I said wickedly, 'Now let's blow on it and drink it out of our saucers, as we do in the north.' To the amazement of the civil servants, and indeed Ingle, that's what we did. Tomlinson became very sympathetic to our educational problems.

Figure 67 Lady Bragg at the opening of Marshall's Airport, Cambridge, in 1946. (Courtesy: Marshall of Cambridge)

[155] Aneurin 'Nye' Bevan (1897–1960) was a Welsh Labour Party politician who was the Minister for Health in the post-war Attlee government from 1945 to 1951.

[156] George Tomlinson (1890–1952) was a Labour politician.

The whole matter of sport presents some difficulty to a woman Mayor but I was determined to do all that a man would. I only drew the line at boxing matches; to them I sent the Deputy Mayor. I faithfully attended football matches, hockey dinners and rowing services, and I bowled against the Mayor of Oxford in our annual match. There were two rather spectacular occasions. One was the opening of Marshall's airport, on a fearfully cold January day.[157] I had not been properly briefed and found that I was expected to fly over the city in a little Tiger Moth, all open to the wind. I told the pilot who was to take me that I must have some proper clothes; so a grubby flying coat and helmet were found for me, and away we went cheered by the crowd. 'Shall we loop the loop?' said the pilot.

'Certainly not', I replied shivering.

'Well, part of the programme is for you to take over the controls.' I did, quite terrified, and it seemed we immediately gave an ominous lurch. Anyway, we landed safely, and the airport was opened.

The other occasion was an American football match, when Ingle had his bit of fun. One day he announced three American officers, whose mouths flew open when they saw a woman, for which Ingle had not prepared them. 'Well, Ma'am,' they said, 'we had come to ask you to kick off at our charity football match, but now we see we can't.' I told them that I would do this. I quickly telephoned my husband for help in finding someone to coach me in kicking. He produced a nice man to come to our garden, and finally I did a fine kick right over our holly bush. Alas! On the day everything went wrong. The American ball is quite a different shape from ours, and the man lay on the ground, holding it in position. 'Don't do that,' I whispered. 'I shall probably kick your head off.' I was so nervous that, in my enthusiasm, I did a tremendous kick, my leg high in the air, and the ball travelled only a yard or two. Very bad; but the papers enjoyed snapping the Mayor in the middle of her kick, with one leg high in the air.

Cambridge was then still very much a man's world, especially within the university. For example, I was not allowed to read a lesson at the famous Christmas carol service, although the Bragg family had processed in the mayoral procession and occupied six choir stalls.

[157] Technically, the airport was first opened on 9 June 1929. The opening by Lady Bragg was the first after it had been closed for the duration of the war and she was in fact the first civilian to take a flight in the UK after the war.

Figure 68 A grainy newspaper photograph from 1946 of Lady Bragg kicking an American football. (Courtesy: Patience Thomson)

It was also the custom to invite the Mayor to college feasts, but this year they invited 'the Mayoress in a black tie', that is, my husband. I bore up under this irritation, until it came to an invitation to dine in Trinity College for the Deputy Mayor (a man) to meet the honorary graduands, of whom I remember Lord Louis Mountbatten was one. I had a tactful letter from the Master, our friend, G. M. Trevelyan saying how sorry he and the Fellows were not to be able to have me, especially as it was my husband's college. It was the only occasion on which I became awkward. I said that if they would not have this Mayor the town would not be represented. Unfortunately, there is no end to this story, as the distinguished guests decided to leave for London before dinner.

On the other hand, there were advantages sometimes. It was annually a burning question whether, on Mayor-Making Sunday, the Vice-Chancellor or the Mayor should leave the university church first. Remember that James I had told the high steward of the day that the

Vice-Chancellor should always take precedence. My councillors urged me not to put up with this. The Vice-Chancellor that year, Sir Henry Thirkill, was the soul of courtesy and would certainly have let me go first; but that seemed a bit mean and, in the event, we pleased everyone by going out exactly together.[158]

The Mayor has to try and take an interest in all the city's activities but he has the chance to boost those especially dear to him. Anything to do with family matters particularly appealed to me. Thus, I was able to start a branch of the National Marriage Guidance Council by taking the chair at an inaugural meeting, which was followed by a mayoral tea, and then becoming its first chairman. I did much the same for the Home Help Service.

At the end of the year, I was completely exhausted; but it had been a wonderful experience, and one I look upon with pride. My father had me painted, but alas it is only good of the mayoral chain. The household returned to normal. I went away for a week's rest, and on my return I found that my husband and Patience had gone out and bought a puppy (something I had always resisted, as the pony, hens and tame rabbits seemed enough) and now I was introduced to this minute creature, Scrap, who was totally un-house-trained.

Cambridge slowly returned to normal, or something approaching normal, after the war. The evacuees left, the colleges filled up again with undergraduates, teaching staff returned from war work and the American air force men went home; but life was not easy; there was still food and clothing rationing for several years, and shortages of many commodities. Most people were very tired, but at least we could take holidays and travel again.

Money was the restriction, as only a limited amount could be taken out of the country, and what with taxation and high costs, most people felt poor. Here scientists were very lucky; not only did international conferences start again, but lecturers, especially in the USA, were in great demand. WLB was an outstandingly good popular lecturer, so there was no lack of invitations, and all expenses were paid

[158] Henry Thirkill (1886–1971) was educated at Cambridge as a physicist but gave up what would certainly have been a distinguished scientific career for a rewarding life in the administrative and educational development of Clare College and University. He was Master of Clare College from 1939 to 1958 and Vice-Chancellor of the University of Cambridge from 1945 to 1947.

for us both. Added to this, in the case of American visits, I would now sometimes be asked to lecture as well, under a sort of umbrella title of 'Women in Public Affairs'. To describe these delightful trips chronologically over the next few years would be tedious, so I shall highlight ones that gave us special pleasure.

Our first was just after the war in 1946, to Sweden, when Stephen was able to come with us. None of our children had ever been abroad before, and it was fun to see his delight in everything, starting with the immense unrationed meals on the Swedish boat. We stayed with the famous physicist Siegbhan and his wife in Stockholm and she enjoyed filling Stephen up with steaks and strawberries and cream. We all three had an orgy of shopping for clothes, watches and the luxuries of which the war had deprived us. We completed Stephen's holiday by taking him to friends, the Eckermanns, who had a big estate in the country, where the lakes had ospreys nesting and in their forests there were elk. Eckermann was much impressed by Stephen as an engineer when Stephen pointed out politely that the powerful motorboat on the lake would not gather speed till the anchor was pulled up!

We also enjoyed our regular trips to Brussels—after the war, the international Solvay Conference was held again every three years, and in 1948 WLB was made president and remained so until 1961. As when I first went with WLB in my twenties, the conference still showered lavish hospitality on its twenty scientific guests and their wives. In our case, we now stayed with a descendant of the Solvay founder, Ernest and his wife, at their chateau outside Brussels, in great luxury. WLB has always hated being waited upon, with breakfast in bed and clothes laid out, but I love it. While the daily scientific meetings were being held, the wives would be driven out to Bruges and to the galleries in Brussels. I really rather liked going alone and seeing my favourite pictures again and again.

The autumn of 1948 saw us embarked on an extensive lecture tour in the USA. Everyone was eager to hear what research was starting again in the Cavendish Laboratory. The Americans seem exceptionally fond of attending lectures and now I found myself asked to give some as well. The volume of voluntary work undertaken was of great interest to them, so they wanted to hear about local government, the office of Mayor and the work of the magistrates' courts. We were based in Pittsburgh, where WLB gave a very intensive course, but first we called in at Boston for lectures at Harvard and MIT, though really that was an excuse to see our Stephen, who had been awarded a Commonwealth Fund Scholarship.

To return to Pittsburgh, a city not unlike Manchester in its industrialism, its music and the warmth of its inhabitants, this is where by accident I made my first attempt at lecturing in the States. The real lecturer had had to cancel a few days before his engagement, so I agreed to present myself before an audience of 1,000 women on Monday morning in his stead. I nearly lost my nerve before I reached the platform, as the chairman, a very large lady, receiving me in a small room, told me how unwelcome I was to her as she was completely pro-German; with these remarks, she handed me a cheque for the talk. I should like to record that I tore it up and stamped on it, but when I saw the amount I could not bear to do so. Instructed by WLB, I did not read my talk but used notes and tried where appropriate to be amusing. However, I gathered American women like considerable formality and a serious tone and are not conducive to jokes. In spite of all this, the audience appeared interested and did sometimes laugh. It was rather a success; even the pro-German chairman spoke appreciatively and became a friend. However, I was unprepared for questions at the end. Here I apparently made history. I was asked how much we spent on our new (this was 1948) Health Service. I had no idea, so I simply said, 'I don't know'. Everyone gasped and the chairman said that she had never before heard anyone give such an answer. 'But,' I said, remembering some advice from Lady Reading, 'I know where to find out; if the questioner will leave a name and address, she shall have the information'; but she never did.

I gained confidence, and at my new assignment at Duke's University (actually founded by the Duke of tobacco fame), I told my undergraduate (girls) audience not to knit, please, as the clicking and dropping of needles disturbed me. They were delightful girls, wearing high heels, as the dean of women told me, to do me honour; but I wonder if I got over our idea of voluntary work. I was explaining that, if it was your day to serve in the magistrates' court, you must be there to do the work. 'But,' said a young thing earnestly, 'what if it's the only time you can have your hair done?' In spite of this attitude, I greatly enjoyed these girls, and at a party the dean had for them to meet me, we talked till late at night.

Another visit in those post-war years I must mention was to Stockholm again. The Nobel foundation had its fiftieth anniversary in December 1950, and all the winners of the prize went to wonderful celebrations. A highlight was the banquet for 1,400 people in the beautiful town hall, when each course was brought in by a regiment of waiters marching to music, down the grand staircase, holding aloft the courses and drinks in great silver salvers—a wonderful sight—especially the ices in vast blocks of ice into which electric lights had been frozen. This was typical of the celebrations, and outside Stockholm it was so beautiful, with deep, hard snow and all the Christmas decorations shining and twinkling in the dark that dropped soon after lunch. It was a memorable occasion, and one at which we met again friends from all over the world.

WLB was working very hard. I could not of course begin to understand what was going on in the various departments of the Cavendish, but it has been told to me that during this time, as head of the laboratory 'he fostered what were to prove the two most important directions in science of the past thirty years'. One was radio astronomy and the other molecular biology. He had also to involve himself in university politics, which he hated. I endeavoured to support him. As well as reading to me any popular lecture he was to give, he liked to go through minutes of meetings which might be difficult. He said that it cleared his head. I would try and plan holidays when he could sail, paint and bird watch, and many of these would be on the Norfolk coast with the family.

At home we were both working very hard. For myself, though no longer Mayor I was very busy as a councillor and as the chairman of both the Secondary Education Committee and the Maternity and Child Welfare Committee, both of which involved visits to schools and clinics, appointment of staff and certainly learning a great deal. I had a growing impression that committee members, and perhaps voluntary workers generally, took themselves too seriously and that a little gaiety would do no harm. I remember a headmaster (whom we actually appointed) told me how nervous he was at his interview, seeing all the serious faces, and that when I smiled encouragingly at him, he felt much better.

Apart from my role in local government, I was also a governor of Addenbrooke's Hospital, and the chairman of both the Marriage Guidance Council and the English-Speaking Union. What gave me most pleasure, I think, was being elected in 1951 as an Associate Fellow of Newnham, my old college. After all, as a student there I was told that my work was not up to standard and that my career there was in jeopardy; I had never considered myself an academic, so I was really thrilled. Added to all these commitments, I was now a magistrate and a member of the juvenile panel of the bench.

Work on the bench entailed visits to the local prison at Bedford, and these I would gladly have missed. I have never liked zoos or circuses or seeing anything deprived of freedom, much less men and women. There were one or two amusing incidents. I was late for my first prison board meeting and, as it was mid-morning and seeing a little plate of food beside me, I started to eat what appeared to be 'elevenses'. This was a mistake; it was a specimen evening meal for the prisoners for us to inspect. The other incident was once, when I was touring the workshops, a prisoner whom I had sentenced for refusing to pay his wife's maintenance called out cheerfully, 'Good morning, Lady Bragg, nice to have you with us.'

So far, all this was work in Cambridge, but if one has been Mayor or held that kind of appointment, I gather that one's name goes on a list at national level. At any rate, one day in 1951 I received a letter from the Minister of Education inviting me to be a member of the 'Central Advisory Council for Education in England' for the next three years. It sounded very grand but, as I had never heard of it, I took it to our Director of Education for advice. This was rather awkward, as it was something he very much wanted to do himself. He wondered why I had been asked, and of course I wondered too, but secretly I thought it might have been the result of that drinking tea out of saucers incident in the Mayor's parlour with the minister. So I found myself once a month going to London, sitting with a group of directors and professors of education, youth leaders and a don or two, researching into why so many boys and girls left school early. I thought to myself that two or three sensible people could have done it in half the time.

More was to come. One morning, the postman, with an enquiring look, handed me an envelope with 'Prime Minister' printed on the back; and there was an invitation from Mr Attlee to serve on the Royal Commission for Marriage and Divorce. We sat for four years; I suppose it followed the usual pattern of royal commissions, but it was all new to me. First, the public were invited to send their views, and it seemed to me (who had to read them all) that every possible organisation, and indeed many individuals, did so. It appeared that for a body not to send in a memorandum meant loss of face. Many asked to come in person and put their case.

The chairman, Lord Morton, was a Lord of Appeal, a very tall, solemn man, an elder of his church and rather easily shocked.[159] He hated smoking, so to avoid doing that, I often knitted socks while waiting to begin our deliberations. On one occasion he came into the anteroom and said, 'You look worried, Lady Bragg. Can I help you?' He was startled when I said, 'Yes, please, I don't quite know if this sock is long enough. If you could put your foot on the table and pull up your trouser leg, I could measure.' He did that, much to the surprise of the pressman strolling by us.

Since my name began with 'B', I sat between a Mr Brown and Sir Russell (later Lord) Brain.[160] Mr Brown was a member because he played rugger for Scotland and this was supposed to represent the ordinary man's point of view, I understood. He used to pass me boiled sweets. Sir Russell Brain was added to our number, being an eminent physician, as I had told the chairman when we had to discuss lesbianism as a ground for divorce that most of us did not know the details of this. He wrote an all-too-clear memorandum on the subject. In the end, the result was the expected one. Half of us wanted easier divorce, and half (including myself) did not, which probably reflected opinion in the country. One of the Scottish judges, Lord Walker, who looked as if he had been thrown by Epstein in stone, put in a minority report which actually anticipated the law as it now stands. Finally, we gave the chairman a Georgian silver salver, had a dinner in the House of Lords and broke up.

At home the scene was a happy one. Our house and garden were a lovely setting for our growing family, relations and friends. I still have vivid memories, pictures in my mind to be brought out like photographs, with scents and sounds added. One was the arrival home of WLB on his bicycle at lunch time every day, his large attaché case balanced in an equally large basket, the scrunch of gravel on the drive before he parked under our great copper beech. Then there was the dining room with the sun pouring in through French windows open to a wide border of dark lavender full of bees humming, the table crowded with undergraduates for Sunday tea. Then there was Margaret's 'animal fete and competition' on the lawn, cats and rabbits

[159] Fergus Dunlop Morton, Baron Morton of Henryton (1887–1973), was a British judge and Law Lord.

[160] Walter Russell Brain, First Baron Brain (1895–1966), was a British neurologist.

arriving in baskets, dogs on leads and a goat tethered among the lilacs, chewing up a competitor's school cap. In a quiet corner of the garden I see two little Latvian girls doing their homework, children of the two tragic refugee widows who looked after us at the time.

We all used to visit my father close by as often as possible, and he always had Sunday lunch with us to hear of all our doings. One Sunday, in 1949, all prepared to come as usual, he quietly died. I was there to hold his hand, as we had planned together.

The children were starting to take wing. Stephen had a research post in Derby with Rolls-Royce. He came home quite often, bringing friends. In 1950 he became engaged to Maureen Roberts, only child of one of my fags at St Leonards School; this gave us the greatest pleasure. The following year they were married, and in 1952 they gave us our first grandchild, Nigel.

David, sad to say, suffered ill health and was only able to work intermittently. Painting was his great love and he went to an art school. He was often at home with us, sometimes bringing friends and 'lame ducks'.

Margaret left school in due course and then startled us by electing to go to Somerville, Oxford. I suppose one always wants one's children to go to one's own college, and I suspect that I had pictured her at Newnham, but in the event it was fun seeing something of Oxford when I used to drive her and her effects there. Now with Margaret, and Patience to follow, the era of the young men started. I found this fascinating and it quite difficult to accept that I had only the most passive, if any, role to play in this.

Patience, in 1948, also went to Downe House school, and finally to Newnham College. Each girl had a year based at home after school, going abroad and having special coaching for university. I should like to record that during that year I taught them domestic skills such as cooking, sewing and cleaning, but my own were so limited that they were not worth passing on. I felt that I had failed them. However, they did meet very interesting people and as a result I like to think that they developed standards of behaviour, conversation and taste.

For example, G. M. Trevelyan, the historian and master of Trinity College, and Janet, his wife, were great friends and asked us to bring the girls to stay with them at their family home in Northumberland. There were also two bachelor dons staying, and we used to roam the moors by day and, at night, we would play paper games or our host would read aloud to us. At meals we had what our family called 'intellectual conversation'. In contrast to this academic atmosphere, we knew Edith Evans, the actress who, when she spent the night with us, remained as it were on stage.[161] We had only to see her powder her nose to realise we were the audience for her. Actually, there was rather an unfortunate episode on this occasion. We decided that she must have a bathroom to herself, so we gave her the children's, carefully emptied of family effects. The lavatory there was thought to be too shabby and the girls gave it a coat of paint, but not in time for that to dry, and Dame Edith had to be warned off.

In 1952 we decided that we must leave our lovely house and buy something smaller. Our present one was leased from a college, and that lease would run out in a year or two; in addition, the place was becoming too expensive to run, especially with WLB's retirement in sight. We bought a pretty house not far away, in a charming garden, but we all felt it a great wrench to leave our home. Had we known that we were only going to be in our new house a year, it would perhaps have softened the blow.

3.17 The Royal Institution: A Difficult Decision

We had only been in our new house for a few months before we were faced with a dramatic change in our life. WLB was invited to go to be head of the Royal Institution. Lord Brabazon, its president, asked to come and discuss this with him as soon as possible. As a result of the appointment and subsequent resignation of Professor Andrade (who was well known to be a very difficult man), the position was vacant. WLB asked me how I would feel about going there. 'I think it would be dreadful', I said at once; the disadvantages mounted up as I thought of them. We should have to leave Cambridge, all our friends, all my work and go to live in the heart of London, with no garden and not a tree in sight. This was bad enough, but there was a much worse aspect. The Royal Institution had fallen on evil days, and there had been a terrible row there, culminating in the director, Professor Andrade, being forced by the members to resign. He brought and won a lawsuit

[161] Edith Mary Evans (1888–1976) was a celebrated English actress.

suing the Royal Institution for wrongful dismissal, and the whole affair had become distastefully public. From what we heard of it, it seemed that everybody involved behaved badly. The affair had alienated the Royal Society, and certain Fellows in particular. WLB hated rows, and I could not bear him to go to a place which now seemed in danger of sinking. However, WLB said that we ought to see Lord Brabazon; and so he came and stayed the night with us. Lord Brabazon, always known as 'Brab', was rather a dear but a dropper of bricks and given to making injudicious remarks; he was however devoted to the Royal Institution and begged WLB to go there and 'save the place'. WLB promised to consider it.

I suggested that we go and consult our old friends, the Adrians. One remembers small details in these events and, when I rang Hester Adrian, my great friend, she said we had better come to dinner as she had just had a pair of chickens sent from Norfolk. After dinner we talked about the Royal Institution and, to this day, I am not sure what Adrian really felt. He was then just finishing his term as President of the Royal Society, with whom the Royal Institution was in such bad odour. He paced the big drawing room and finally he said, 'I think you'll have to go there with all your family connections with the place. Certainly, if you don't, no one else will.' After this, WLB accepted the post, but I felt very unhappy.

My anxiety increased when we both went to London to see the officers of the Royal Institution, not a very distinguished trio. My impression was that, since Professor Andrade's resignation some months ago, they had run the place and were not determined to relinquish powers they had gained in favour of another possibly dictatorial 'Director'. They wanted a well-known scientist who would lecture and do research there but was not to do the traditional entertaining or deal with the members. Finally, there appeared to be no money for anything. I remained silent till the end of the meeting, when I suddenly said a great deal, so that the treasurer declared afterwards apparently, 'Beware of the female of the species'.

Thus, in the new year, we moved into the beautiful official flat at the Royal Institution and for the next few years were beset with troubles inside and out, before we came into a triumphant calm and an atmosphere of general goodwill. Inside, there was a lack of co-operation; in their fear of WLB taking too much power, managers and officers too often thwarted him and did not back him up properly. However, on the research side, with which they had nothing to do, WLB collected the necessary money and gathered together a research team. In this he would always say he owed everything to Professor Max Perutz and Sir John Kendrew, who, while remaining in Cambridge, undertook to build up protein research with him and a small team.

Meanwhile, the breach with the Royal Society continued. A group of chemists especially among its Fellows boycotted the Royal Institution, refusing all invitations to come there and showing their displeasure in many ways, now best forgotten. Actually, I am glad to say that most of this deplorable behaviour was lost on WLB. Half the time he really did not notice and when occasionally he did, he found satisfactory explanations for it. I saw it all, and I vowed that one day in the future I would say what I thought about it, probably to Sir Robert Robinson, who might, I suppose, be called the doyen of the leading chemists; and I did, as I shall recount later. How WLB laughed when we were discussing it all over our early morning tea and I said, 'You know, I am one of the children of this world, and you are one of the children of light in the way you make allowances for all this sad behaviour.' Anyway, in the scientific world, we had many loyal friends who never failed, especially Sir George Thomson and Sir Alfred Egerton,[162] Sir Charles Darwin, and of course Hester Adrian, who used our flat as her London spare room and was the only person in whom I confided our difficulties. I think we never spoke of them to anyone else at the time, not even the family.

Meanwhile, I was extremely busy organising the domestic side of our lives. Our official flat must have been one of the largest in London but it was rambling and very hard to run. There was room for all the family and guests. During all the thirteen years we were there, I managed to get someone or other to live in and cook though, to be on the safe side, I took a few lessons in cooking at the Cordon Bleu. This was essential, as we had a traditional entertaining programme. Every Friday night in term time there has been, since Faraday's time, a 'Discourse' or popular lecture for Royal Institution members and their friends, with an audience of anything up to 500. We had to give a dinner party for the lecturer beforehand and put him and his wife (if he had one) up for the night. For the first year or two, my cleaner

[162] Alfred Charles Glyn Egerton (1886–1959) was a British chemist and Physical Secretary of the Royal Society.

from Cambridge, Mrs Stubbings, came up for the evening to wait at dinner, returning on the late theatre train. The lecturer and our dinner guests wore white tie and tails, and the women dressed to match. This led to some funny scenes. Scientists are notoriously vague, and often the lecturer would forget part of his wardrobe. WLB collected 'spares' of everything possible: ties, shoe laces, collars and studs; we could produce almost everything. I remember the Astronomer Royal of the day wandering in his underwear in search of help, having forgotten his shirt. No wonder that I greeted one lecturer, arriving for dinner, with the words 'Oh! Professor So-and-So, how nice to see you in all your clothes.'

To which he replied, 'Alas! I left my braces behind, but your kind head porter downstairs sent out for a pair.'

In inviting guests to dinner, we tried to get people who did not know the Royal Institution and would be interested and perhaps become members; judges, politicians and industrialists, whether we knew them personally or not, they generally came, since the Royal Institution was unique.

At Christmas, again from Faraday's time on, a set of six famous lectures for children are given, and after each one we would have the lecturer and about twenty of the juvenile audience to tea upstairs in the flat. It was the only time in my life that our Christmas cake was always finished to the last crumb. We had a big Christmas tree in our hall, and the flat was all decorated. We also gave an annual sherry party for new members, and lunches for the research people and official visitors to the place. It was hard work, but WLB and I enjoyed it all, though we should have been glad of inside support; but the managers at that time were not very socially competent.

The reason that the Director had this large and beautiful flat was so that it should be used for entertaining; it also had to be our home. During the time that we lived there, it was the base for our children and grandchildren. Stephen had now a home of his own in Derby, but he and Maureen and, in due course, his three small sons often visited us. There was a delightful occasion when the managers invited him to give a Friday evening discourse, and we gave the dinner party beforehand for him and Maureen. It was home for David, who did not marry till after we had left.

We did not have Margaret for long. After taking her degree at Oxford, she had a year at the London School of Economics, obtaining a Social Science Diploma, at the end of which she became engaged to Mark Heath.[163] This gave us enormous pleasure, the only sadness being that, as he was in the Foreign Office and posted abroad, they left for Indonesia soon after the wedding (September 1954). It would be hard to imagine a better setting for a wedding reception than the Royal Institution or, indeed, a more beautiful church than St James' Piccadilly for the marriage. Although they left for the other side of the world, they returned to us on leave from their various postings—Denmark, Sofia and Canada—and besides visiting them, we often took responsibility for their children, two boys and a girl, when they came to the Royal Institution.

Patience left Cambridge with a degree in languages; this degree enabled her to take a job translating in the Foreign Office library, and she lived with us until she too married in 1959. Thus, there was another lovely wedding at the Royal Institution. This marriage was a very scientific alliance, as her husband David was the second son of our great friend Sir George Thomson. I remember how the two fathers congratulated each other by telephone in a series of ecstatic grunts, typical of the way in which scientists communicate. David was a merchant banker, and they lived in London and so were easily accessible. To round off the story, they had two boys and two girls.

Besides the immediate family, our own relations and friends came and stayed; the spare room seemed always occupied. There were my two brothers Harry and Philip, and close friends like Hester Adrian and George Thomson, who had almost regular bookings. My dear Enid was living in London so we were able to meet constantly, and Willie's sister Gwendy Caroe was close by. No wonder I expected to live a domestic life!

Certainly, I had thought sadly to give up public work but that turned out to be far from the case. To my surprise and delight, the Lord Chancellor allowed me to stay on the Cambridge Bench (which I did until I reached retiring age) on condition that I paid my own expenses and attended for the accepted number of times. I used to catch the breakfast train and sit in court for two consecutive days, generally staying at Trinity with the Adrians, but I had a 'bed-list' of old

[163] Mark Evelyn Heath (1927–2005) was a British diplomat who was Her Britannic Majesty's Envoy Extraordinary and also the Minister Plenipotentiary to the Holy See from 1980 to 1982. He became the first ambassador to the Holy See, from 1982 to 1985.

friends and had great fun seeing them all. When we were in residence at Waldringfield, our house in Suffolk, of which more later, I drove over for the day starting at 8 a.m. For some years I was on the Council of the Magistrates Association but that of course met in London.

In 1955, just as the Royal Commission for Marriage and Divorce ended, I was invited to serve on the Lord Chancellor's Advisory Committee for Legal Aid. It met in the House of Lords and was chaired by a splendid peer who managed to bring hunting terms even into legal aid. My contribution was an earthy one. I would not take assumptions for granted. For instance, when we were assured that everyone knew about legal aid and that it was well advertised, I asked a number of people in buses, shops and trains and assured myself that hardly anyone knew anything about it. I did once succeed in a getting a change in the law on a very small point, but I was rather proud of that. The Lord Chancellor, at the beginning of the legal year in October, gave a pre-lunch party for the judges after they had processed to their church service in full regalia. To this, members of my committee were invited, so I decided to go and found it alarming. I appeared to be the only woman in the crowd on the staircase of the House of Lords. In fact, the policeman at the top asked me politely if I was not in the wrong place until he saw my invitation card. I am not a 'women's lib' person and, had I not been squashed in by judges and senior members of the Bar, I would have gone home. However, once arrived, I found below the wigs some friendly faces that were known to me, and I was soon being plied with most potent hot punch and sausage rolls. I remained on that Legal Aid Committee with great pleasure and interest for fourteen years.

The other appointment that I valued most highly was to the chairmanship of the National Marriage Guidance Council when Lord Merthyr retired.[164] I was told that this would mean taking the chair at the annual meeting and a 'general interest'. It has never seemed to me much good being a chairman unless you really know your subject. You can so easily make an ass of yourself by exposing your ignorance. In order to understand, I stumped the country getting to know local Marriage Guidance Councils; it might be Leeds, Cardiff, Truro or St

[164] William Brereton Couchman Lewis, Third Baron Merthyr (1901–1977), known as The Honourable William Lewis between 1914 and 1932, was a British barrister and politician.

Helen's, as opportunity offered, but I chose times when WLB was away lecturing. He was himself very interested in this venture and sometimes came with me to the annual conference. We agreed that people working in this field, though sometimes temperamental, were among the nicest we had ever known. I did this work for nearly twenty years.

One thing leads to another, and I was also a governor of Moorfields Eye Hospital, involved with the Mental Health Association and the Health Visitors and, finally, a member of a committee on the Succession Rights of Illegitimate Persons. Although when I was asked by the Lord Chancellor he wrote, 'the work should prove both interesting and important', it did not in the event seem to me either; but I think the truth was that it was really beyond my competence.

The interest that really enlivened those first rather dreary years at the Royal Institution was writing feature articles for *The Guardian*. I hold my bicycle responsible for starting me on this. Like everyone else, I bicycled everywhere in Cambridge, and I insisted on bringing my bicycle to London, where I used to ride around Hyde Park. My first article included some of my cycling adventures. Anyway, the then editor (Mr Wadsworth[165]) accepted it and told me that I could send him more. I did so regularly until he died, and after that I think I was not to his successor's taste; but I was pleased to have my article on my Aunt Mary and her policemen included in a volume of *The Bedside Guardian*.

There is no doubt that these outside interests were of great value to me, as they enabled me in an atmosphere of frustration and even intrigue to keep a sense of proportion and detachment.

The end of the year 1960 saw the turning of the tide in the life of the Royal Institution. To begin with, we must take relations with the Royal Society; it was the society's tercentenary, to be marked with many lavish celebrations. In these, the Royal Institution would normally have played its part, but all offers of help or co-operation were rejected. The President of the Royal Society at the time was a chemist, Sir Cyril Hinshelwood.[166] He left no impression upon me whatsoever;

[165] Alfred Powell Wadsworth (1891–1956) became Editor of *The Manchester Guardian* newspaper in 1944.

[166] Cyril Norman Hinshelwood (1897–1967) was an English physical chemist. He shared the 1956 Nobel Prize in Chemistry for his research into the mechanisms of chemical reactions.

I only know that he was extremely anti the Royal Institution and, I suppose, exerted considerable influence in that direction. However, at the end of the year he and several officials were due to retire, and I think one could date a welcome change of attitude with the coming of successors friendly to the Royal Institution, and a happy future seemed assured.

I satisfied my own feelings in the matter some years later. Earlier I recorded that one day, if the opportunity came, I would 'have it out' with a leading chemist and tell how we had felt. The opportunity came one night at a Royal Society dinner at which I found myself sitting next to Sir Robert Robinson, doyen of this country's chemists. When we were young, he and WLB had been professors at Manchester, and his wife and I used to pram-push our babies together. Now at dinner, we reminisced happily until he said warmly, 'You and I have always been friends, haven't we.'

'No,' I said, 'we have not.'

He looked startled, and I knew that my moment had arrived. I asked him if he could deny that he and some of his fellow chemists had wanted to see the end of the Royal Institution after the 'row'. He could not. I itemised one or two of the occasions on which I knew that he had, as it were blackballed WLB. 'But,' he kept assuring me, 'it was nothing personal, only that the RI was in such bad odour. Do tell him that.' I asked him to tell WLB that himself as soon as possible. It was now about time for the after-dinner speeches; our neighbours must have been surprised to see us shake hands solemnly, as I sank back and sipped my cognac. At the end of the dinner, I saw Sir Robert seek out WLB, but I did not hear what they said as, overcome with emotion, the heat of the room and the excellence of the dinner, I fainted!

At the same time, the fortunes of the Royal Institution were reviving. In fact, the years now until his retirement in 1966 were among the happiest of WLB's life and so of course of mine. He had a good staff: Ronald King, who had been there in Professor Andrade's time, was Assistant Director of Research; a master of detail and minutia, he relieved WLB of many 'chores'.[167] There was a good and forward-looking librarian in Kenneth Vernon; the Royal Institution has a fine

[167] Ronald King was Professor of Metal Physics at the Royal Institution from 1957 to 1978.

library in which WLB took a great interest.[168] Bill Coates, known now to thousands of scientifically-minded TV watchers, was Lecturer's Assistant.[169] Good and supportive managers accepted appointment. One of the happiest events was the appearance of our friend Admiral Sir William Davis, who WLB persuaded to become treasurer and rescue us from the 'red'.[170]

This is not the place to tell about the research that was going on in the Royal Institution lab, even if I could. I can only say that it was going remarkably well and I enjoyed the people who were doing it, especially David Phillips, who arrived from Canada to work and became a warm friend.

I suppose the greatest contribution WLB made in bringing the Royal Institution before the public eye was the introduction of regular popular scientific lectures to schoolchildren. This was the implementation of a suggestion by his sister Gwendy Caroe, who felt it a pity that the famous 'Christmas lectures for a juvenile audience' were the monopoly of members' children. These new 'Schools Lectures' were an instant success. I endeavoured to go to many of WLB's 'popular' lectures, although I did not really understand them, but the Schools Lectures I attended whenever possible were for sheer pleasure. The lecture theatre would be crammed with 500 children gripped by a series of wonderful experiments mounted by Bill Coates and explained by an inspired lecturer obviously enjoying himself as much as his audience. There is now a plaque at the Royal Institution commemorating WLB's achievement in lecturing to 200,000 school children.[171]

Our anxiety was during the next few years not the Royal Institution but WLB's health. Several times he had pneumonia very badly, and twice he underwent operations, one in 1962 so serious that he was in hospital for five weeks. It was during those anxious weeks that

[168] Kenneth Denis Cecil Vernon (1917–) was Keeper of the Library at the Royal Institution from 1950 to 1966.

[169] The wonderful William Albert Coates (1919–1993) was a famous lecturer's assistant for many years at the Royal Institution. See <http://richannel.org/ri-discourse-bill-coates--1948-1984-my-years-between> for a video of his retirement lecture.

[170] William Wellclose Davis (1901–1987) was a Royal Navy officer who went on to be Vice Chief of the Naval Staff.

[171] This plaque was made by the sculptor John William Mills (1933–) and can be seen at the entrance to the Royal Institution lecture theatre.

he heard that four of his 'Cavendish boys', Perutz, Kendrew, Crick and Watson, had been awarded the Nobel Prize.[172, 173] This so excited and delighted him that the surgeon told me, half laughing, that he thought it would kill or cure him. He recovered and went to our house in Suffolk for a long convalescence and then abroad.

We had started soon after we came to London to look for a place in the country where we could spend weekends and holidays and above all have a garden. We felt starved of fresh air and exercise and needed somewhere where we could relax. Actually, we had found a tennis court and even a derelict garden on a bomb site at the edge of Green Park, and we could sometimes be seen crossing Piccadilly either with our tennis rackets or pushing a wheelbarrow with plants and garden tools. Though hardly a form of exercise, but certainly of relaxation, I often dropped into St James's Piccadilly, that beautiful Wren church, to meditate and soothe myself.

After a long search, we found a place in Suffolk which delighted us. It was at Waldringfield on the River Deben. The house itself was not exactly to our taste; it was in fact made from the parts of a heavily timbered old inn, and the ceilings of the house had the original Suffolk pargeting work. But there is no doubt that it only just missed the gnomes in the garden, and flights of china ducks up the inside walls. However the site, up a farm lane, close to the river and with a coppice and a lovely garden, left nothing to be desired. For more than twenty years we revelled in it, and so did our family. WLB and I had a chubby dinghy on the river, and the grandchildren learnt to sail there. Every summer we had the research people for the day and had great fun with treasure hunts, a long walk by the river and lunch and tea in the garden. Patience and David Thomson and their family had a luxurious caravan in the wood. We were happy too in making great friends, the Tysons and Lubbocks, in the village and, in Woodbridge, close by, the

[172] Francis Harry Compton Crick (1916–2004) was an English molecular biologist, biophysicist and neuroscientist, most noted for being a co-discoverer of the structure of the DNA molecule in 1953 with James Watson and Maurice Wilkins, who were awarded the 1962 Nobel Prize in Physiology or Medicine.

[173] James Dewey Watson (1928–) is an American molecular biologist, geneticist and zoologist, best known as a co-discoverer of the structure of DNA in 1953 with Francis Crick. Author of one of the most popular scientific books ever printed, *The Double Helix*.

Ian Jacobs.[174] Altogether we loved it and could hardly tear ourselves away and go back to London to work, when the time came.

I am always told that I am gregarious, and I confess I loved the social side of our life in London. I liked dressing up and I had some lovely evening dresses (not really much needed in Cambridge). We went to private views, Wimbledon, banquets, theatres and parties. Of parties, I remember two vividly. One was a dinner with Sir Alexander Fleming and his fascinating Greek wife.[175] He rang me up one day and invited us, saying to my astonishment that it was to meet Marlene Dietrich, who had expressed a wish to meet one or two Nobel Prize winners, and WLB was the most accessible.[176] It was quite a large party and, when Marlene Dietrich arrived, rather late, we all stood up. There she was, with her marvellous legs and azure blue eyelids. One associates her with the singing of sexy songs in a husky voice, but at dinner, between two Nobel Prize winners, she discussed education very seriously. The other dinner party I especially remember was at Sir Stephen Runciman's.[177] Royalty was to be there: Princess Marina and her sister Princess Olga. Incidentally, they were two of the most beautiful women I ever saw. We had quite a long wait, and I was rather squashed on a sofa by a very stout lady; not roped in anywhere, she overflowed me and made me very hot. I asked her about herself and she told me that her husband was an archaeologist and she went with him on his digs and cooked. I was to find out later that this was very much a half-truth. After supper (lobster, which I love) in one of those silences that the presence of royalty engenders, someone told WLB that Agatha Christie was a fellow guest.[178] He, a passionate devotee

[174] Lieutenant-General Sir Edward Ian Claud Jacob (1899–1993), known as Ian Jacob, was the Military Assistant Secretary to Winston Churchill's war cabinet and later a distinguished broadcasting executive, serving as Director-General of the BBC from 1952 to 1960.

[175] Alexander Fleming (1881–1955) was a Scottish biologist, pharmacologist and botanist who discovered the enzyme lysozyme in 1923 and the antibiotic penicillin in 1928. He shared the 1945 Nobel Prize for Physiology or Medicine with Howard Florey and Ernst Boris Chain.

[176] Marie Magdalene 'Marlene' Dietrich (1901–1992) was a German actress, singer and film star.

[177] James Cochran Stevenson Runciman (1903–2000) was an English historian known for his work on the Middle Ages. He spoke many different languages.

[178] Agatha Mary Clarissa Christie (1890–1976) was an English crime novelist, short-story writer and playwright.

Figure 69 Porch in Woodbridge, 1964. (Sketch by WLB; Courtesy: the Royal Institution London)

of detective stories, absolutely shouted, 'Agatha Christie, lead me to her.' The princesses laughed, and in a few minutes WLB was talking and declaring admiration to the fat lady who had sat next to me on the sofa earlier in the evening.

At intervals during our time at the Royal Institution, we had wonderful trips abroad. We had holidays during our normal vacations, when we visited Margaret and Mark in their various postings to Canada, Denmark and Sofia as well as at a villa lent to them in Greece. We took Patience to Spain and to take part in the Brussels Exposition, where her father was in charge of the British scientific side. We had also more official tours, and the Royal Institution managers were very good in allowing us time for these in term time. A detailed description of these trips would be boring since so many scientists have trodden the same paths, but I keep a store of memories which, like snapshots, I often return to look at in my mind's eye. We did in fact travel in every continent, except South America. We went to South Africa, where I have pictures of watching elephants, giraffe and lions in a national park, the glories of the botanical gardens in Cape Town, and a trip to watch a colony of night herons in the trees while on the ground beneath were wild arum lilies, each with a minute frog at its heart. I remember standing in a post office and realising the unpleasantness of apartheid when I was told very roughly that I could not be served as I was in the non-European queue.

We went on a lecture trip to India, based in Madras, where we visited Delhi, Benares, Bangalore and Calcutta, seeing the sights that all visitors see, so that I must not describe them but only take a few special memories from my store. There was Christmas Day in the south, in Trivandrum, where we went to service in a church packed with communicants and dazzling with sunshine, with great butterflies flying in and out and the clergy bare-footed. I recall an expedition where we were to be received by a maharajah on his estate, where on his lakes there were thousands of duck for WLB to watch. Alas, he was at prayer when we arrived and kept us waiting for an hour, when we were longing to be off to the marshes. I was buoyed up by the thought of finally seeing an Indian prince covered in rubies and emeralds. When he at last turned up, he was in an old tweed jacket.

Now I see myself in our high commissioner's exquisitely tended garden in Delhi, lecturing to a crowd of Indian women dressed in beautiful saris and politely listening to my talk on marriage guidance, though their marriages were arranged and they could not be divorced. Horrid memories also come of burning ghats along the Ganges, emaciated sacred cows wandering in the streets and the drive into Calcutta from the airport, as shocking scenes of life and death were enacted along the packed road.

Finally, there was in 1960 a trip round the world. WLB was invited to give the Rutherford Memorial Lecture in Nelson, New Zealand, where Rutherford was born and brought up. WLB insisted in taking one of the protein models with him to illustrate his lecture. It was packed in a light container the size of a large hatbox and became the cause of many amusing scenes. The customs officers were, in several places, convinced that it was a bird or animal. WLB refused to be parted from it, and it travelled as hand luggage. Once the pilot wanted to see for himself, and WLB opened it and gave a short talk on protein research on the plane! Arrived in New Zealand, inevitably we visited universities, but we were given wonderful opportunities of seeing the country from Auckland to Christchurch. There followed a month in Australia, a great pleasure to us both, as WLB was able to show me in Adelaide his old home (now a residential club) covered with a great vine, which he, when a boy, had planted as a twig, and his old school, St Peter's, where his portrait hangs, and we met those of his friends who were still around. He had not been back for fifty years.

After a few days in Hong Kong, where we bought masses of presents, and WLB bought me a lovely embroidered coat, we flew home, receiving surprisingly VIP treatment. We had special food and, wherever we touched down, all the passengers had to wait for us to go first. We were rather puzzled; when WLB said that perhaps science had come into its own, I replied to tease that it was for the Chairman of the National Marriage Guidance Council! At Heathrow we were told that, of course, we did not need to go through customs. Suddenly, up bustled some official and explained that, through an error in coding, we had been confused with the Governor of Hong Kong and his wife on leave; apparently, they were to come on the next plane. This unexpected episode rounded off our splendid tour.

The last half of our time at the Royal Institution was happy. It was clear that the place was returning to its former glory under WLB's direction. He himself had honours, degrees and medals; the protein research work was forging ahead and he was lecturing to his heart's content. Many distinguished visitors came, and our flat was a lovely place to entertain them. We acquired a very good cook, who stayed with us till WLB retired. She made our menus for the day in French, even when I was alone. She only came because 'she had never worked in a professor's home before'! We sometimes put her in the gallery to watch the children's lectures so that she should have the full academic flavour. I really was cushioned from all the domestic care, for we had not only two splendid cleaners, 'Bonny' and Mrs Jiggins (our Gladys), who carried on with our successor, but devoted help when needed from porters and workshop personnel.

I was doing a great deal of work; as membership of one committee ended, a new one seemed to take its place. I also had many speaking engagements, mostly on marriage guidance or magistrate's work. I imagine that it was this that led to the managers, to my astonishment, inviting me to give a Friday evening discourse entitled 'Changing Patterns in Marriage and Divorce'. I felt this a great honour and went into retirement for a few days in Newnham College, where a don kindly lent me her rooms. All went well on the night, except that WLB had inveigled me into having some graphs made on divorce figures. I had to explain them in due course, and I did this badly. One of WLB's research men, whom I knew well, told me afterwards that it was clear that I was unfamiliar with graphs.

During this time, we shared in the drama of Jim Watson's book *The Double Helix*, which was to prove a bestseller both in the States and in this country. It will be remembered that the author was one of the Cavendish Laboratory people to receive the Nobel Prize in 1962 when WLB was so ill. WLB was attached to this young American (known in our home as 'the Bush Baby' because of his protruding eyes) and encouraged him to write a popular account of his work on DNA. My feelings towards him were mixed. I admired him for being himself, but it did not seem to me to be a very nice self. I have known and had friends among many scientists, and they have a respect for each other, and a certain code of behaviour which Jim Watson did not appear to share. I saw the first draft of the book when it was shown to WLB. It was originally called *Honest Jim*. I was appalled; so many of those people that he had met in his host country, and especially in Cambridge, were held up for criticism and ridicule, and none more than his professor, WLB. We were in the States about that time and had been invited to see the people at Harvard University Press who were considering publishing *Honest Jim*. Pressure was put on them not to do so by interested parties, and I understand that it was finally decided that it was not suitable for a university press. Subsequently, it was published in England under the title *The Double Helix* and, with regard to personalities, in a somewhat milder form. Jim Watson asked WLB to write the introduction and to my amazement, he accepted. I recall standing with the two of them in the hall of our flat, and Jim saying that he hoped that I realised that there was no one he admired more than my husband. I was very annoyed. 'No one could possibly think that after all the horrid remarks you make about him in your wretched book', I said angrily. The introduction was duly written and I was pleased at the many messages WLB received paying tribute to his magnanimity. He had written gently at the end: 'Those who figure in this book must read it in a very forgiving spirit.'

I had not seen the last of the Bush Baby. After he had gone back to the States, he returned to England for a day or two, on one of which he hired a car and drove down to Suffolk where we happened to be, to show us his newly acquired fiancée. I found this endearing and we made them very welcome. WLB said to her with a twinkle, 'I hope you know what you have taken on'.

Now, in 1965, followed some delightful events for us. WLB had won the Nobel Prize in 1915, half a century before, and the BBC decided to make a film of his life, called *Fifty Years a Winner*. I found the making of it great fun, though now I realise WLB and I, in different ways, led Philip Daly, the producer, a dance.[179] WLB would get tired and then wanted to direct the production. I tried to insist that if any of our children and grandchildren were taking part, all of them must, in case feelings were hurt. TV vans appeared in Waldringfield, thrilling our village, as well as Albemarle Street. There were shots of WLB gardening and sketching, and our gardener, Charlie Barker, rushed to the local pub to brandish a ten-pound note which he had earned for picking an apple with the picker WLB had invented. Somebody was sent to Canada to film an interview with our enchanting little granddaughter saying what she thought of her grandfather. This of course was the domestic side of the film; it was mostly scientific. We all felt it was a great success, and Philip Daly won an award for it.[180]

[179] Philip Daly was a BBC producer, well known for his programmes on science.

[180] This film can be viewed at the Royal Institution by arrangement.

This film was shown in Sweden the night before WLB and I arrived in early December for the annual Nobel Celebrations, when the current winners of the prize received their awards from the King in Stockholm. We were especially invited and received very special treatment. As it is bitterly cold and snowy then in Sweden, I hired a lovely fur coat with a blond mink collar. WLB liked it so much that on our return home he bought it for me. Sweden is very proud of its Nobel Prize winners, and everyone had seen the film the night before we arrived. When we went out shopping next day in Stockholm, men stopped us to shake our hands and take off their fur hats, and women blew kisses; in the restaurant where we had coffee, everyone stood up and clapped. Our rooms were full of flowers.

Traditionally a highlight of the celebration is a banquet at the palace, and after dinner the King had a long talk with WLB and persuaded him to end our trip by flying up to Umeå, a university in the far north, in which the King took a great personal interest. It was the coldest expedition I have ever known. When we left the plane at lunch time, it was almost dark and it was so cold that the hairs in my nostrils tinkled with ice. Anyway that's what it felt like. The people there are very cut off, and everyone at the university, so we were told, came to WLB's lecture as it was such an event.

At home the Royal Institution gave us a wonderful evening party, to which twenty of the twenty-eight British Nobel Prize winners came. The BBC was upset that they were not allowed to make a TV programme on this unique occasion, but the managers wanted it to be a private one. Among our many relations and friends present, we were delighted to have Margarethe Bohr, widow of the famous Danish physicist, as she happened to be in London. Our president, Lord Fleck, presided;[181] the Lord Chancellor Lord Gardiner was present and made a speech;[182] but it was left to WLB's oldest friend, Sir George Thomson, to make the presentation of an illuminated address signed by the officers of the Royal Institution and all the Nobel Prize winners, on behalf of the members. The Lord Chancellor made a kind reference to me, saying that I was well known to him and his

[181] Alexander Fleck, First Baron Fleck (1889–1968), was a British industrial chemist.

[182] Gerald Austin Gardiner, Baron Gardiner (1900–1990), was a British Labour politician.

Figure 70 Learning to draw geese. (Sketch by WLB; Courtesy: Patience Thomson)

department as a member of various committees and as a magistrate. I might add that this was the only time I had the fun of wearing a diamond necklace, lent to me for this special occasion by our friend Richard Ogden, the jeweller.[183]

[183] Richard Ogden (1919–2005) was a well-known jeweller in Burlington Arcade, London.

The following year, in 1966, at the age of seventy-six, WLB retired. Sir George Porter succeeded him, and that gave WLB great pleasure, as he felt that George and Stella, George's wife, would be just right for the place; and so indeed has been the case.[184] We were given a formal but most imaginative farewell by the officers and members; during this occasion, Lord Adrian unveiled a plaque commemorating WLB's work. WLB replied to his speech, and I was allowed also to speak and to say goodbye. We were presented with our portraits and other lovely presents and, after this ceremony, there were on show exhibits relating to WLB's work, his manuscripts and papers. What especially pleased us was an exhibition of his own watercolours and pencil portraits.

We had five happy years of retirement living in Suffolk, where we could garden and WLB could paint. Although his health was failing, he continued his lectures to schoolchildren and was writing a final book, which he was just finishing at the time of his death in 1971.

After that there is no more to be told. I have the joy of children, grandchildren and all our friends, and I have come home to Cambridge, where I feel that I belong. I look back with gratitude, knowing that I have had the greatest of human experiences, that of loving and being loved.

Lady Alice Grace Jenny Bragg, née Hopkinson, born 1899, died in 1989.

[184] George Hornidge Porter, Baron Porter of Luddenham (1920–2002), was a British chemist. He was awarded the Nobel Prize in Chemistry in 1967.

Suggestions for Further Reading

Andrade, E. N. da C. and Lonsdale, K. William Henry Bragg 1862–1942. *Biographical Memoirs of Fellows of the Royal Society* **4**: 276–300 (1943).

Authier, A. *Early Days of X-Ray Crystallography*. (Oxford University Press, 2013).

Bragg, W. L. *The Development of X-Ray Analysis*. (Bell, 1975).

Brown, A. *J.D. Bernal. The Sage of Science*. (Oxford University Press, 2007).

Caroe, G. M. *William Henry Bragg, 1862–1942, Man and Scientist*. (Cambridge University Press, 1978).

Ferry, G. *Dorothy Hodgkin. A Life*. (Granta, 1998).

Ferry, G. *Max Perutz and the Secret of Life*. (Pimlico, 2008).

Glazer, A.M. The first paper by WL Bragg—what and when? *Crystallography Reviews* **19**: 117–124 (2013).

Glusker, J. P., Patterson, B. K. and Rossi, M. *Patterson and Pattersons: Fifty Years of the Patterson Function*. (Oxford University Press, 1987).

Glynn, J. *My Sister Rosalind Franklin*. (Oxford University Press, 2012).

Greig, J. *John Hopkinson: Electrical Engineer*. (Science Museum, London, 1970).

Hall, K. T. *The Man in the Monkey Nut Coat: William Astbury and the Forgotten Road to the Double Helix*. (Oxford University Press, 2014).

Hunter, G. K. *Light is a Messenger: The Life and Science of William Lawrence Bragg*. (Oxford University Press, 2004).

Jenkin, J. *William and Lawrence Bragg, Father and Son: The Most Extraordinary Collaboration in Science*. (Oxford University Press, 2008).

Julian, M. 'Women in crystallography', In Kass-Simon, G. and Farnes, P., eds, *Women of Science, Righting the Record* (Indiana University Press, 1993), pp. 335–384.

Keen, D. A. Crystallography and physics. *Physica Scripta* **89**: 128003 (2014).

Maddox, B. *Rosalind Franklin: The Dark Lady of DNA*. (Harper Collins, 2003).

Perutz, M. F. Sir Lawrence Bragg. *Nature (London)* 233: 74–76 (1979), reprinted *Acta Crystallographica* **A69**: 8–9 (2013).

Perutz, M. F. *I Wish I'd Made You Angry Earlier. Essays on Science and Scientists*. (Oxford University Press, 1998).

Phillips, D. William Lawrence Bragg. 31 March 1890–1 July 1971. *Biographical Memoirs of Fellows of the Royal Society* **25**: 74–143 (1979).

Thomas, J. M. and Phillips, D., eds. *Selections and Reflections: The Legacy of Sir Lawrence Bragg*. (Science Reviews Ltd, 1990).

Thomson, P. M. A tribute to W. L. Bragg by his younger daughter. *Acta Crystallographica* **A69**: 5–7 (2013).

Watson, J. D. *The Double Helix*. (Weidenfeld & Nicholson, 1968); annotated version, Gann, A. and Witkowski, J., eds. (Simon and Schuster, 2012).

Name Index